Safe and Sustainable Use of Arsenic-Contaminated Aquifers in the Gangetic Plain

AL. Ramanathan • Scott Johnston
Abhijit Mukherjee • Bibhash Nath
Editors

Safe and Sustainable Use of Arsenic-Contaminated Aquifers in the Gangetic Plain

A Multidisciplinary Approach

Editors
AL. Ramanathan
School of Environmental Sciences
Jawaharlal Nehru University
New Delhi, India

Scott Johnston
Southern Cross GeoScience
Southern Cross University
East Lismore, New South Wales, Australia

Abhijit Mukherjee
Department of Geology and Geophysics
Indian Institute of Technology
Kharagpur, West Bengal, India

Bibhash Nath
School of Geosciences
The University of Sydney
Sydney, Australia

Co-published by Springer International Publishing, Cham, Switzerland, with Capital Publishing Company, New Delhi, India.

Sold and distributed in North, Central and South America by Springer, 233 Spring Street, New York 10013, USA.

In all other countries, except SAARC countries—Afghanistan, Bangladesh, Bhutan, India, Maldives, Nepal, Pakistan and Sri Lanka— sold and distributed by Springer, Haberstrasse 7, D-69126 Heidelberg, Germany.

In SAARC countries—Afghanistan, Bangladesh, Bhutan, India, Maldives, Nepal, Pakistan and Sri Lanka—printed book sold and distributed by Capital Publishing Company, 7/28, Mahaveer Street, Ansari Road, Daryaganj, New Delhi 110 002, India.

ISBN 978-3-319-16123-5   ISBN 978-3-319-16124-2   (eBook)
DOI 10.1007/978-3-319-16124-2
Springer Cham Heidelberg New York Dordrecht London

Library of Congress Control Number: 2015940544

© Capital Publishing Company 2015
This work is subject to copyright. All rights are reserved by Capital Publishing Company, whether the whole or part of the material is concerned, specifically the rights of translation, reprinting, reuse of illustrations, recitation, broadcasting, reproduction on microfilms or in any other physical way, and transmission or information storage and retrieval, electronic adaptation, computer software, or by similar or dissimilar methodology now known or hereafter developed. Exempted from this legal reservation are brief excerpts in connection with reviews or scholarly analysis or material supplied specifically for the purpose of being entered and executed on a computer system, for exclusive use by the purchaser of the work. Duplication of this publication or parts thereof is permitted only under the provisions of the Copyright Law of the Publisher's location, in its current version, and permission for use must always be obtained from Capital Publishing Company. Permissions for use may be obtained through Capital Publishing Company. Violations are liable to prosecution under the respective Copyright Law.
The use of general descriptive names, registered names, trademarks, service marks, etc. in this publication does not imply, even in the absence of a specific statement, that such names are exempt from the relevant protective laws and regulations and therefore free for general use.
While the advice and information in this book are believed to be true and accurate at the date of publication, neither the authors nor the editors nor the publisher can accept any legal responsibility for any errors or omissions that may be made. The publishers make no warranty, express or implied, with respect to the material contained herein.

Printed on acid-free paper

Springer is part of Springer Science+Business Media (www.springer.com)

Foreword

There is a lack of detailed understanding on the primary and secondary sources controlling the spatial variability of arsenic in alluvial aquifers in the Gangetic plain. Further the factors controlling the temporal changes in arsenic concentrations are of great concern now since arsenic toxicity creates an impediment to water usage for agriculture and drinking water purposes. These knowledge gaps critically affect our scientific understanding and ability to develop informed policy to mitigate and manage arsenic contamination. The book thus addresses the extent to which the reductive dissolution of As-bearing Fe(III) oxides hypothesis is applicable, sustainability of the deeper aquifer system(s) as an alternate arsenic-free water source, vertical connectivity between the shallow and deep aquifer system(s) and the lateral heterogeneity of aquitards/semi-confining layers, type of groundwater extraction strategies required to maintain the sustainability of the deeper aquifer system(s) and institutional approach to address and mitigate arsenic issues. In order to resolve these extant knowledge gaps, this book has been brought out as a research frontier to address this public health problem faced by millions of people, through highly integrated, multi-disciplinary hydrogeochemical research. Further it attempts to integrate quantification of arsenic mobilization, sustainable abstraction of arsenic-safe drinking water, predication of the long-term viability of As-safe groundwater, quantifying reactive transport processes of arsenic and simultaneous and comprehensive documentation of As-contaminated shallow aquifer systems.

Department of Sustainable Development,
   Environmental Science and Technology,
   KTH-International Groundwater
   Arsenic Research Group,
Teknikringen 76, SE-100 44 Stockholm, Sweden

Gunnar Jacks
Prosun Bhattacharya Ph.D.

Preface

The Arsenic problem in the groundwater of the Gangetic fluvial plains has direct consequences for human health and is presumed to be one of the biggest natural groundwater calamities that mankind is facing. The Gangetic basin is affected by arsenic contamination in the groundwater that is above the permissible limit of 10 µg/L. People are chronically being exposed to drinking arsenic-contaminated hand tube-wells water from Holocene aquifers (recent alluvial sediments). The arsenic distribution in this region has routes originating from the Himalayan region, and its ingestion through the food chain has far-reaching consequences including health hazards and socio-economic impacts. The rapid spreading of arsenic toxicity in these ground waters is due to the slow to moderate movement and over-exploitation.

Despite taking a number of precautionary measures by various agencies, the arsenic contamination in ground water continues to be an unsolved social problem. To resolve this problem in an adequate manner it desperately needs the bonding of strategic scientific research. This book thus addresses numerous scientific investigations in the Central Gangetic plain and comes out with a number of findings, and alternative propositions, which are collectively presented in this book reflecting exclusively the authors' view and diverse perspectives from their intensive research outputs. The chapters are thus aimed at highlighting the state-of-affairs of arsenic in ground water with scientific narration done in different states. These chapters also discuss the knowledge gaps and areas in which further actions are to be taken up along with their scope.

Section I covers chapters on the role of fluvial geomorphology, quaternary stratigraphy and sedimentology in arsenic distribution using hydrogeochemical evolution and resistivity survey approaches in the different shallow aquifer of Central Gangetic plain. Section II has four chapters covering groundwater topography and aquifer characterization and risk assessment in public groundwater wells along with the existing challenges in managing arsenic contamination.

The chapters in Section III deal with arsenic mobilization and distribution processes and their temporal and spatial variations. Spatio-vertical heterogeneity and surface generated organic matter act as a driver in arsenic mobilization in these basins including delta and coastal regions are also discussed in detail. Section IV emphasizes "arsenic and health" and its impact on the food chain. This section also encompasses innovative new remediation techniques including the low-cost arsenic removal method.

Thus, this book offers a meaningful and practicable guidance for the better management of arsenic problems in the ground water of the Gangetic plain with contributions from distinguished scientists from both academic and research institutions from all over the world, who have been actively working in this area. We would like to thank all the authors for their contributions and the publishers for bringing out this volume successfully. This book will be of great treasure for those working and planning to work on ground water, hydrogeology, hydrogeochemistry, water quality and other issues related to drinking water quality and the remediation techniques. This book is recommended for all libraries of universities, colleges and other institutions working on water and will be an invaluable reference work for planners who are working on water supply and sanitation.

New Delhi, India	AL. Ramanathan
East Lismore, Australia	Scott Johnston
Kharagpur, India	Abhijit Mukherjee
Sydney, Australia	Bibhash Nath

Contents

Section I Role of Fluvial Geomorphology and Sedimentology in Arsenic Distribution

1 Hydrogeochemical Evolution in the Different Shallow Aquifers of Central Gangetic Plain and Kosi Alluvial Fan and Their Implications for the Distribution of Groundwater Arsenic 3
Abhijit Mukherjee

2 Assessment of Subsurface Lithology by Resistivity Survey Coupled with Hydrochemical Study to Identify Arsenic Distribution Pattern in Central Gangetic Plain: A Case Study of Bhagalpur District, Bihar, India 17
Pankaj Kumar, Ram Avtar, Alok Kumar, Chander Kumar Singh, and AL. Ramanathan

3 Arsenic Contamination in Groundwater in the Middle Gangetic Plain, India: Its Relations to Fluvial Geomorphology and Quaternary Stratigraphy 33
Babar Ali Shah

Section II Groundwater Arsenic Characterisations and Risk Assessments

4 Preliminary Assessment of Arsenic Distribution in Brahmaputra River Basin of India Based on Examination of 56,180 Public Groundwater Wells 57
Chandan Mahanta, Runti Choudhury, Somnath Basu, Rushabh Hemani, Abhijit Dutta, Partha Pratim Barua, Pronob Jyoti Borah, Milanjit Bhattacharya, Krisaloy Bhattacharya, Wazir Alam, Lalit Saikia, Abhijit Mukherjee, and Prosun Bhattacharya

5   **Problem, Perspective and Challenges of Arsenic Contamination in the Groundwater of Brahmaputra Flood Plains and Barak Valley Regions of Assam, India**.................... 65
Nilotpal Das, Latu Khanikar, Rajesh Shah, Aparna Das, Ritusmita Goswami, Manish Kumar, and Kali Prasad Sarma

6   **Arsenic Contamination of Groundwater in Barak Valley, Assam, India: Topography-Based Analysis and Risk Assessment**.................. 81
Abhik Gupta, Dibyajyoti Bhattacharjee, Pronob Borah, Tushar Debkanungo, and Chandan Paulchoudhury

7   **Hydrogeochemistry and Arsenic Distribution in the Gorakhpur District in the Middle Gangetic Plain, India**.. 97
Hariom Kumar, Rajesh Kumar Ranjan, Shailesh Yadav, Alok Kumar, and AL. Ramanathan

Section III   Arsenic Hydrogeochemistry and Processes

8   **Arsenic Distribution and Mobilization: A Case Study of Three Districts of Uttar Pradesh and Bihar (India)**...................... 111
Manoj Kumar, Mukesh Kumar, Alok Kumar, Virendra Bahadur Singh, Senthil Kumar, AL. Ramanathan, and Prosun Bhattacharya

9   **Understanding Hydrogeochemical Processes Governing Arsenic Contamination and Seasonal Variation in the Groundwater of Buxar District, Bihar, India**.......... 125
Kushagra, Manish Kumar, AL. Ramanathan, and Jyoti Prakash Deka

10 **Chemical Characteristics of Arsenic Contaminated Groundwater in Parts of Middle-Gangetic Plain (MGP) in Bihar, India**... 143
Sanjay Kumar Sharma, AL. Ramanathan, and V. Subramanian

11 **An Insight into the Spatio-vertical Heterogeneity of Dissolved Arsenic in Part of the Bengal Delta Plain Aquifer in West Bengal (India)**.................................. 161
Santanu Majumder, Ashis Biswas, Harald Neidhardt, Simita Sarkar, Zsolt Berner, Subhamoy Bhowmick, Aishwarya Mukherjee, Debankur Chatterjee, Sudipta Chakraborty, Bibhash Nath, and Debashis Chatterjee

12 **Surface Generated Organic Matter: An Important Driver for Arsenic Mobilization in Bengal Delta Plain**...................... 179
S.H. Farooq and D. Chandrasekharam

Contents

13 A Comparative Study on the Arsenic Levels
    in Groundwaters of Gangetic Alluvium and Coastal
    Aquifers in India 197
    S. Chidambaram, R. Thilagavathi, C. Thivya, M.V. Prasanna,
    N. Ganesh, and U. Karmegam

Section IV Arsenic in Food Chain, Health and Its Remediation

14 Groundwater Arsenic Contamination in Bengal Delta
    and Its Health Effects 215
    Mohammad Mahmudur Rahman, Khitish Chandra Saha,
    Subhas Chandra Mukherjee, Shyamapada Pati, Rathindra Nath Dutta,
    Shibtosh Roy, Quazi Quamruzzaman, Mahmuder Rahman,
    and Dipankar Chakraborti

15 Impact of Arsenic Contaminated Irrigation Water
    on Some Edible Crops in the Fluvial Plains of Bihar 255
    N. Bose, A.K. Ghosh, R. Kumar, and A. Singh

16 A Greenhouse Pot Experiment to Study Arsenic
    Accumulation in Rice Varieties Selected from Gangetic
    Bengal, India... 265
    Piyal Bhattacharya, Alok C. Samal, and Subhas C. Santra

17 Status of Arsenic Contamination Along the Gangetic
    Plain of Ballia and Kanpur Districts, Uttar Pradesh,
    India and Possible Remedial Measures... 275
    N. Sankararamakrishnan, A. Gupta, and V.S. Chauhan

18 A Low-Cost Arsenic Removal Method for Application
    in the Brahmaputra-Ganga Plains: Arsiron Nilogon 289
    Shreemoyee Bordoloi, Sweety Gogoi, and Robin K. Dutta

Index 299

About the Editors

AL. Ramanathan Professor in the School of Environmental Sciences, Jawaharlal Nehru University, New Delhi, India, has research interests in the fields of hydrogeochemistry of inland and coastal surface and ground waters and their resource management. He has taught at various universities in India and abroad and has guided a number of Ph.D. scholars. Professor Ramanathan has published more than 100 articles in reputed referred journals and authored six books. He is also continuing his research work on groundwater for Indian and international agencies.

Scott Johnston an Australian Research Council Future Fellow is based at Southern Cross University, Australia. He has received his Ph.D. in 2005 (UNE) on hydrology and biogeochemistry in sulfidic wetlands. His main work, including more than 45 peer-reviewed manuscripts, stresses on how to unravel the physical and hydrological characteristics of floodplain sedimentary environments. He has worked closely with the Australian industry, state and local governments to develop and refine techniques for remediating degraded floodplain wetlands.

Abhijit Mukherjee Assistant Professor, Geology and Geophysics, IIT Kharagpur, India, did his Ph.D. from the University of Kentucky and post-doctorate from University of Texas on hydrogeology and arsenic geochemistry of Indo-Gangetic plain. He is serving as an Associate Editor, *Journal of Hydrology and Applied Geochemistry*. He has done extensive work on elevated arsenic in deeper groundwater of western Bengal basin, India. His work, including more than 30 peer-reviewed articles, also relates to arsenic and other toxic elements in natural water systems, groundwater recharge in natural dune systems, and agricultural ecosystems in the desert region.

Bibhash Nath Research Associate at the School of Geosciences, The University of Sydney, is an 'Environmental Geochemist' with specialization in 'Arsenic Hydrogeochemistry'. Dr. Nath researched in Taiwan on water quality issues, especially hydrogeology and geochemistry of arsenic contaminated coastal aquifer in SW Taiwan. He is also a recipient of the prestigious DAAD fellowship (2003).

Dr. Nath has co-authored 52 peer-reviewed journal articles, three book chapters, two technical reports, one editorial and more than 30 conference abstracts/proceedings which have been highly cited. He is an Associate Editor of the *Journal of Hydrology* and a Guest Editor (special issue on 'Arsenic Ecotoxicology') of the *Journal of Hazardous Materials*.

Section I
Role of Fluvial Geomorphology and Sedimentology in Arsenic Distribution

Chapter 1
Hydrogeochemical Evolution in the Different Shallow Aquifers of Central Gangetic Plain and Kosi Alluvial Fan and Their Implications for the Distribution of Groundwater Arsenic

Abhijit Mukherjee

1.1 Introduction

For a long time it was understood that the extent of the arsenic (As) enriched groundwater is confined within the lower Ganges plain and delta in eastern India. However, during the last few years, reports of elevated As in groundwater of different parts of the middle portions of the Gangetic plain, upstream from West Bengal, in U.P. (Ramanathan et al. 2006) and Bihar, got published. Subsequently, As contamination was detected in foothills of Himalayas in Nepal (Shrestha et al. 2003), which is a sediment provenance for many of the tributaries of the Ganges, and also in the Indus alluvial plains in Punjab and Sindh provinces of Pakistan (Nickson et al. 2005). In spite of some surveys on As distribution, there is a lack of hydrogeochemical knowledge about the distribution, extent, severity, source and cause of the contamination in these areas. However, initial estimates show that the poisoning might be widespread and several million people may be at risk.

The groundwater chemistry of the Gangetic aquifers and its associated tributary basin aquifers e.g. the river Kosi fan is suggested to be controlled by the presence of carbonates, the composition of silicates, and the oxidation of sulfides (Galy and France-Lanord 1999). According to these authors, in scarcity of Ca-plagioclase in the Himalayas, weathering of the alkaline Himalayan silicates releases Na and K as the dominant cations. Dissolution of hydrobiotite, vermiculite, and smectite of the sediments derived from the Himalaya (Baumler and Zech 1994; Grout 1995), and introduced from weathering of biotite, are the source of Mg in the groundwater (Galy and France-Lanord 1999). However, Dowling et al. (2003) suggested that the groundwater chemistry is indicative of dissolution of detrital carbonates in the upper

---

A. Mukherjee (✉)
Department of Geology and Geophysics, Indian Institute of Technology-Kharagpur, Kharagpur, West Bengal 721302, India
e-mail: amukh2@gmail.com

reaches of the river leading to dominance of carbonate weathering. However, proponent of the silicate weathering hypothesis suggested that the chemical evolution would have been dominated by the deposition of silt-dominated sediments in the foreland basin from the rising Himalayas (Burbank 1992; Derry and France-Lanord 1996; Galy and France-Lanord 1999; Kumar et al. 2006). Nevertheless, it is believed that most of the weathering in basin is caused by H_2CO_3 liberated by degradation of organic matter in the soil, and <10 % of the weathering is caused by H_2SO_4 derived from sulfide oxidation. The groundwater has concentrations of NO_3^- (<1–120 mg/L) and SO_4^{2-} (15–379 mg/L), suggesting a relatively different redox environment than the Bengal basin groundwater. The observed Eh ranges from 0.12 to 0.7 V, indicating oxic to slightly iron reducing conditions. Such a redox condition might not be conducive of and would limit the microbial reduction of NO_3^-, SO_4^{2-} and Fe(III) (Kumar et al. 2010; Seyler and Martin 1989). A detailed hydrogeochemical characterization of the study area has been provided in Mukherjee et al. (2012).

Shallow groundwater were sampled during the present study along few regional transects, which covered the majority of the aquifers encountered in the central Gangetic plain, comprising the states of Bihar and Jharkhand. The detailed hydrogeology of the area, till date, is largely un-comprehended. The objective of this study is to decipher the regional hydrochemical conditions and trends in various aquifers of the study area, and their influence on the distribution of the groundwater As. The importance of the study is in the fact that the study area is hydrologically and sedimentologically just upstream of Bengal basin (Mukherjee et al. 2009a), and thus the study results can provide immense insight on the conducive environment, conditions and mechanisms for groundwater As mobilization in various geologic-geomorphic terrains, which are obscure in the predominantly fluvio-deltaic Bengal basin.

1.2 Study Area

The study area extends over the flood plain deposits of the river Ganges (also known as the river Ganga) and its tributaries in the states of Bihar and Jharkhand in eastern parts of India (Fig. 1.1). The area is stretched from Indian cratonic outcrop rocks, representing Peninsular India, and extending up to the Terrai plains in north, located in the piedmonts of the Himalayas near the Indo-Nepal international boundary. The southern bank lithology of the river Ganges are composed mostly of Pre-Cenozoic metamorphics, along with some igneous rocks, mostly as volcanics of Rajmahal flood basalt, overlain by a layer of alluvial flood plain deposits. In the northern bank of the river Ganges, the fluvial sedimentation are much more pronounced, and together with the alluvial deposits of river Kosi (or river Kaushiki in Sanskrit), one of the primary tributaries of the Ganges forms the extensive alluvial deposits of the Kosi alluvial fan, which, with its thick coarse to medium grained sand aquifer systems, is sometimes regarded as the "bread basket" for Northern India.

The alluvial deposits in the northern bank of the river Ganges and the Kosi fan is made up of thick pile of unconsolidated Quaternary Alluvium (Fig. 1.2) with several

1 Hydrogeochemical Evolution in the Different Shallow Aquifers

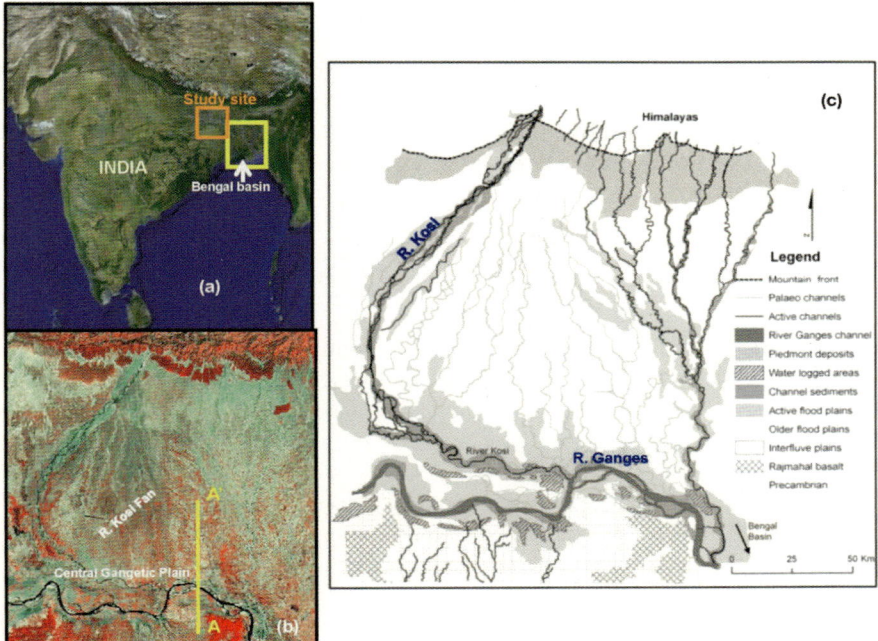

Fig. 1.1 Maps of the study area showing (**a**) disposition of the study area in the Indian-subcontinent, and with reference to the Bengal basin; (**b**) a FCC (false colour coded) imagery of the study area highlighting the geological and hydrological features; and (**c**) delineated geomorphologic features within the study area (Modified from Mukherjee et al. 2012)

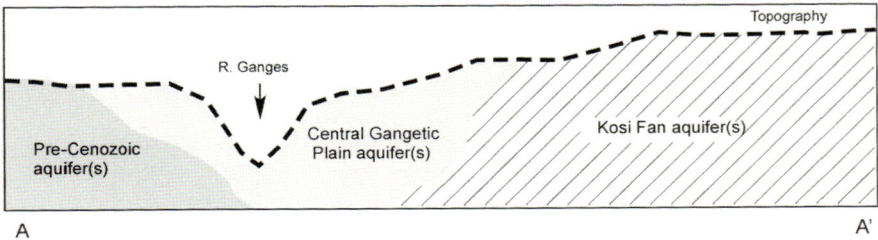

Fig. 1.2 A conceptual cross-section along the line AA′ shown in Fig. 1.1b, showing the probable disposition of the aquifers within the study area

cycles of fining upward sequence. Seismic studies have indicated the thickness of Quaternary Alluvium in the range of 300–400 m. The beds have gentle dip towards south. Within a depth of 100 m, the beds steeply dip toward south in the northern part of this fan deposit, and the layers are mostly composed of coarse sand and pebble with localized thin layers of clay. From north to south laterally continuous aquifers are present within a depth of 80 m bgl (Fig. 1.2). These are made up of coarse sand and gravels. The water levels observed during pre-monsoon ranged from 2.5 to 7.8 m bgl (mean 4.2 m bgl). During mid-monsoon the difference between maximum (2.6 m bgl) and minimum (0.9 m bgl) water levels were found to be least. Mean water level during August is near surface (~1.5 m bgl) indicating a large part of post-August rainfall may be considered as rejected recharge. Average water level rise from monsoon recharge is ~3 m. Between August and November, i.e. monsoon period ~37 % of the recharged water has been calculated to be seeping out as base-flow to rivers.

1.3 Methodology

Groundwater samples were collected from two transects (Fig. 1.3): (1) A 135 km long north-south transect extending from southern bank of the river Ganges up to the Indo-Nepal border in the north, through the river Kosi fan (groundwater sample $n=38$); (2) A 90 km long east-west transect, running parallel to, and along the southern bank of the river Ganges. The groundwater sample collection and field measurements were done following standard procedures (e.g. Woods 1981). The groundwater samples are collected from hand-pumped tube wells and public water supply wells. The total depth intervals of the samples vary between 8 and 137 m.

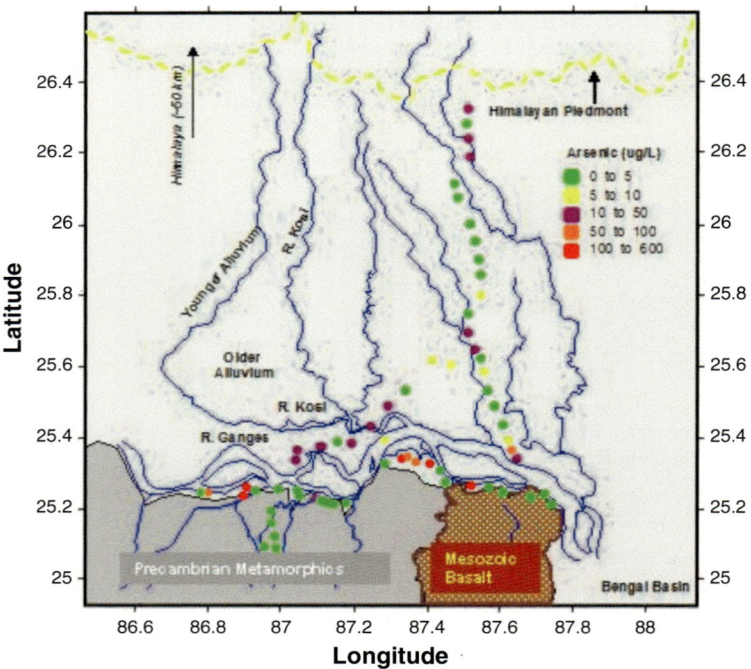

Fig. 1.3 Map showing the groundwater samples collected from the study area, classed by the arsenic concentrations detected in those samples

1 Hydrogeochemical Evolution in the Different Shallow Aquifers

The groundwater sample locations in this study have been described following field information and other secondary information e.g. remote sensing. Accordingly, the samples have been classified into four geologic-geomorphic groups: aquifers composed of Pre-Cenozoic aged geologic media (comprising Precambrian metasediments and Rajmahal basalts, $n=20$ [PC]), Alluvial sediments in the valleys of the river Ganges and its tributaries including Older Alluviums (the older flood plains and interfluve plains, $n=17$ [OA]) and Younger Alluvium (comprise the channel sediments and active flood plain deposits of rivers Ganges, river and other tributary systems, $n=25$ [YA]), and Piedmont aquifers at the foothills of the Himalayas, irrespective of age or deposition agents, $n=4$ [PD] (Figs. 1.3 and 1.4). Hence, the PD aquifer group does comprise characteristics of both younger and older alluvium of the numerous Himalayan first and second order tributaries of the river Kosi system.

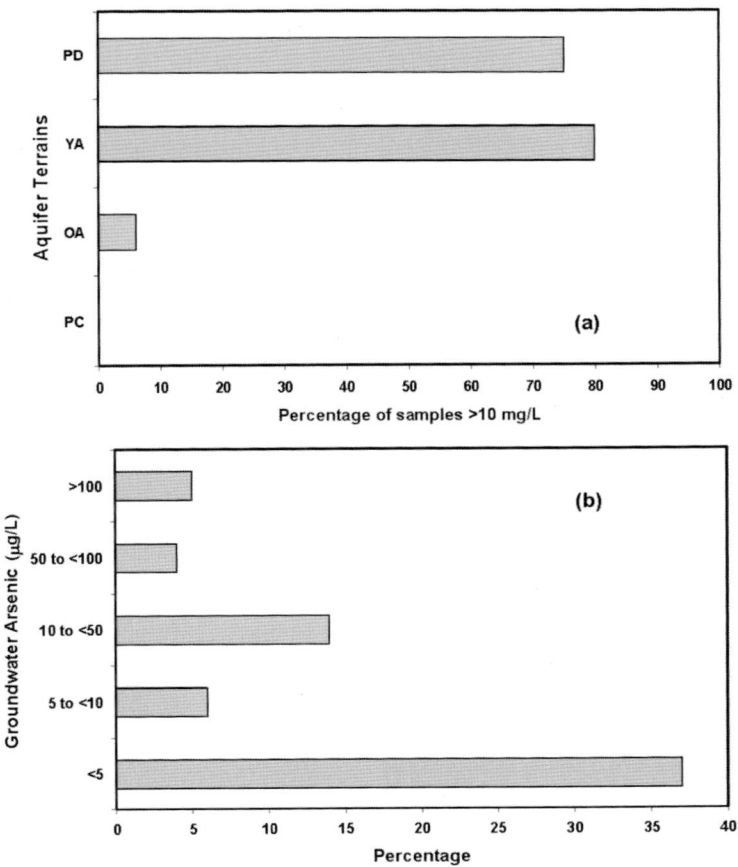

Fig. 1.4 Plots showing the (**a**) percentage of the samples having groundwater arsenic concentrations ≥ 10 µg/L in each of the aquifer terrains, and (**b**) percentage of samples within each concentration classes

1.4 Groundwater Chemistry

Here, we evaluate the different sources and mechanisms, which have led to the present composition of shallow groundwater for each of the terrains of the Central Gangetic Plain. The cations and anions in the groundwaters in the study area are probably sourced from silicate weathering, carbonate dissolution, and/or reactions like cation exchange (Figs. 1.5 and 1.6). Because of the known tropical, humid condition of the study area, presence of evaporite minerals and its dissolution seems to be impractical.

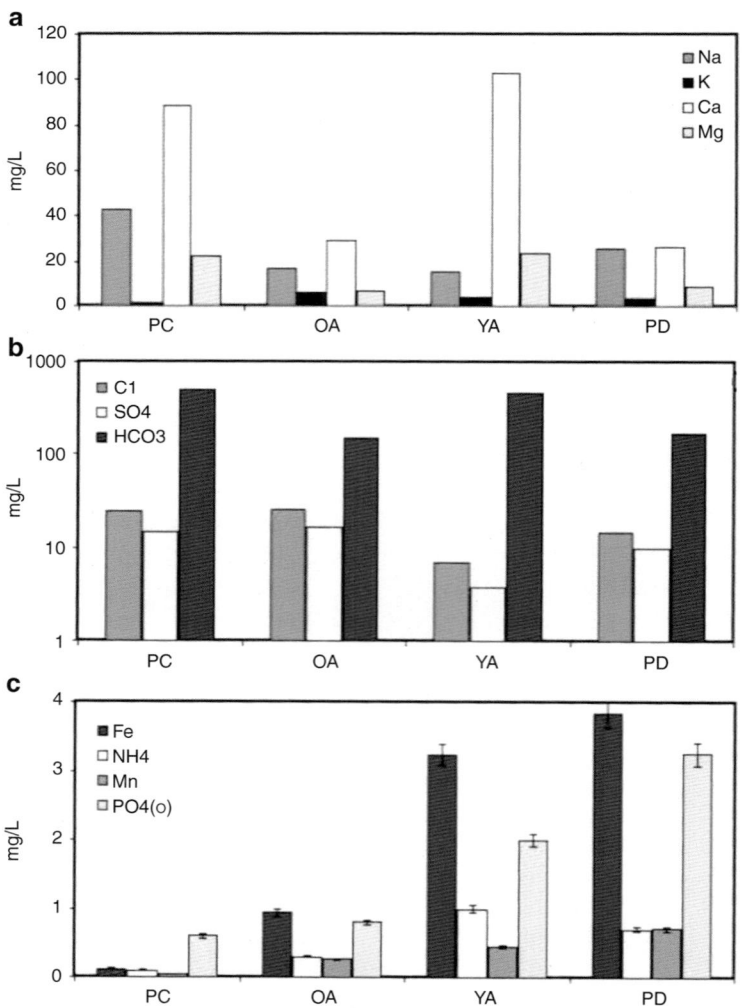

Fig. 1.5 Plots showing median concentrations of (**a**) selected major cations, (**b**) selected major anions, and (**c**) selected minor ions present in the different aquifer terrains

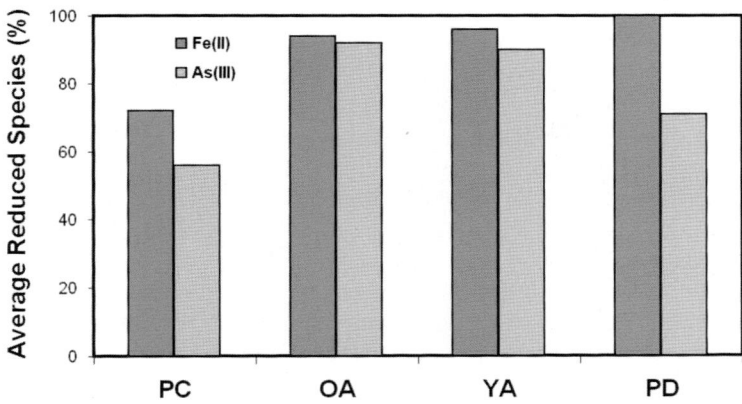

Fig. 1.6 Plot showing the percentage of the Fe(II) and As(III) redox species present within the total Fe and As concentrations detected in the groundwater samples of the different terrains

Groundwaters in all of the aquifers are mostly of Ca–HCO$_3$ facies with variants ranging from Ca–Na-HCO$_3$–Cl to Na-HCO$_3$ types, suggesting evolution from water-rock interaction. To ascertain the sources of the major solutes Ca+Mg vs. total cations, were plotted to show distinct trends between the alluvial and non-alluvial aquifers (PC terrain). However, although the ratio of Ca+Mg to total cations tends to be higher for the YA group (medians of ratios ~0.7) than the other type of aquifers in the study area (the ratios are 0.34 for PD, 0.45 for OA and 0.57 for PC), the ratio for Na+K to total cations tends are lowest for YA (median ratio=0.17 for YA), and median ratio ranges from 0.25 to 0.4 for other three groups. If the bivalent cations (e.g. Ca and Mg) in the YA groundwaters are derived from the silicate weathering, the provenance, i.e. the Himalayan source rocks would have been enriched in alkaline earth silicates e.g. Ca-feldspars, which are in contrary more enriched in orthoclase silicates, providing the flux of monovalent Na and K to the groundwater systems by silicate dissolution (Sarin et al. 1989; Galy and France-Lanord 1999; Mukherjee et al. 2009b).

The concentration of the total anion present in the PC and YA aquifer groundwater samples (>8,000 μM) are much more than of the total anion concentrations of the other groups, although in all of the aquifers, HCO$_3^-$ predominates the total anion budget. However, the concentrations of the anion in PC and YA (median >7,000 μM) are more than three times that of the OA and PD samples (median <3,000 μM). The median values for the molar ratio of total anion concentrations to HCO$_3^-$ in OA samples (<0.8) are less than that of YA samples (>0.9) suggesting that carbon cycling from primary mineral and/or secondary organic sources are a more dominant process in the YA aquifers than other aquifer groups. The median ratio of PC and PD aquifers are between that of OA and YA aquifer types. The HCO$_3^-$ versus Ca+Mg and Na+K bivariate plots suggest that the carbonate weathering has contributed to the HCO$_3^-$ in the YA and silicate dissolution contributing that to the PC aquifer groundwater. Generally, in the PC and OA aquifers, the Cl and SO$_4^{2-}$ concentrations are quite enriched than the YA and PD groundwaters.

Bivalent cations (Ca^{2+} and Mg^{2+}) may have mobilized sorbed monovalent cations like Na^+ and K^+ from the aquifer matrix by cation exchange reactions for long residing groundwater in the study area, and the influence of the process can be estimated by subtracting equivalent concentrations of co-introduced anions (HCO_3^- and SO_4^{2-}) from other processes of solute introduction like water-rock reactions. The Na^+ leaching from the aquifer sediments are the residual from the subtraction of the equivalent concentrations of meteoric originated Cl from the Na (McLean and Jankowski 2000). A negative slope in bivariate plot (i.e. $y=-x$) would suggest a possibility of active ion exchange reactions in the study area, as being observed in the case of PC samples, that have a slope of about −0.8. The proximity to the $y=-x$ line might indicate that the long residing groundwater in the PC aquifers probably would have been exposed extensively to cation exchanges. On the contrary, the groundwater in the alluvial or piedmont aquifers can be conceptualized to have shorter groundwater residence time, along with dominance of shallow, local scale flow path in the shallow aquifers in higher gradient terrains of the study area, thus possibly translated to low to very low negative slope value, as indicator of lack of cation exchanges.

1.5 Groundwater Arsenic Distribution and Fate

The regional signature of the extent of groundwater As in the study area is not as pervasive as observed in the downstream fluvio-deltaic plains of the Bengal basin, and only ~35 % of the groundwater samples collected from the study area, irrespective of the aquifers, were detected to have dissolved As concentrations ≥10 μg/L (WHO standard for safe drinking water), and only 14 % were found to have As ≥0.05 mg/L. Groundwater As detection ranges from <1 to 520 μg/L (mean 30 μg/L and median <5 μg/L), if all the aquifers of the study area are taken together (Fig. 1.4). However, when classified by the separate terrains observed in the study from where the samples have been collected, even 75 % samples for the recent alluviums and foothill deposits were detected to have As ≥10 μg/L. In contrast there is none to negligible As detected in the groundwater collected from the aquifers of the PC and OA terrains. The YA aquifers are found to be most enriched in groundwater arsenic with an arithmetic mean of 73 μg/L and median 27 μg/L.

It is expected that because of the proximity, and similarity in lithology and groundwater systems, the As fate in the present study area existing in the upstream in the Ganges alluvial system should have some similarity with the downstream Bengal basin in terms of chemical characteristics and the spatial distribution of the As, along with its fate and transport. In order to understand the As fate, multivariate relationship have been evaluated in the study area, non-parametric Spearman correlations were calculated for As and other potentially influencing groundwater parameters of the samples suggested some interesting insight. These calculations indicate that different processes control the As fate in different terrains. For aquifers of all terrain, redox dependent mobilization of As is demonstrated by strong negative

1 Hydrogeochemical Evolution in the Different Shallow Aquifers

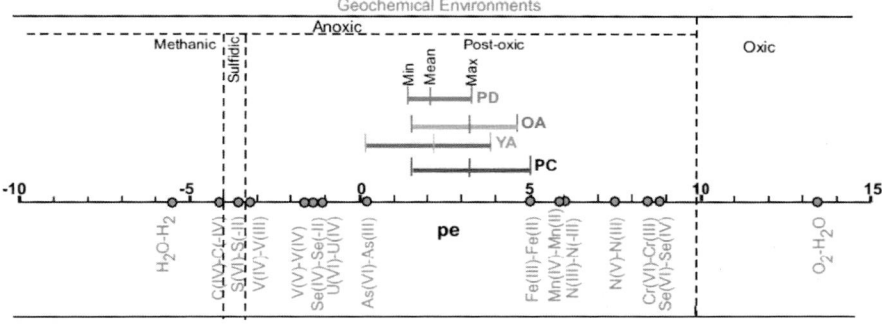

Fig. 1.7 Range of p_e values for the groundwater samples collected from each of the terrains along with their relative position in the geochemical environment (Modified after Mukherjee et al. 2012)

correlation with Eh ($\rho=-0.44$ to -0.76), as observable in reductive (Fe-Mn)OOH dissolution (Fig. 1.7). Such reduction dependent mobilization is also reinforced by correlation of As with reduction indicators like DO, SO_4 and NH_4^+ (Acharyya et al. 1999). It is expected that reductive dissolution of FeOOH onto which As is adsorbed, would lead to moderate to strong correlation between As and Fe. Such relationships have been advocated by workers like Nickson et al. (1998), Dowling et al. (2003), McArthur et al. (2004) and Stüben et al. (2003). However, in many other studies, such good correlation have not been observed (e.g. Swartz et al. 2004; Mukherjee and Fryar 2008; Mukherjee et al. 2008), probably suggesting the heterogeneity in the inter-dependence of the processes that might be controlling the co-existence of these elements. In the samples of this study, Fe correlates moderately to strongly with As with $\rho=0.38$ for YA and 0.66 for OA. However, for PD, Fe and As have very weak negative correlation ($\rho=-0.12$). Similarly, while Mn and As show moderate to strong correlation for YA and OA ($\rho=0.32-0.47$), PD shows a negative correlation ($\rho=-0.57$). In the contrary, competitive anions to As like HCO_3 and ortho-PO_4 show very strong (for PD, $\rho>0.8$) or strong positive correlationship (for OA, $\rho=0.53$ and 0.49, respectively), while they are moderate to insignificant for YA ($\rho=0.32$ and 0.14, respectively).

Such differences in correlations between the OA, YA and PD sites suggest that As is mobilized by multiple processes, in-situ or along the flow path. The YA and PD also have some correlation for As with pH ($\rho=0.48$ and 0.78), indicating probability of pH-dependent sorption reactions playing an influential role in regulating As mobility in the study area, which has not been suggested to have significant role in the Bengal basin (Mukherjee and Fryar 2008; Mukherjee et al. 2011). In summary it might be stated that the enrichment processes of dissolved As in the YA and PD might be influenced by considerably different mechanisms. While in YA, which are located mostly at the discharge ends of the flow paths, mobilization by metal-reductive dissolution is a potential dominant mechanism, enrichment in the recharge areas, i.e. PD might be more influenced by competitive anion exchange between adsorbed As and ion introduced from mineral dissolution processes or competitive ionic exchanges by agricultural processes (e.g. ortho-PO_4).

1.6 Conclusion and Synthesis

It can be concluded from the present study and in earlier work (Mukherjee et al. 2012) that the water chemistry and elevated As distribution in the previously not-much-studied Central Gangetic plain and Kosi fan aquifers have some similarity, but more dissimilarity to that of the downstream, extensively studied, Bengal basin. The present study area is located in the eastern parts of the India subcontinent, stretching from the northern edge of the cratonic India to Himalayan foothills. Tectonically, thus the aquifers become part of the Himalayan foreland basin, mostly of stable shelf composition. The aquifers considered in the present study may be considered to be one of the best yielding aquifers of the subcontinent, and probably among some of the most prolific aquifers across the globe. Hence, the aquifers are most extensively agriculturally exploited. However, the concern about the reported As enriched groundwater from geogenic and non-point source has raised serious alarm in the huge population dependent on groundwater for their day to day needs. Moreover, the proximity of the immediately downstream, highly groundwater As enriched Bengal basin pose an intriguing question about the connection of the source and sedimentary history of the two adjoining basins. In this study, the objective was to delineate the As-enriched/prone aquifers of the Central Gangetic basin (including the Central Gangetic plain and Kosi fan). The hydrogeological techniques used in this study have shown that the groundwater in the aquifers within different host rocks in this basin suggests hydrochemical compositions that they have inherited from their geological and geomorphical evolutionary history. The aquifers in the study area can be generalized to be composed of the geologic material of Pre-Cenozoic aged rocks that comprise Precambrian metasediments and Cretaceous flood basalts, alluvial sediments in the valleys of the River Ganges and its tributaries including older alluviums in the older flood plains and interfluve plains, and younger alluviums that are comprised of the channel sediments and active flood plain deposits of rivers Ganges, Kosi and other tributary systems, and Piedmont aquifers at the foothills of the Himalayas.

In general, the groundwaters existing in all of the aquifers are mostly of Ca–HCO_3 hydrochemical facies, with compositions ranging from Ca–Na-HCO_3–Cl to Na-HCO_3 types. Like groundwater in many of the modern alluvial aquifers with wide water-rock interaction and organic load, HCO_3 is found to be the dominant anion. However, concentrations of anion in the PC and YA group are more than double that of the OA and PD groups, probably for totally different type of geochemical reactions. Much of the YA solutes in the groundwater are introduced from carbonate weathering reactions. However, the PD groundwater and many of the PC and OA samples are largely affected by silicate weathering by incongruent leaching of the argillaceous metamorphics and volcanics in the Himalayas and the Pre-Cenozoic deposits, which act as a major solute provenance for groundwaters residing in these aquifers. Further, the potentially long residence time of groundwater in the PC terrain has resulted in to suspected cation exchange reactions, whereby monovalent cations got introduced from the host aquifer matrix. The PD samples are found to be more similar to OA in terms of major solutes composition but are more similar to YA groundwater in term of minor solutes.

The presence of redox sensitive species like As(III), Fe(II), $NH_{3(dis)}$, and elevated HS^- indicates the occurrence of a reductive, post-oxic redox environment, with most groundwater samples being plotted in the metal reducing hydrogeochemical condition. Dissolved Fe concentrations in predominantly Fe(II) in groundwater for all terrains and As exist mostly as As(III). Reduction is more dominant in groundwaters in YA and PD aquifers, with signature of influence of agricultural recirculated water. Redox conditions (oxic to methanic, dominated by metal-reduction) are highly spatially variable, with no systematic depth-dependence. Almost 35 % of all of the collected groundwater from all of the four aquifers of the study area were cumulatively found to have dissolved As concentrations ≥ 10 µg/L, of which, a major percentage (~45 %) have concentrations ≥ 50 µg/L. However, when classified by the aquifer terrains, while the PC and OA groundwaters have none or negligible As, almost 75 % or more YA and PD groundwaters are enriched by As ≥ 10 µg/L. The YA groundwaters are most enriched with As, with detected concentrations up to 520 µg/L. Analyses of correlation between As and other redox parameters suggest that most of the As liberation and mobilization in the YA and OA aquifers are influenced by redox related mobilization, with suggested reductive dissolution of metal (Fe-Mn)-oxyhydroxides leading to reasonably good correlation of As with Fe. Interestingly, the piedmont groundwaters strongly correlate competitive anions that are probably influenced from anthropogenic activities like nutrients introduced by agricultural processes. Hence, the As fate in the regional groundwater recharge areas near the Himalayan foothills are influenced by mechanistically different geochemical methods than the groundwater residing in the discharge areas typified by extensive alluvial deposits in vicinity of the modern river channels. Hence, in the present study area of Central Gangetic Basin that is located upstream of the Bengal basin or the Lower Gangetic plain and delta, hydrochemistry are much more variable in the solute distribution and fate as a function of differently evolved geological and geomorphic terrains. In comparison, the Bengal basin aquifers have overlapping facies aquifer systems with obscure terrains divisions. Consequent to the huge active sedimentation in that basin, most of the aquifers are geochemically similar to composition of the younger alluviums, with similar distribution and fate of As like the YA aquifers of the present study area.

Acknowledgement The author acknowledges the help and support received from the collaborators Bridget Scanlon (University of Texas at Austin, USA), Alan Fryar (University of Kentucky, USA), Dipankar Saha (Central Ground Water Board, India), Ashok Ghosh (A.N. College, India), and Sunil Chaudhuri and Ranjan Mishra (T.M. Bhagalpur University). Funding for fieldwork for the study was obtained from the Jackson School of Geoscience Initiative fund at the University of Texas at Austin. Analytical and computational help for the study were provided by the University of Texas at Austin, University of Kentucky and University of Arizona. The author acknowledges the help provided by Barindra Lal Mukherjee, Murali Singh and Kareya Lal for field sampling, and Dr. Phil Bennet (UT Austin) for gas sample analyses.

References

Acharyya SK, Lahiri S, Raymahashay BC, Bhowmik A (1999) Arsenic poisoning in the Ganges delta. Nature 401:545

Baumler R, Zech W (1994) Soils of the high mountain region of Eastern Nepal: classification, distribution and soil forming processes. Catena 22:85–103

Burbank DW (1992) Causes of recent Himalayan uplift deduced from deposited patterns in the Ganges basin. Nature 357:680–683

Derry LA, France-Lanord C (1996) Neogene Himalayan weathering history and river $^{87}Sr/^{86}Sr$: impact on the marine Sr record. Earth Planet Sci Lett 142:59–74

Dowling CB, Poreda RJ, Basu AR (2003) The groundwater geochemistry of the Bengal Basin: weathering, chemisorption, and trace metal flux to the oceans. Geochim Cosmochim Acta 67(12):2117–2136

Galy A, France-Lanord C (1999) Weathering processes in the Ganges-Brahmaputra basin and the riverine alkalinity budget. Chem Geol 159:31–60

Grout H (1995) Characterization physique, mineralogique, chimique et signification de la charge particulaire et colloýdale derivieres de la zone subtropicale. Unpublished PhD thesis, Aix-Marseille, France

Kumar K, Ramanathan AL, Rao MS, Kumar B (2006) Identification and evaluation of hydrogeochemical processes in the groundwater environment of Delhi, India. Environ Geol 50:1025–1039

Kumar P, Kumar M, Ramanathan AL, Tsujimura M (2010) Tracing the factors responsible for arsenic enrichment in groundwater of the middle Gangetic Plain, India: a source identification perspective. Environ Geochem Health 32:129–146

McArthur JM, Banerjee DM, Hudson-Edwards KA, Mishra R, Purohit R, Ravenscroft P, Cronin A, Howarth RJ, Chatterjee A, Talukder T, Lowry D, Houghton S, Chadha DK (2004) Natural organic matter in sedimentary basins and its relation to arsenic in anoxic ground water: the example of West Bengal and its worldwide implications. Appl Geochem 19(8):1255–1293

McLean W, Jankowski J (2000) Groundwater quality and sustainability in an alluvial aquifer, Australia. In: Sililo A (ed) XXX IAH congress on groundwater: past achievements and future challenges. A.A. Balkema, Rotterdam, Cape Town

Mukherjee A, Fryar AE (2008) Deeper groundwater chemistry and geochemical modeling of the arsenic affected western Bengal basin, West Bengal, India. Appl Geochem 23:863–892

Mukherjee A, Scanlon BR, Chaudhary S, Misra R, Ghosh A, Fryar AE, Ramanathan AL (2007) Regional hydrogeochemical study of groundwater arsenic contamination along transects from the Himalayan alluvial deposits to the Indian shield, Central Gangetic Basin, India. Geol Soc Am Program Abstr 39(6):519

Mukherjee A, von Brömssen M, Scanlon BR, Bhattacharya P, Fryar AE, Hasan MA, Ahmed KM, Jacks G, Chatterjee D, Sracek O (2008) Hydrogeochemical comparison and effects of overlapping redox zones on groundwater arsenic near the western (Bhagirathi sub-basin, India) and eastern (Meghna sub-basin, Bangladesh) of the Bengal basin. J Contam Hydrol 99:31–48

Mukherjee A, Fryar AE, Thomas WA (2009a) Geologic, geomorphic and hydrologic framework and evolution of the Bengal basin, India. J Asian Earth Sci 34:227–244

Mukherjee A, Bhattacharya P, Shi F, Fryar AE, Mukherjee AB, Xie ZM, Sracek O, Jacks G, Bundschuh J (2009b) Chemical evolution in high arsenic groundwater in Huhhot basin, Inner Mongolia, P.R. China and its difference from Western Bengal basin, India. Appl Geochem 24:1835–1851

Mukherjee A, Fryar AE, Scanlon BR, Bhattacharya P, Bhattacharya A (2011) Elevated arsenic in deeper groundwater of western Bengal basin, India: extents and controls from regional to local scale. Appl Geochem 26:600–613

Mukherjee A, Scanlon BR, Fryar AE, Saha D, Ghosh A, Chaudhari S, Mishra R (2012) Solute chemistry and fate of arsenic in the aquifers between the Himalayan foothills and Indian craton: influence of geology and geomorphology. Geochim Cosmochim Acta 90:283–302

Nickson RT, McArthur JM, Burgess WG, Ahmed KM, Ravenscroft P, Rahman M (1998) Arsenic poisoning of Bangladesh groundwater. Nature 395:338

Nickson RT, McArthur JM, Shrestha B, Kyaw-Myint TO, Lowry D (2005) Arsenic and other drinking water quality issues, Muzaffargarh District, Pakistan. Appl Geochem 20(1):55–68

Ramanathan AL, Bhattacharya P, Tripathi P (2006) Arsenic in groundwater of the aquifers of the central Gangetic plain of Uttar Pradesh, India. Geol Soc Am Program Abstr 38(7):241

Sarin MM, Krishnaswami S, Dilli K, Somayajulu BLK, Moore WS (1989) Major ion chemistry of the Ganga-Brahmaputra river system: weathering processes and fluxes to the Bay of Bengal. Geochim Cosmochim Acta 53(5):997–1009

Seyler P, Martin JM (1989) Biogeochemical processes affecting arsenic species distribution in a permanently stratified lake. Environ Sci Technol 23:1258–1263

Shrestha RR, Shrestha MP, Upadhyay NP, Pradhan R, Khadka R, Maskey A, Maharajan M, Tuladhar S, Dahal BM, Shrestha K (2003) Groundwater arsenic contamination, its health impact and mitigation program in Nepal. J Environ Sci Health Part A 38(1):185–200

Stüben D, Berner Z, Chandrasekharam D, Karmakar J (2003) Arsenic enrichment in groundwater of West Bengal, India: geochemical evidence for mobilization of As under reducing conditions. Appl Geochem 18(9):1417–1434

Swartz CH, Blute NK, Badruzzaman B, Ali A, Brabander D, Jay J, Besancon J, Islam S, Hemond HF, Harvey CF (2004) Mobility of arsenic in a Bangladesh aquifer: inferences from geochemical profiles, leaching data, and mineralogical characterization. Geochim Cosmochim Acta 66:4539–4557

Woods WW (1981) Guidelines for collection and field analysis of ground-water samples for selected unstable constituents. U.S. Geol. Surv. Techniques Water-Resour. Invest. Book 1 (Chapter D2)

Chapter 2
Assessment of Subsurface Lithology by Resistivity Survey Coupled with Hydrochemical Study to Identify Arsenic Distribution Pattern in Central Gangetic Plain: A Case Study of Bhagalpur District, Bihar, India

Pankaj Kumar, Ram Avtar, Alok Kumar, Chander Kumar Singh, and AL. Ramanathan

2.1 Introduction

The occurrence of arsenic (As)-rich alluvial groundwater is a worldwide problem (Kumar et al. 2010). Most studies of As pollution have focused on the predominance of As poisoning in the groundwater of West Bengal (India) and Bangladesh and thought to be limited to the Ganges delta i.e. the lower Gangetic plain (Bhattacharya et al. 1997; Ahmed et al. 2004; Ben et al. 2003). Some states as Uttar Pradesh and Bihar reported the presence of elevated concentrations of arsenic in drinking water wells sporadically (Acharyya and Shah 2004; Chakraborti et al. 2004; Acharyya 2005; Chauhan et al. 2009; Sankararamakrishnan et al. 2008; Srivastava et al. 2008; Kumar et al. 2010). Several authors suggested that the reductive dissolution of Fe (III)-oxyhydroxides in strongly reducing conditions in the young alluvial Holocene sediments is the cause for arsenic mobilization (Harvey et al. 2002; Nickson et al. 1998; Nickson et al. 2000). Groundwater quality is controlled by various factors viz. composition of recharging water, the mineralogy and

---

P. Kumar (✉)
Institute of Science and Technology for Advance Studies and Research (ISTAR), Vallabh Vidyanagar, Gujarat 388120, India

School of Environmental Sciences, Jawaharlal Nehru University, New Delhi, India
e-mail: pankajenvsci@gmail.com

R. Avtar
Institute of Industrial Science, The University of Tokyo,
Ce 509, 4-6-1, Komaba, Meguro-Ku, Tokyo 153-8505, Japan

A. Kumar • C.K. Singh • AL. Ramanathan
School of Environmental Sciences, Jawaharlal Nehru University, New Delhi, India

© Capital Publishing Company 2015
AL. Ramanathan et al. (eds.), *Safe and Sustainable Use of Arsenic-Contaminated Aquifers in the Gangetic Plain*, DOI 10.1007/978-3-319-16124-2_2

reactivity of the geological formations in the region of aquifer recharges, the impact of human activities and the environmental parameters that may control the geochemical mobility of redox (oxidation and reduction potential as varies from 169 mV to −134 mV respectively in this case) sensitive elements in the groundwater environment (Bhattacharya et al. 2009). The arsenic contaminated aquifers are persistent within lowland organic rich, clayey deltaic sediments in the Bengal basin and locally within similar facies in narrow, entrenched river valleys within the Gangetic alluvial plain (Acharyya and Shah 2004; Acharyya 2005).

The current drinking water quality guideline for arsenic is 0.01 mg/L (WHO 2004). Contaminated drinking water (i.e. water samples having arsenic in excess of above mentioned permissible limit/concentration level) is the main source of chronic human intoxication (Gabel 2000; Smith et al. 2000), which results in skin ailments such as hyper pigmentation and keratosis, and leads progressively to cancer and ultimately death (WHO 2004; National Academy Press 2001). In south Asia, enrichment of arsenic (As) in groundwater possesses a serious health threat not only limited to human being (e.g., circulatory disease, neurological effects, black foot disease and carcinogenicity) but also accumulates in plants and fish (Nath et al. 2010).

Electrical resistivity is primarily a function of porosity, pore fluid resistivity (salinity), mineral content, texture, moisture content, fissures and fractures of geological formations, temperature and clay content. The resistivity value of rocks varies depending upon the presence of secondary porosity such as weathered, fractured and joints (Yadav and Singh 2007; Raju and Reddy 1998). The resistivity value depends more on water content and quality in a porous aquifer as compared to hard rock aquifer. For aquifers composed of unconsolidated materials, the resistivity decreases with the degree of saturation and the salinity of groundwater (Khan et al. 2008). Clay minerals conduct current through their matrix, therefore clay formations tend to display lower resistivity than do permeable alluvial aquifers (Mukherjee et al. 2007). Resistivity methods are intensively used by geophysicists for evaluation of subsurface features (Rhoades et al. 1990). Several methods such as Schlumberger, Wenner, pole-dipole and dipole-dipole have been used in several studies for delineating subsurface features. The probes are applied on the soil surface as well as in bore-hole logging (Rhoades and Schilfgaarde 1976). Vertical electrical sounding was applied to estimate hydraulic conductivity (Mazac et al. 1990) and texture (Banton et al. 1997) of the stratified soils and sediments.

Though there are few studies on resistivity application to locate arsenic 'hotspots' through lithological characterization in lower Gangetic plain (Charlet et al. 2007; Aziz et al. 2008; Hoque et al. 2009; Nath et al. 2010), the idea is fairly new in case of middle Gangetic plain. Because of patchy distribution of As, drilling and geochemical study at point scale can only help to target As-safe aquifers; however these are economically not viable in developing countries like India and Bangladesh. Therefore this study strives to show the hydro-geochemical signature along with spatial distribution of As and Fe in Bhagalpur district in middle Gangetic plain. We used the above findings as a validation tool while targeting the As-safe aquifer through geophysical survey which is cost effective. It will definitely help in sustainable use of the As-low aquifer which can reduce the future casualty due to its contamination.

2.2 Study Area

Bhagalpur district is located in the eastern part of Bihar from 25°1329′ to 25°608′ N latitude and from 86°6408′ to 87°542′ E longitude (Fig. 2.1). The district is a peneplain, intersected by numerous streams. Surface levels varied due to high banks of the Ganga, Koshi (Ghugri), Chanari and Chandan. The geomorphology of the area is monotonously flat and featureless plain with gradient towards the river Ganges.

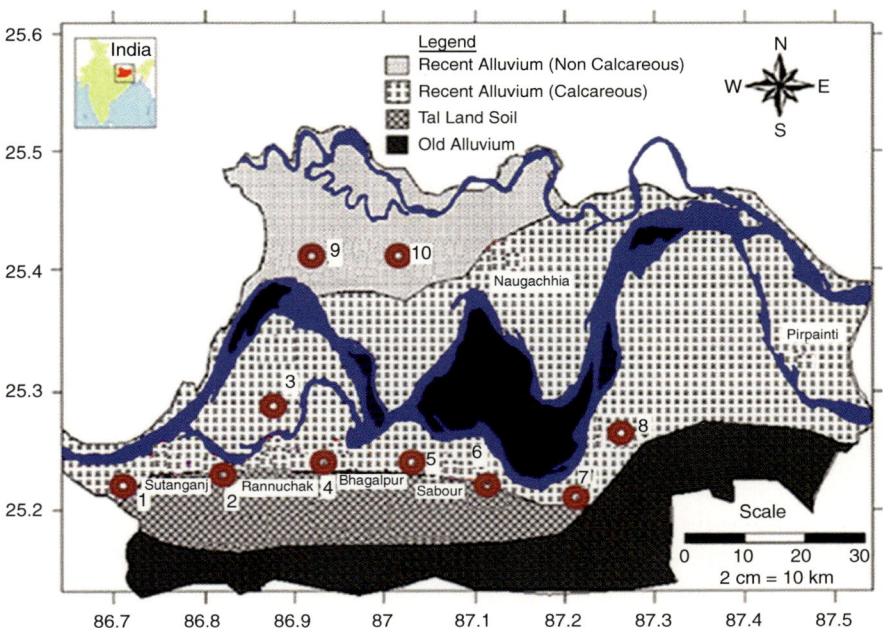

Fig. 2.1 General geomorphology of Bhagalpur district, Bihar (India) (Kumar et al. 2010) (Resistivity surveys were conducted along the points shown in brown colour)

2.2.1 Geology and Hydrogeology

Geologically, the area is represented by alluvial deposits of Quaternary age. The area has been divided into four different zones: Recent alluvium (non-calcareous), Recent alluvium (calcareous), Tal land soils and Older alluvium. The large-scale features of the Gangetic plain correspond to major climate changes in the late Quaternary (Singh 2004). The geomorphic surfaces identified in the regional mapping of the Quaternary deposits of the Gangetic plain are upland interfluvial surface (T_2), marginal fan upland surface (MP), mega fan surface (MF), piedmont fan surface (PF), river valley terrace surface (T_1) and active flood plain surface (T_0) (Singh 2004; Shah 2008) as shown in Fig. 2.2. A significant aspect of these surfaces is that all of them are depositional surfaces, having a succession of overlying sediments. The Bihar Gangetic plain shows prominent distinction between T_0, T_1 and T_2 surfaces. The Holocene aggradations, mostly due to rising base level and climate-driven sediment supply, are pronounced as much as 5–10 m thick. XRD studies on soil samples of arsenic-safe older alluvial and arsenic-contaminated newer alluvium from Middle Gangetic plain reveals mineralogical assemblage of quartz, muscovite, chlorite, kaolinite feldspar, amphibole and goethite (Shah 2008). The depth of well in uppermost aquifer varies from 6 m bgl (below ground level) to 13 m bgl. The multiple aquifer system of this region has variable hydraulic conductivity and water quality. This district has two types of irrigation system: (a) canals and (b) tube wells.

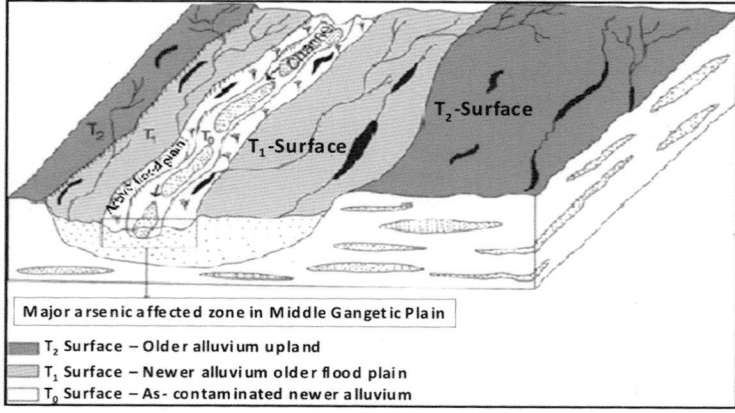

Fig. 2.2 Schematic model shows major As-contaminated zone in Newer Alluvium. (T_0-Surface) entrenched channels and floodplains and As-free zone in Older Alluvium upland terraces (T_2-Surface) in Middle Ganga Plain (Bihar and UP). Newer Alluvium older floodplains (T_1-Surface) are locally As-contaminated (Singh 2004; Shah 2008)

2.3 Materials and Methods

2.3.1 Sampling Phase I: Groundwater and Core Sampling and Analysis

Thirty-six groundwater samples (both from hand pumps/tube wells) were collected in polypropylene bottles (Tarsons). Care was taken while sampling that a number of times sufficient volume of groundwater has to be purged prior to collecting the groundwater sample for chemical analysis so that sufficient inflow is induced from the subsurface geological formation of the well and to minimize the impact of iron pipes through which water was pumped out. Further these samples were stored below 4 °C in a portable ice-box to minimize chemical alteration. In-situ measurements—mainly electrical conductivity (EC), pH and oxidation-reduction potential (ORP)—were measured using a portable Orion Thermo water analyzing kit (Model Beverly, MA, USA; 01915) with a precision level of 1 %. Total arsenic was determined with the help of a Digital Arsenator (Wagtech, UK) and arsenic speciation was performed in the field with disposable cartridges (Metal Soft Center, PA, USA) with precision level of 5 %. Here, mode of operation is simple as it absorbs As (V), but allows As (III) to pass through and thus help to identify different species of arsenic.

H_3BO_3 was used as preservative for nitrate analysis. Fe and Mn were analyzed in the laboratory using an atomic absorption spectrophotometer (AAS) (Shimadzu AA-6800) with a precision level of 1 %. Concentrations of total arsenic were cross-checked on acidified samples using graphite furnace (GF) AAS (Shimadzu AA-6800) in absorption mode using chemical standards, with the detection limit of 2 µg/L. Major cations analysis was carried out by use of an EEL flame photometer with an error percentage of <2 % using duplicates (APHA 1995). The concentration of HCO_3^- was measured by acid titration. Other anions F^-, Cl^-, NO_3^-, SO_4^{2-} and PO_4^{3-} were analyzed by use of Dionex DX-120 ion chromatograph with a precision level of 1 %. Other characteristics, for example SiO_2 and NH_4^+, were analyzed with a Jenway model 6505 dual beam spectrophotometer. Oxygen and hydrogen isotopes were analyzed by mass spectrometer (Model MAT 252, Thermo Finnigan Inc.) at university of Tsukuba. The results for both isotopes are expressed through deviation from the VSMOW (Vienna Standard Mean Ocean Water) standard using the δ-scale according to the Eq. (2.1) and unit is per mil

$$\delta\text{\textperthousand} = \left[\frac{R_{sample} - R_{VSMOW}}{R_{VSMOW}}\right] \times 1000 \qquad (2.1)$$

where R is the isotopic ratio (i.e. $^2H/^1H$ and $^{18}O/^{16}O$) for the sample and standard. For the sample preparation before isotopic measurement, we adopted hydrogen gas equilibrium method using platinum catalyst with 6 h for δD and carbon dioxide gas equilibrium method with 9 h for $\delta^{18}O$. Analytical precisions of stable isotopes were better than 0.1‰ for $\delta^{18}O$ and 1.0‰ for δD. High-purity reagents (Merck) and

Milli-Q water (Model Milli-Q, Biocel) were used for all the analyses. The analytical precision for ions measurement was determined by calculating the ionic balance error which was in between ±5 % (Kumar et al. 2011). One core sediment (up to 21-m depth from surface) was taken through borehole drilled with the help of local drillers using hand percussion technique (Kumar et al. 2010). Mineralogical study for clay at different lithological units is done with the help of XRD (PANalytical) to confirm arsenic enrichment.

2.3.2 Sampling Phase II: Geophysical Survey

In order to get insight of deep lithology, electrical resistivity surveys were carried out to determine the lithology, weathered, fractured pattern, depth to basement and resistivity variations in the study area. Ten vertical electrical soundings (VES) were taken at different locations within the study area. The Schlumberger method was used, in which the distance between the two current electrodes (AB) is successively expanded, while the distance between the two potential electrodes (MN) is kept at a minimum (MN $\leq$1/5 AB). Resistance was calculated, with the known value of position of current and potential electrodes.

The apparent resistivity is given by equation (2.2):

$$R_a = \pi \left[\frac{S^2}{A} - \frac{A}{4} \right] \frac{V}{I} \tag{2.2}$$

where S and A are current and potential electrode spacing, V is voltage difference between the potential electrodes, and I is apparent current.

Increase in the distance between the electrodes leads to the greater depth penetration and gives information about more sub-surface properties. The measured apparent resistivity values were plotted against half the distance between current electrodes (AB/2) on a log-log graph to calculate the vertical electrical sounding curves, which were interpreted using IPI2 win software to determine the apparent resistivity and the thickness of subsurface layer of different composition.

2.4 Results and Discussion

The value of resistivity has a direct relationship between porosity of subsurface material. Subsurface material with high porosity like sandstone generally leads to increase in resistivity value. So the variation in subsurface lithology of the aquifer can be clearly defined on the basis of vertical electrical sounding (VES). The statistical distribution of apparent resistivity for the study area is shown in Fig. 2.3, whereas depth wise resistivity profile for different locations of the study area is shown in Fig. 2.4.

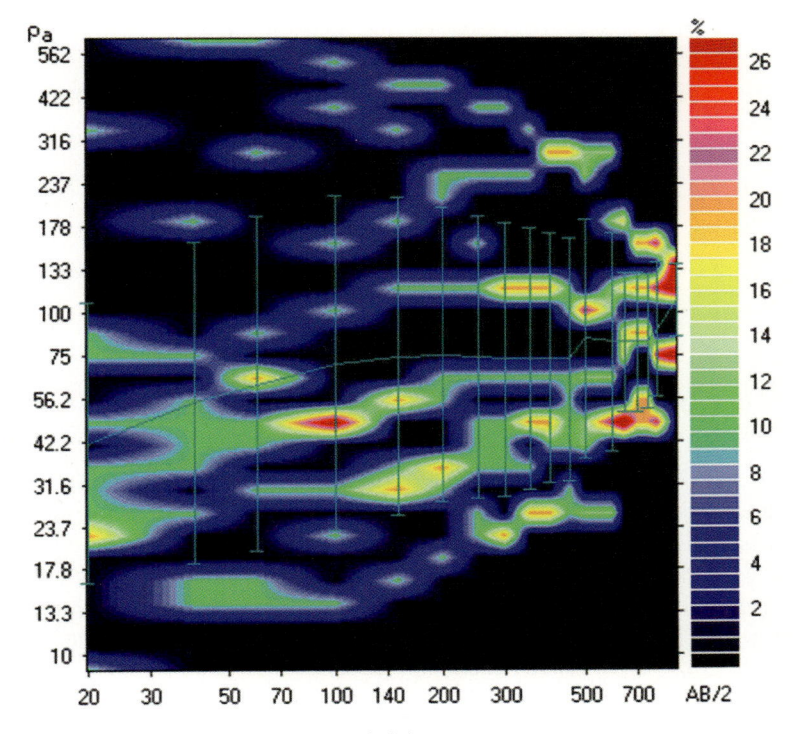

Fig. 2.3 Statistic distribution of apparent resistivity

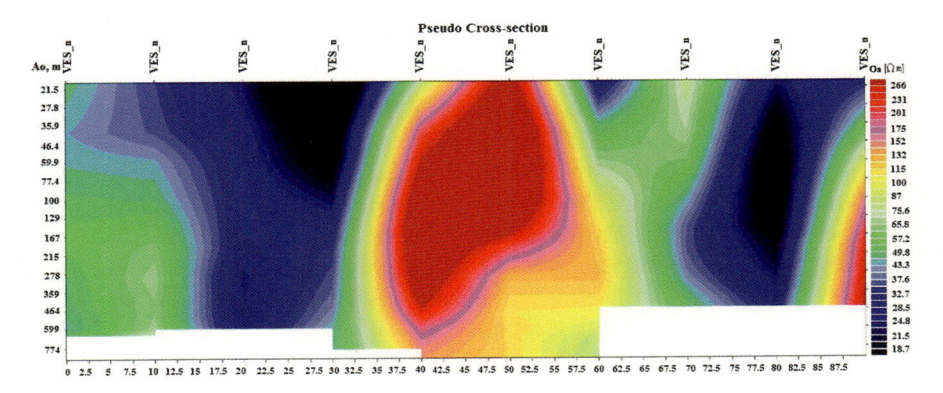

Fig. 2.4 Profile showing resistivity change with the depth in different location of the sounding of the study area (where horizontal and lateral distances are given in metres)

As most of the study area lies in Gangetic alluvial plain and the top horizon of soil has relatively high humus content with high cations exchange capacity; therefore it provides high density of mobile electric charges and thus the resistivity value of this region is very low (Wilding et al. 1983). Points 1 and 2 show lower value of resistivity in the top-most layers but it increases with depth. Here it is suggested that

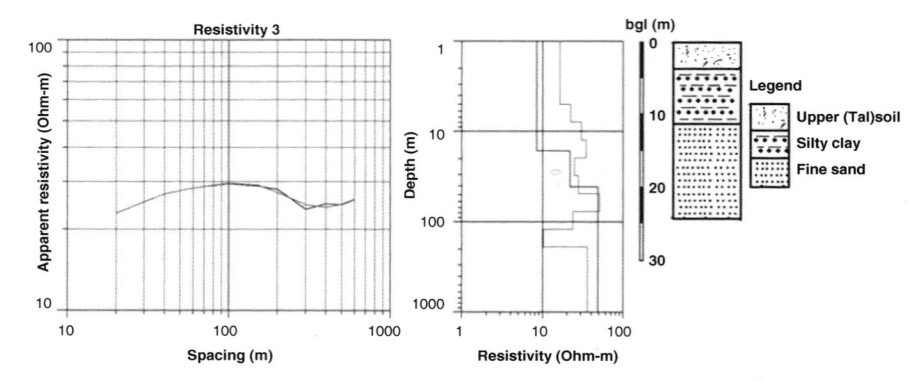

Fig. 2.5 Comparison between apparent and interpretive resistivity with the lithology of the borehole near to the point VES 3. *Thick black lines* denoting true/calculated resistivity value

surface layer soil has higher salinity because of secondary salts like chlorides and sulphates coming from organic matter breakdown and agricultural runoff carrying unutilized sulphates which ultimately result in to lower value of resistivity. On the other hand, subsurface lithology is dominated with sand and silt with high porosity resulting in to high resistivity. For points 3 and 4, since both of them are located at meandering of river i.e. active flood plain, they show low value of resistivity because of the presence of clay mineral in shallow subsurface layer (which is ranging 6–15 m below ground level in this case). To validate this result, borehole lithology (taken near point 3) is presented along with the resistivity curve (Fig. 2.5). It was found that VES curve is consistent with the subsurface structure. Resistivity value was very low (>10 Ω m) up to a depth of 16 m approximately due to presence of humus dominated top surface layer followed by thick layer of clay. From 16-m downward there is a huge deposition of homogenous fine sand layer which can be shown by increasing trend of resistivity value (Paranis 1997) with an exception of sudden decrease in resistivity value which might be because of thin layer of old alluvium which forms perched aquifer for few metres. Points 5 and 6 show sudden lateral increase in the value of resistivity mainly because of presence of potential aquifer and presence of sandstone. However, for point 6, through the examination for depth profile of resistivity, it was found that value of resistivity decreases after a certain depth showing presence of clay layer. Lithological evidence indicates the presence of pyrites and dolomites in the sandstones which can contribute arsenic to groundwater.

For point 9, because of bare soil with very high rate of evaporation, upper horizon of soil has high salinity, which was well supported by relatively lower value of resistivity. The pH of the groundwater varied in the range at 8.1 ± 0.1 (Table 2.1). It indicates that when iron is more the arsenic is retained on the iron compound and during favourable redox conditions ferric arsenate gets aerated and releases arsenic in water.

Table 2.1 Statistical analysis for water samples of Bhagalpur

Parameters	Unit	Minimum	Maximum	Average	St. Dev.	P.L. WHO (2004)
Depth	(ft)	30.0	95.0	55.7	34.6	NA
pH		7.8	8.3	8.1	0.1	9.2
ORP	mv	−134.0	169.0	8.9	95.2	NA
EC	μs/cm	250.0	980.0	599.2	175	1,400
TDS	mg/L	191.2	710.0	455.2	133	NA
Na^+	mg/L	5.7	41.9	18.7	8.54	200
K^+	mg/L	0.4	3.2	1.4	0.7	150
Ca^{+2}	mg/L	20.7	126.0	64.8	30.3	200
Mg^{+2}	mg/L	7.8	18.2	10.7	2.2	150
HCO_3^-	mg/L	15.2	214.5	104.8	42.2	NA
F^-	mg/L	0.1	4.9	0.8	0.9	1.5
Cl^-	mg/L	6.6	219.9	69.7	50.8	600
NO_3^-	mg/L	1.0	39.0	23.7	11.6	45
SO_4^{2-}	mg/L	3.8	72.8	32.9	20.2	400
PO_4^{3-}	mg/L	2.7	6.4	4.0	0.9	150
SiO_2	mg/L	14.0	49.3	30.0	8.1	NA
NH_4^+	mg/L	0.6	3.1	1.4	0.8	NA
Fe	mg/L	0.7	7.6	3.2	2.2	0.3
Mn	mg/L	0.01	1.8	0.7	0.6	0.1
As(tot)	μg/L	19.1	118.0	51.2	27.6	10
As^{3+}	μg/L	10.9	81.1	34.4	21.0	10
As^{5+}	μg/L	7.1	55.2	16.8	10.3	10
$\delta^{18}O$	Per mil	−9.61	−7.15	−8.36	0.62	NA
δ^2H	Per mil	−65.47	−49.89	−56.27	4.07	NA

Note: P.L. Permissible limit, *NA* Not applicable

2.4.1 Arsenic and Its Speciation

The statistics of general water geochemistry is given in Table 2.1. Total arsenic concentrations in the groundwater varied from 19.1 μg/L to 118 μg/L with an average value of 51.23 μg/L (Table 2.1). Distribution pattern of arsenic and iron in this area is shown in Figs 2.6 and 2.7 respectively, from which it is observed that higher concentration of both the elements is clustered near river Ganges and its tributary Koshi (Ghugri). From XRD study of the soil samples, it was found that goethite, dolomite, calcite, quartz and feldspar are the major minerals for most of the samples. Speciation modeling (calculated with the help of PHREEQC code) results for selected minerals in specified water samples is shown in Table 2.2. The result shows that the value for log PCO_2 is low for most of samples i.e. all selected samples are undersaturated with respect to it. Most samples are saturated with respect to calcite and dolomite, except for some samples from Golgamma, Panchrukhi, Marwa and

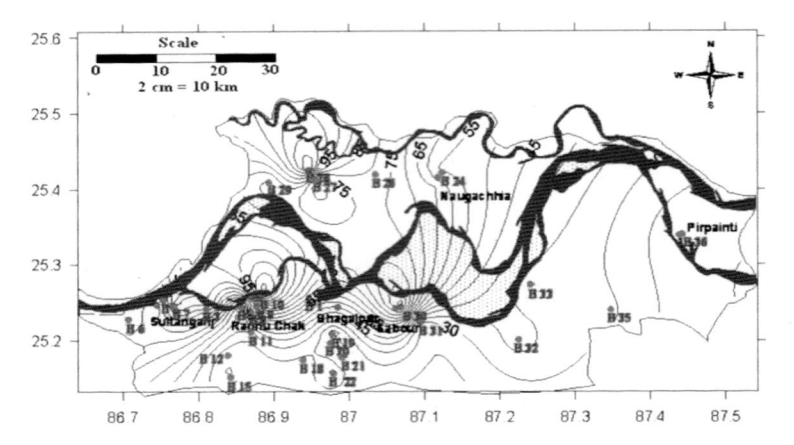

Fig. 2.6 Contour diagram showing spatial distribution of arsenic concentration (μg/L) (Kumar et al. 2010)

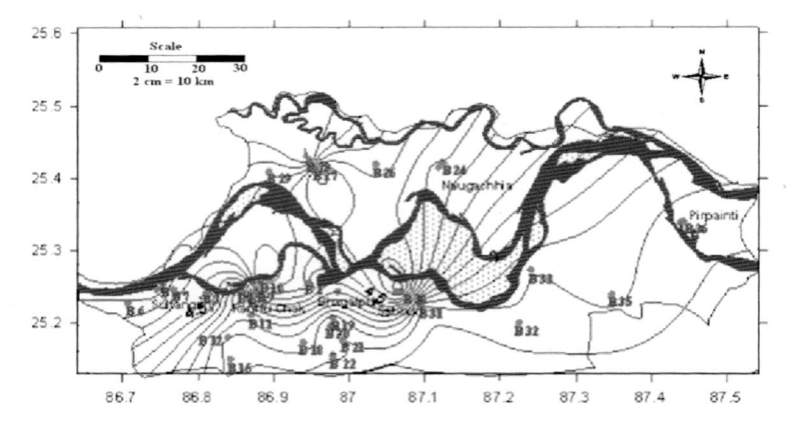

Fig. 2.7 Contour diagram showing spatial distribution of iron concentration (mg/L)

Mahesitilakpur. Groundwater is undersaturated with respect to poorly crystalline minerals like $FeCO_3$ and $Fe(OH)_3$, but saturated with respect to goethite mineral. Hence goethite is suggested as the sink for dissolved iron minerals which precipitates during oxidation and may result in the de-coupling of As and Fe (Mukherjee et al. 2008). Here redox coupling (As (III)/As (V)) of As associated with Fe is assumed to govern arsenic mobilization in groundwater. Oxyhydroxide reduction mechanism driven by microbial activity degrades the arsenate to arsenite and triggers the mobilization.

So presence of iron mineral associated with clay results in high conductance and thus low resistivity reaffirm the conclusion made for relatively lower resistivity value in case of clay layer at shallow depth for points 3 and 4. Most of the arsenic compounds are present in the form of sulfides or as arsenide of iron, copper, nickel

Table 2.2 Speciation modelling result for selected minerals in specified water samples

Sample/SI	PCO$_2$	Calcite	Dolomite	FeCO$_3$	Goethite	Fe(OH)$_3$
Golgamma	−1.447	−0.332	−1.611	−2.557	−0.172	−0.053
Panchrukhi	−1.898	−0.182	−0.735	−1.608	−0.436	−1.708
Merwa	−2.457	−0.277	−0.223	−2.263	−0.300	−1.788
Mahesitiakpur	−0.681	−0.006	−0.630	−0.770	−0.116	−1.818
Sultanganj	−0.614	−2.415	−1.240	−0.599	−0.619	−2.048
Bhagalpur	−1.858	−1.648	−0.458	−0.382	−0.530	−1.917
Sabour	−3.406	−0.845	−0.476	−3.371	−2.071	−2.186
Rannuchak	−2.146	−1.441	−2.462	−1.316	−1.430	−1.598
Pirpainti	−2.475	−1.761	−0.308	−0.360	−1.470	−2.135
Naugachhia	−1.708	−1.542	−0.521	−0.925	−1.869	−2.137

and cobalt ores, into the thick sedimentary deposit such as clay and their mobilization is being governed by the active terminal electron accepting processes (TEAPs) as reported by Bhattacharya et al. (2009).

2.4.2 Relationship Between VES and Hydrochemical Signature of the Aquifer

When the resistivity profile at point 3 was compared with the contour for both As and Fe, a good correlation was found which suggests that VES values can be used to delineate As-contaminated aquifer zone. The low resistivity which represents very fine silt and clay soil in subsurface is believed to be deposited during meandering in a floodplain environment and thus has a strong anoxic condition to result in the negative Eh value along with high As and Fe concentrations in groundwater. Thus such clay layer at shallow depth with abundant iron minerals along with high concentration of arsenic can be detected by geophysical as well as geochemical techniques.

Combined with the geochemical composition, this study reveals that the controlling factor of arsenic in groundwater is redox condition and iron content of the aquifer and release of arsenic may depend on oxidation followed by reduction due to vertical movement of water. Finally it can be concluded that "oxyhydroxide reduction theory" is responsible for release of arsenic in the aquifer of Bhagalpur.

2.4.3 Stable Isotopic Chemistry

Relationship between $\delta^{18}O$ and δD values for groundwater samples is shown in Fig. 2.8. Value for $\delta^{18}O$ ranges from −9.61 to −7.15‰ with a median value of −8.36‰, whereas for δD it ranges from −65.47 to −49.89‰ with a median value of

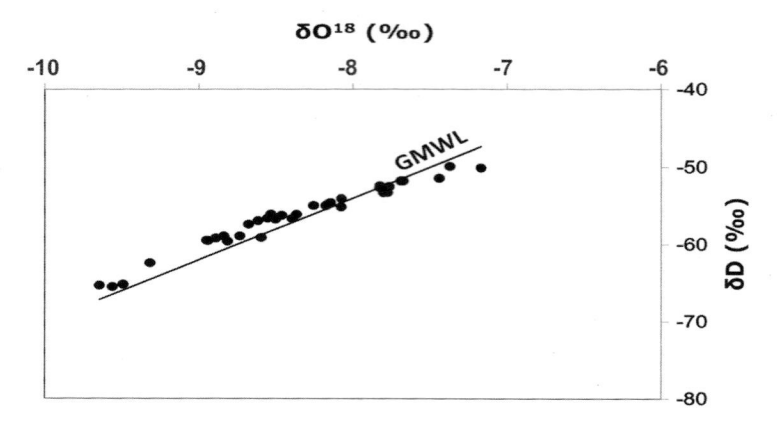

Fig. 2.8 Bivariate plot of $\delta^{18}O$ and δD values for groundwater samples

−56.27‰. Most of the result points were plotted near the GMWL (drawn from the equation $\delta^2H = 8\delta^{18}O + 10$ (Craig 1961)) except fewer samples but the value range is huge. Groundwater samples falling along the global meteoric water line (GMWL) indicates that origin of ground water is meteoric and non-evaporated water rapidly infiltrated to the saturated zone. High range of value may be due to continental effects. However point away from the GMWL might be because of non-equilibrium fractionation during evaporation and exchange with carbonaceous rock minerals as in case of younger flood plain (Tsujimura et al. 2007). Groundwater samples with highest $\delta^{18}O$ values ranging between −7.00‰ and −8.00‰ are mainly representing samples from calcareous young alluvial plain region. Therefore, the high As concentration in the groundwater is mainly because of infiltration of river water through contaminated river bed sediment in this region. Finally it can be concluded that depositional environment and geological age are also very important factors in controlling arsenic mobilization.

2.5 Conclusion

This study reveals that resistivity survey is useful for predicting subsurface geological formation. It also gives information about soil salinity, water content, humus and texture especially presence of clay, silt and sand stone content. It is observed that at shallow depth i.e. low resistivity zone (>10 Ω m), redox reactions support the release of As from arseniferous iron oxyhydroxides in the groundwater and the given facts were also validated through geochemical characterization of water samples (Kumar et al. 2010) and the borehole lithology (silty clay). For observation site, though the subsurface distribution pattern of clay layer is very irregular but in general after 40-m depth there is a thick layer of fine sand which can be labelled as potentially safe zone for groundwater extraction. This important finding can help the local driller/local administration in many ways to target arsenic safe aquifer if resistivity survey will be done on extensive spatial scale. Result for stable isotopic composition shows that most of sample points plotted near the global meteoric water line (GMWL) i.e. origin of ground water is meteoric in principle; however point away from the LMWL might favour exchange with rock minerals and evaporation processes. Younger alluvial plain water samples with distinguished isotopic values also supporting the idea that depositional environment and aquifer structure are main driving force behind arsenic mobilization.

Acknowledgement Authors would like to thank Indian Council of Medical Research (ICMR), Government of India for giving fellowship and grant for this research work. Without fail, we are also grateful to School of Environmental Sciences, Jawaharlal Nehru University, New Delhi, India for providing the Central Instrumentation Facility to complete all analytical works.

References

Acharyya SK (2005) Arsenic levels in groundwater from Quaternary alluvium in the Ganga plain and the Bengal basin, Indian subcontinent: insight into influence of stratigraphy. Gondwana Res 8:1–12

Acharyya SK, Shah BA (2004) Risk of arsenic contamination in groundwater affecting Ganga Alluvial Plain, India. Environ Health Perspect 112:A19–A20

Ahmed KM, Bhattacharya P, Hasan MA, Akhter SH, Alam MA, Bhuyian H, Imam MB, Khan AA, Sracek O (2004) Arsenic enrichment in groundwater of the alluvial aquifers in Bangladesh: an overview. Appl Geochem 19:181–200

American Public Health Association (APHA) (1995) Standard methods for the examination of water and wastewater, 19th edn. American Public Health Association, Washington, DC

Aziz Z, van Geen A, Stute M, Versteeg R, Horneman A, Zheng Y, Goodbred S, Steckler M, Weinman B, Gavrieli I, Hoque MA, Shamsudduha M, Ahmed KM (2008) Impact of local recharge on arsenic concentrations in shallow aquifers inferred from the electromagnetic conductivity of soils in Araihazar. Bangladesh Water Resour Res 44, W07416. doi:10.1029/2007WR006000

Banton O, Seguin MK, Cimon MA (1997) Mapping field-scale physical properties of soil with electrical resistivity. Soil Sci Soc Am J 61:1010–1017

Ben DS, Berner Z, Chandrasekharam D, Karmakar J (2003) Arsenic enrichment in groundwater of West Bengal, India: geochemical evidence for mobilization of As under reducing conditions. Appl Geochem 18:1417–1434

Bhattacharya P, Chatterjee D, Jacks G (1997) Occurrence of arsenic contamination of groundwater in alluvial aquifers from Delta Plain, Eastern India: option for safe drinking supply. Int J Water Res Dev 13:79–92

Bhattacharya P, Hasan MA, Sracek O, Smith E, Ahmed KM, von Brömssen M, Huq SMI, Naidu R (2009) Groundwater chemistry and arsenic mobilization in the Holocene flood plains in south-central Bangladesh. Environ Geochem Health 31:23–44

Chakraborti D, Samanta MK, Rahman MM, Ahmed S, Chowdhury UK, Hossain MA, Mukherjee SC, Pati S, Saha KC, Dutta RN, Quamruzzaman Q (2004) Groundwater arsenic contamination and its health effects in the Ganga-Meghna-Brahmaputra plain. J Environ Monit 6:74N–83N

Charlet L, Chakraborty S, Appelo CAJ, Roman-Ross G, Nath B, Ansari AA, Lanson M, Chatterjee D, Mallik B (2007) Chemodynamics of an arsenic "hotspot" in a West Bengal aquifer: a field and reactive transport modeling study. Appl Geochem 22:1273–1292

Chauhan D, Nickson R, Iyengar L, Sankararamakrishnan N (2009) Groundwater geochemistry and mechanism of mobilization of arsenic into the ground water of Ballia District, U.P. India. Chemosphere 75(1):83–89

Craig H (1961) Isotopic variations in meteoric waters. Science 133:1702–1703

Gabel T (2000) Confounding variables in the environmental toxicology of Arsenic. Toxicology 144:155–162

Harvey C, Swartz CH, Badruzzaman ABM, Keon-Blute NE, Yu W, Ashraf AM, Jay J, Beckie R, Niedam V, Brabander DJ, Oates PM, Ashfaque KN, Islam S, Hemond HF, Ahmed MF (2002) Arsenic mobility and groundwater extraction in Bangladesh. Science 298:1602–1606

Hoque MA, Khan AA, Shamsudduha M, Hossain MS, Islam T, Chowdhury SH (2009) Near surface lithology and spatial variation of arsenic in the shallow groundwater: Southeastern Bangladesh. Environ Geol 56:1687–1695

Khan S, Rana T, Hanjra MA (2008) A cross disciplinary framework for linking farms with regional groundwater and salinity management targets. Agric Water Manag 95:35–47

Kumar P, Kumar M, Ramanathan AL, Tsujimura M (2010) Tracing the factors responsible for arsenic enrichment in groundwater of the middle Gangetic Plain, India: a source identification perspective. Environ Geochem Health 32:129–146

Kumar P, Iwagami S, Yaping L, Mikita M, Tanaka T, Yamanaka T (2011) Multivariate approach for surface water quality mapping with special reference to nitrate enrichment in Sugadaira, Nagano Prefecture (Japan). Environmentalist 31:358–363

Mazac O, Cislerova M, Kelly WE, Landa I, Venhodova D (1990) Geotechnical and environmental geophysics. In: Ward SH (ed) Geotechnical and environmental geophysics. vol. 2. Environmental and groundwater applications. Society of Exploration Geophysicists, Tulsa

Mukherjee S, Sashtri S, Gupta M, Pant MK, Singh C, Singh SK, Srivastava PK, Sharma KK (2007) Integrated water resource management using remote sensing and geophysical techniques: Aravali quartzite, Delhi, India. J Environ Hydrol 15(10):1–10

Mukherjee A, von Brömssen M, Scanlon BR, Bhattacharya P, Fryar AE, Hasan MA, Ahmed KM, Chatterjee D, Jacks G, Sracek O (2008) Hydrogeochemical comparison and effects of overlapping redox zones on groundwater arsenic near the Western (Bhagirathi sub-basin, India) and Eastern margins (Meghna sub-basin, Bangladesh) of the Bengal Basin. J Contam Hydrol 99(1-4):31–48

Nath B, Mallik SB, Stuben D, Chatterjee D, Charlet L (2010) Electrical resistivity investigation of the arsenic affected alluvial aquifers in West Bengal, India: usefulness in identifying the areas of low and high groundwater arsenic. Environ Earth Sci 60:873–884

National Academy Press (2001) Arsenic in drinking water: 2001 update. In: Goyer R (ed) Sub-committee to update the 1999 Arsenic in drinking water report. National Academy Press, Washington, DC

Nickson RT, McArthur JM, Burgess WG, Ahmed KM, Ravenscroft P, Rahman M (1998) Arsenic poisoning of Bangladesh groundwater. Nature 395:338

Nickson RT, McArthur JM, Ravenscroft P, Burgess WG, Ahmed KM (2000) Mechanism of arsenic release to groundwater, Bangladesh and West Bengal. Appl Geochem 15:403–413

Paranis DS (1997) Principles of applied geophysics. Chapman & Hall, London

Raju NJ, Reddy TVK (1998) Fracture pattern and electric resistivity studies for groundwater exploration. Environ Geol 34:175–183

Rhoades JD, Schilfgaarde JV (1976) An electrical conductivity probe for determining soil salinity. Soil Sci Soc Am J 40:647–650

Rhoades JD, Shouse PG, Alves WJ, Manteghi NA, Lesch SM (1990) Determining soil salinity from soil conductivity using different models and estimates. Soil Sci Soc Am J 54:46–54

Sankararamakrishnan N, Chauhan D, Nickson RT, Iyengar L (2008) Evaluation of two commercial field test kits used for screening of groundwater for arsenic in Northern India. Sci Total Environ 401:162–167

Shah BA (2008) Role of Quaternary stratigraphy on arsenic-contaminated groundwater from parts of Middle Ganga Plain, UP–Bihar, India. Environ Geol 35:1553–1561

Singh IB (2004) Late Quaternary history of the Ganga Plain. J Geol Soc India 64:431–454

Smith AH, Lingas EO, Rahman M (2000) Contamination of drinking water by arsenic in Bangladesh: a public health emergency. Bull World Health Organ 83:177–186

Srivastava AK, Govil PC, Tripathi RM, Shukla RS, Srivastava RS, Vaish DP, Nickson RT (2008) Initial data on arsenic in groundwater and development of a state action plan, Uttar Pradesh, India. In: Bhattacharya P, Ramanathan AL, Mukherjee AB, Bundschuh J, Chandrasekharam D, Keshari AK (eds) Groundwater for sustainable development: problems, perspectives and challenges. Taylor and Francis/A. A, Balkema

Tsujimura M, Abe Y, Tanaka T, Shimada J, Higuchi S, Yamanaka T, Davaa G, Oyunbaatar D (2007) Stable isotopic and geochemical characteristics of groundwater in Kherlen River basin, a semi-arid region in eastern Mongolia. J Hydrol 333(1):47–57

WHO (World Health Organization) (2004) Guidelines for drinking water quality: training pack. WHO, Geneva

Wilding LP, Smeck NE, Hall GF (1983) Pedogenesis and soil taxonomy. I. Concepts and interactions, Developments in soil science, 11A. Elsevier, Amsterdam/Oxford/New York

Yadav GS, Singh SK (2007) Integrated resistivity surveys for delineation of fractures for ground water exploration in hard rock areas. J Appl Geophys 62(3):301–312

Chapter 3
Arsenic Contamination in Groundwater in the Middle Gangetic Plain, India: Its Relations to Fluvial Geomorphology and Quaternary Stratigraphy

Babar Ali Shah

3.1 Introduction

Pollution of groundwater by naturally occurring arsenic (As) is found in sedimentary aquifers worldwide and health problems associated with groundwater As have been documented in many parts of the world such as Bangladesh, India, Pakistan, Nepal, China, Hungary, Vietnam, Thailand, Cambodia, Taiwan, Inner Mongolia, Ghana, Egypt, Japan, Argentina, Mexico, USA and Chile (Mandal and Suzuki 2002; Mukherjee et al. 2006; Ravenscroft et al. 2009). Long-term intake of As-contaminated groundwater above 50 µg/L has caused skin diseases (pigmentation, dermal hyperkeratosis, skin cancer), cardiovascular, neurological, hematological, renal and respiratory diseases, as well as lung, bladder, liver, kidney and prostate cancers (Smith et al. 1992, 1998). The mode of occurrence, origin and mobility of As in sedimentary aquifers are influenced by local geology, geomorphology, hydrogeology and geochemistry of sediments (Bhattacharya et al. 1997; Nickson et al. 1998; Acharyya et al. 2000; Kinniburgh and Smedley 2001). The upper permissible limit of As in drinking water is 10 µg/L as per WHO guideline (WHO 1993), which has been endorsed by Bureau of Indian Standards (BIS 2003). India is the second most populated country in the world, where a large percentage of world's population (17.5 %) is living on limited land area (2.4 %). The Gangetic plain is one of the vast Quaternary alluvium track in Asia, and many cities, towns, villages, and hamlets are located on the bank of the Ganga and Ghaghara rivers. The Ghaghara river has originated from Matsatung glacier in the Himalayas and travelled a distance of about 1,080 km in NW–SE direction, to join the Ganga River. The topography of the Gangetic plains are heterogeneous, varying between upland surfaces, plain areas and low lying small natural bodies, viz., swamp and ponds. Recently, groundwater As contamination has

---

B.A. Shah (✉)
Department of Geological Sciences, Jadavpur University,
Kolkata 700032, West Bengal, India
e-mail: bashahju@yahoo.com

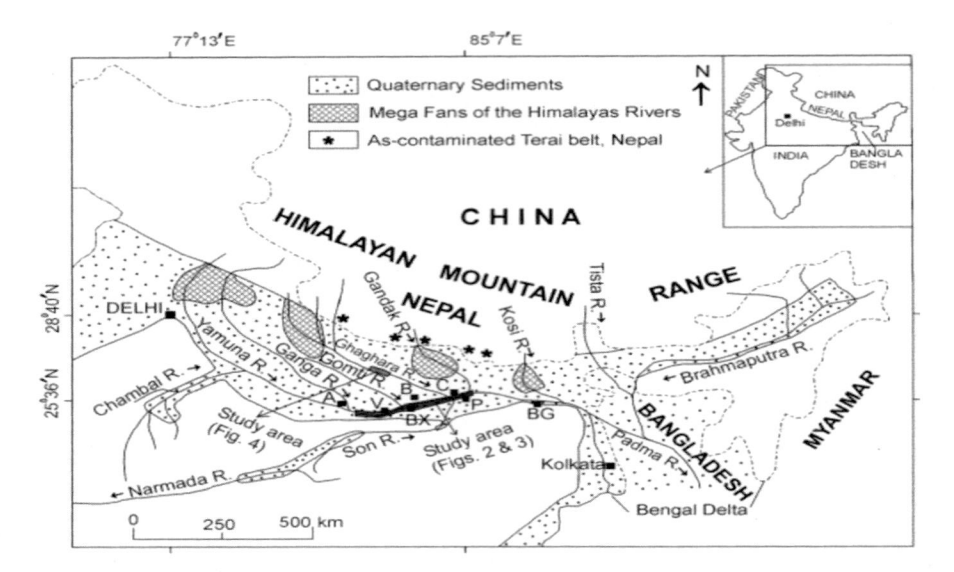

Fig. 3.1 Quaternary sediments in the Indo-Ganga foredeep and Bengal Basin. The study areas from parts of the Middle Gangetic Basin and the Ghaghara Basin are shown in Figs. 3.2, 3.3, and 3.4. Abbreviations: *A* Allahabad, *V* Varanasi, *BX* Buxar, *B* Ballia, *C* Chhapra, *P* Patna, *BG* Bhagalpur

been reported in the states of Uttar Pradesh (UP), Bihar and Jharkhand (Chakraborti et al. 2003; Acharyya and Shah 2004; Bhattacharjee et al. 2005; Ahamed et al. 2006; Shah 2008). In this study, a survey on arsenic content in groundwater was carried out in Mirzapur, Varanasi, Ghazipur, Ballia, Buxar, Bhojpur, Patna, Vaishali, Faizabad, Gonda and Basti districts of UP and Bihar in the Gangetic plain (Fig. 3.1).

Moreover, groundwater As in tubewells were tested within the Holocene Newer Alluvium aquifers, as well as the Pleistocene Older Alluvium aquifers. The main objective of study is to investigate the distribution of groundwater As in entrenched channels and flood plains of the Ganga and Ghaghara rivers system under Quaternary geomorphologic setting.

3.2 Study Area

The study areas in the entrenched channels and floodplains of the Ganga and Ghaghara rivers system are shown in three parts viz., Buxar, Ghazipur, Varanasi and Mirzapur districts (Fig. 3.2), Vaishali, Patna, Bhojpur and Ballia districts (Fig. 3.3), and Faizabad, Gonda and Basti districts (Fig. 3.4). The geomorphologic and Quaternary morphostratigraphic maps in the Middle Gangetic Basin (85° 21′ 11.80″ E to 82° 27′ 23.56″ E) and the Ghaghara Basin (82° 00′ E to 82° 30′ E) were prepared based on Survey of India topographic sheets of 1:50,000 scale. The geomorphic features of the Gangetic plains show differences in their elevations, spatial

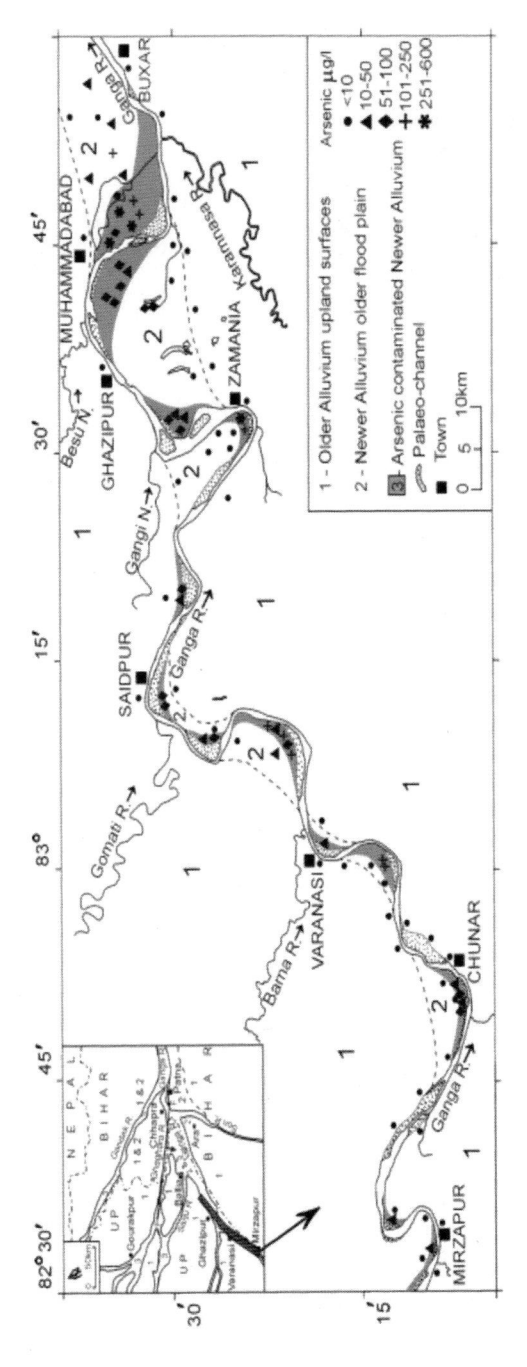

Fig. 3.2 Groundwater As distribution in entrenched channels and floodplains of the Ganga River from Buxar to Mirzapur towns

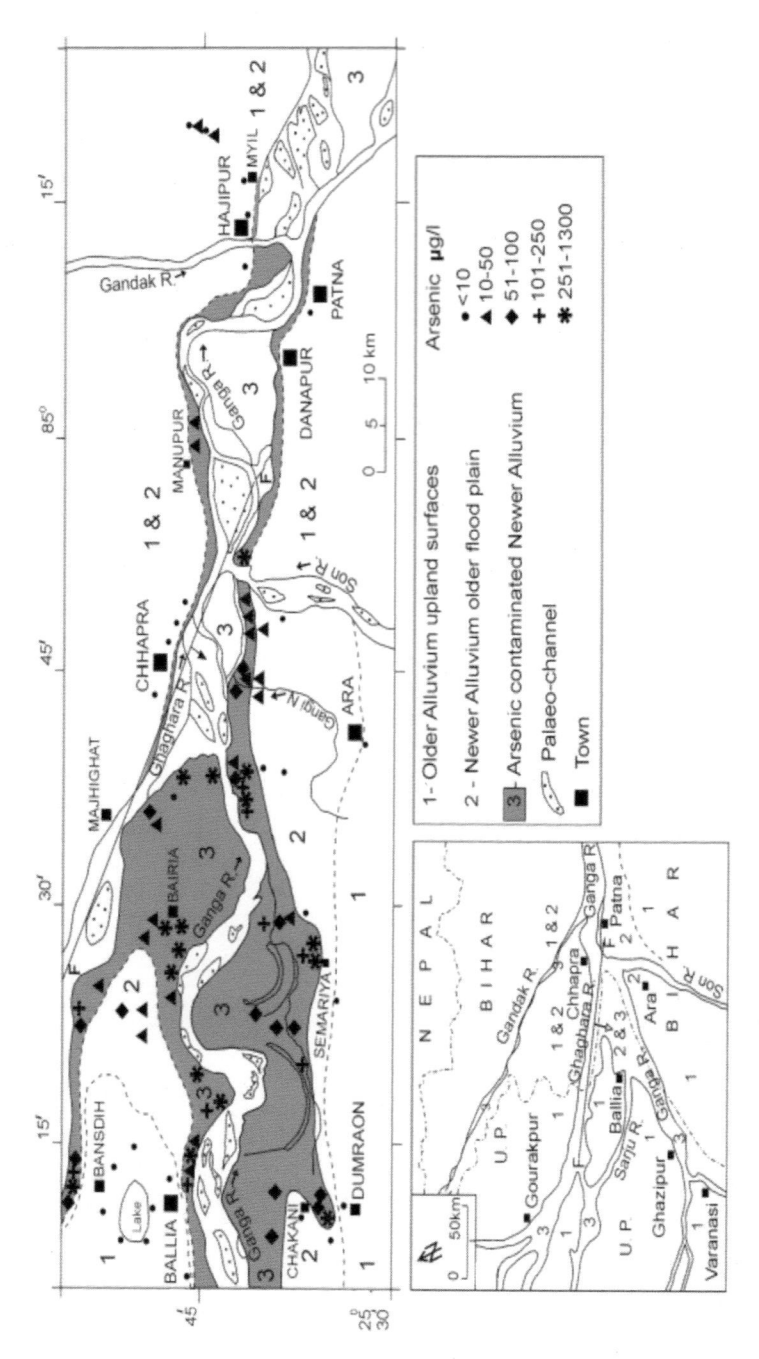

Fig. 3.3 Groundwater As distribution in entrenched channels and floodplains of the Ganga and Ghaghara rivers system from Patna to Ballia towns

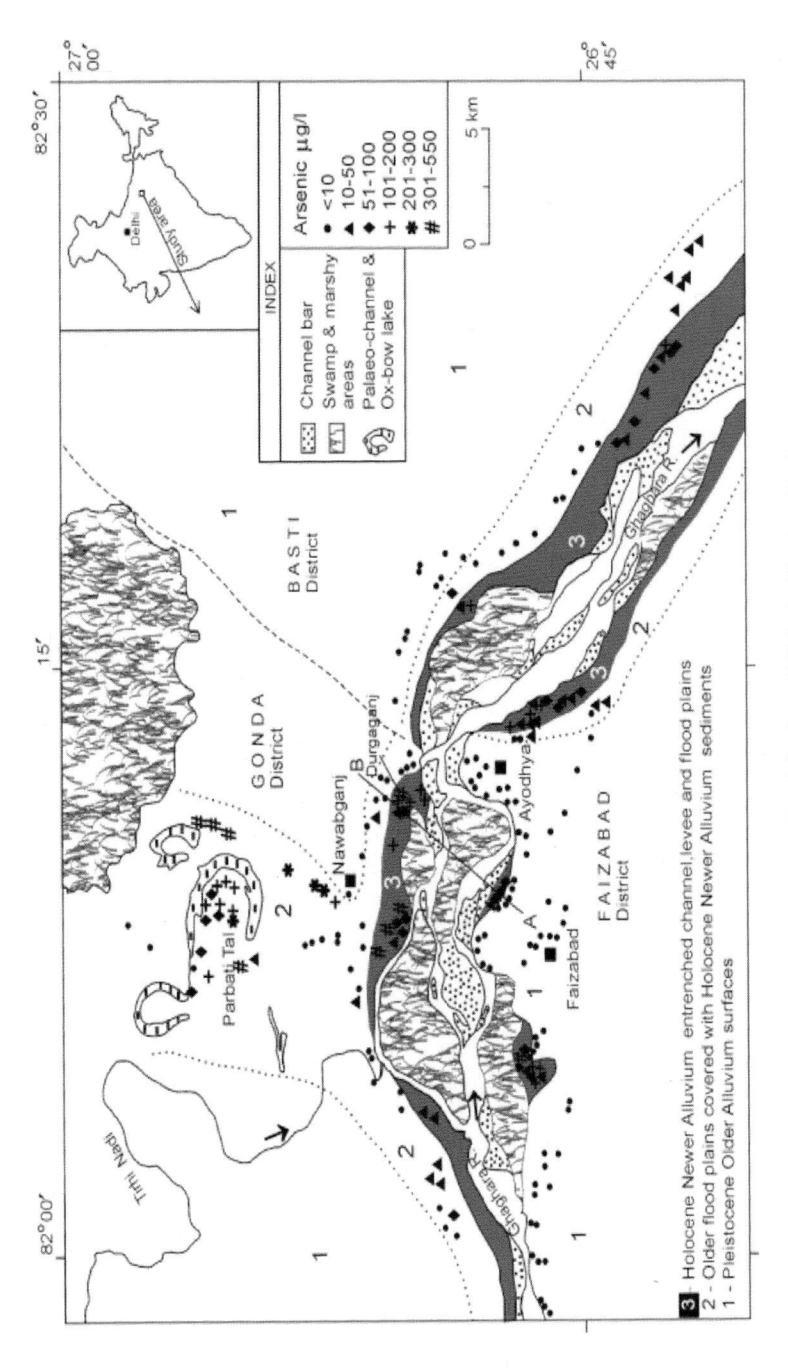

Fig. 3.4 Groundwater As distribution in Faizabad, Gonda and Basti districts of UP in the Ghaghara Basin

distribution and nature of sediment, indicating that their formation must have occurred at different times, under different climates, water budget and sediment supply during the Pleistocene-Holocene period (Singh 2004). The Gangetic plains are subdivided into three zones i.e., the Pleistocene Older Alluvium upland Inter-flue surface (47 ± 12 ka), the Holocene Newer Alluvium river valley terrace surface (3 ± 1 ka), and the Holocene to Recent active channels and flood plains (Srivastava et al. 2003).

The major parts of the Gangetic plains consist of inter-fluve upland surfaces of the Older Alluvium. These Pleistocene Older Alluvium inter-fluve upland surfaces are characterised by yellow-brown coloured sediments with profuse calcareous and ferruginous concretions, and is exposed or occurs under shallow cover of the Holocene sediments. The Holocene Newer Alluvium surfaces are characterised by organic rich grey to black coloured argillaceous sediments in entrenched channels and floodplains of the Gangetic rivers. In this study, approximately 6,500 km^2 (Figs. 3.2 and 3.3) in the Middle Gangetic Basin, and approximately 1,500 km^2 (Fig. 3.4) in the Ghaghara Basin have been mapped to delineate groundwater As-contaminated and As-safe areas. This work is mainly compilation of earlier data from the Middle Gangetic Basin and the Ghaghara Basin under Quaternary geomorphologic setting.

3.3 Geological Setting

The Quaternary geology of the Indo-Gangetic plain has been discussed by Pascoe (1964) who sub-divided the sediments into Older Alluvium and Newer Alluvium. This classification was revised by Pathak et al. (1978) as Upper Siwalik (Upper Pliocene to Lower Pleistocene), Older Alluvium (Middle to Upper Pleistocene), and Newer Alluvium (Upper Pleistocene to Recent) in order of superposition. The first cycle of Quaternary sedimentation in the post-Middle Siwalik basin probably started during Upper Pliocene/Lower Pleistocene with sediment supply in the southern part coming from the peninsular provenance leading to the deposition of variegated clays followed by clastic sediments overlying Chitrakoot Formation (Dwivedi and Sharma 1992). The Banda Group is the oldest Quaternary lithostratigraphic unit in the southern part of the Indo-Gangetic Plain. The sediments formed Bandra Older Alluvium, which is characterized by Siwalik boulder conglomerate, sandstone, siltstone, quartzite that are embedded with yellowish to brownish coloured matrix (Pathak et al. 1978; Kumar et al. 1996).

A new basin was evolved due to Middle Pleistocene Orogenic movement with upliftment and folding in the northern part of the basin and development of the Himalayan foothill faults. This basin gradually shifted towards south due to subsidence and received sediments supply from the northern Himalayan provenance. The second cycle of sedimentation during Middle to Early Upper Pleistocene period formed under warm, humid and dry climate (Dwivedi and Sharma 1992) has been referred here as the Varanasi Alluvium. The Varanasi Older Alluvium constitutes upland surfaces, which are occupying major parts of the Gangetic Plain. The Varanasi Older Alluvium consists dominantly of multiple fill polycyclic sequence of sand, silt and clay. The Pleistocene Older Alluvium sediments are recognized by yellow-brown coloured sediments with profuse calcareous and ferruginous concretions (Kumar et al. 1996). The river Ganga draining the area has developed a second geomorphic unit, the low-lying floodplains. The second-generation sediments in the Holocene period were deposited within the floodplain domain of each river defined by their palaeobank. These sediments constitute the Newer Alluvium. The main geomorphic features are giant channel bar, natural levee, and floodplains. It is the youngest geomorphic surface present on all other surfaces of the Gangetic plain. The Holocene Newer Alluvium comprise grey to black coloured organic-rich argillaceous sediments in entrenched channels and floodplains of rivers in the Gangetic plain. The top thin layer of silt-clay has been deposited in a lacustrine condition towards the close of sedimentation. The Gangetic plain exhibits variety of landforms, namely incised river valleys, abandoned channels, palaeo-channels, alluvial ridges, ponds, lakes etc. (Singh 2004).

The valley of Ghaghara river is very wide (5–15 km) with narrow channel (100–200 m during summer). The Ghaghara Basin consists of thick pile of Quaternary sediments. Geomorphologically, the Older Alluvium surfaces i.e. the Varanasi Older Alluvium constitute upland surfaces, which are also recognised as the Bhangar surface and are occupying major parts of the area. In the Ghaghara Basin, the top 3 m part is of silt-clay facies corresponding to suspension load of flood plain environment underlain by arenaceous facies corresponding to channel-fill palaeoenvironment. The Older Alluvium and Newer Alluvium deposits in the Gangetic plain has been shown in Table 3.1.

3.4 Materials and Methods

In this study, hand tubewell water samples were collected in acid pre-washed 10 ml plastic bottles. Immediately after collection, one drop of dilute nitric acid (1:1) GR Grade was added as preservative. The geographic locations of all samples were recorded by a hand-held Global Positioning System (Garmin eTrex Vista). The information of tubewell depth was acquired from owner of the tubewell. The Older and Newer Alluvium sediment samples were collected from vertical sections of the Ganga river and its tributaries. All samples were collected from 15 to 20 cm inside of vertical sections, and GPS locations of all samples were recorded. Each sediment sample was dried in an oven at about 50 °C for approximately 24 h. Sediment samples were ground homogenously and crushed with a mortar and pestle. Analysis of As and iron (Fe) were carried out both from tubewell water and sediment samples. Sediment sample weighing 0.5 g was taken in a Teflon bomb and 2 ml nitric acid was added with it. It was digested at 120 °C temperature for 8 h in hot air oven. After cooling, 5 ml distilled de-ionised water was added in digested sample and it was filtered through 0.45 μm millipore filter and prepared volume 10 ml. During

Table 3.1 Classification of Quaternary deposits in the Indo-Gangetic Basin

Age	Lithounit	Distribution/Lithology
Holocene	Channel alluvium	Active channel deposits of light grey, fine to medium grained micaceous sand and pebbles-cobbles (Himalayan provenance) and red quartzo-feldspathic medium to coarse sand along with drapes of silt and clays (Peninsular provenance)
	Terrace alluvium	Older flood plain deposits (Terraces) composed of grey fine grained micaceous sand with silt beds in the Himalayan rivers
	Alluvium fan deposits	Occur bordering the Himalayan hill front and composed of sediments ranging from cobbles-pebbles to fine sand and silt
		Disconformity
Pleistocene	Varanasi alluvium	Polycyclic sequence of clay, silt and micaceous sand of the Himalayan provenance with *Kankar* and ferruginous concretions
		Unconformity
	Banda alluvium	(a) Chitrakoot Formation: medium to coarse quartzo-feldspathic and silt, clay of the Peninsular provenance with minor *Kankar*
		(b) Variegated clays
		Unconformity
Pre-Quaternary	Vindhyan Supergroup	Upper Siwalik Boulder Conglomerate
		Bundelkhand Gneissic Complex in south and (? Marwar Supergroup)
	Aravalli group	Lower and Middle Siwalik groups in north

Modified after Khan et al. (1996)

analysis, 5 ml acidified sample was added with 10 ml buffer, 5 ml hydroxylamine hydrochloride, and 2.5 ml phenanthroline and a final volume 25 ml was prepared. The prepared solution was used for As and Fe analyses.

Iron was analyzed by 1,10 phenanthroline method by the use of UV spectrophotometer. Arsenic analysis was done at the laboratory through flow injection hydride generation atomic absorption spectrometry (FI-HG-AAS) system. (A Perkin-Elmer Model 3,100 atomic absorption spectrometer equipped with a Hewlett-Packard Vector Computer with GEM software.) The minimum detection limit with 95 % confidence level was 3 µg/L As (Samanta et al. 1999). A solution of 1.25 % $NaBH_4$ (Merck, Germany) was prepared in 0.5 % NaOH (Merck, India). A 5.0 M solution of HCl (Merck, India) was used. All reagents are Analar grade. All these solutions were prepared using distilled de-ionised water. The flow rate for both tetrahydroborate and hydrochloride acid was 1 ml/min. Blank was prepared and measured under the same conditions. Details of the reagents and glassware were given elsewhere (Samanta et al. 1999). The accuracy of analytical method using FI-HG-AAS was verified by analyzing Standard Reference Materials CRM (BND 301) NPL, Indian water (certified value 990 ± 20 µg/L; observed 960 ± 40 µg/L); Water SRM (quality control sample for trace metal analysis) from USEPA Environmental Monitoring and Support Laboratory, Cincinnati, Ohio, USA (certified value 17.6 ± 2.21 μg/L; observed 16 ± 3.5 μg/L). The X-ray Diffractometry (XRD) with Cu target was used (Model Philips APD-15) to determine the constituent minerals in sediment sample.

3.5 Results and Discussion

3.5.1 Distribution of Groundwater Arsenic in Middle Gangetic Basin

A result of As concentrations of 144 tubewells from Buxar, Ghazipur, Varanasi and Mirzapur districts in the Middle Gangetic Plain (Fig. 3.2) has been shown in Table 3.2.

About 66 % of tubewells have As concentrations above 10 μg/L (WHO guideline), and 36 % of tubewells have As above 50 μg/L. Maximum concentrations of As and Fe in tubewell water are 550 μg/L and 9.3 mg/L, respectively at Chyan Chappra village (25° 43.28′N : 84° 16.24′E), UP. About 87 % of tubewell water samples have higher concentrations of Fe beyond its permissible limit of 1 mg/L (WHO guideline). The Fe content in tubewell waters varies from 0.2 to as much as 9.3 mg/L (Shah 2014). Arsenic contaminated areas in and around Buxar, Muhammadabad, Ghazipur, Zamania, Varanasi, Chunar and Mirzapur towns are mainly confined in the meandering belt of the Ganga river (Fig. 3.2). Maximum As concentration at Karkatpur village is 450 μg/L, which is located in the entrenched channels and floodplains of the Ganga river. However, tubewells located in its opposite side at Zamania town are As-safe in groundwater (<10 g/L), as the areas belong to the Pleistocene Older Alluvium surfaces. Few As-contaminated areas are also located in the Newer Alluvium older floodplains at the northeast of Zamania and east of Muhammadabad towns. These areas are shallow cover of the Newer Alluvium with As-level ≤100 μg/L. However, the Pleistocene Older Alluvium upland surfaces in and around Zamania and Muhammadabad towns exposing yellow-brown oxidized clay with calcareous and ferruginous concretions are As-safe in groundwater (Shah 2014). Moreover, tubewells in the Older Alluvium floodplains covered with the Holocene sediments on top are locally As-contaminated. Tubewells located in Buxar, Muhammadabad, Ghazipur, Saidpur, Varanasi, Chunar and Mirzapur towns are As-safe in groundwater (<10 g/L), because of their positions on the Older Alluvium upland surfaces (Fig. 3.2).

Table 3.2 Distribution of tubewell water As (μg/L) in percentage in the Middle Gangetic Basin

Study areas	Samples	As>10	As>50	As>100	As>250	Max. As (μg/L)
Buxar to Mirzapur towns (Fig. 3.2)	144	66	36	20	8	550
Patna to Ballia towns (Fig. 3.3)	80	89	50	29	18	1,300

A result of As concentrations of 80 tubewells from Vaishali, Patna, Bhojpur and Ballia districts in the Middle Gangetic plain (Fig. 3.3) are shown in Table 3.2, where 89 % of tubewells have As concentrations above 10 µg/L, and 50 % of tubewells have As above 50 µg/L. Maximum concentrations of As and Fe in tubewell waters are 1,300 µg/L and 12.93 mg/L at Semariya Ojjhapatti (25° 36.97′N : 84° 25.71′E) and Pandey Tolla (25° 41.27′N : 84° 36.97′E) villages, respectively. About 85 % of tubewell water samples have higher concentrations of Fe beyond its permissible limit of 1 mg/L. The Fe content in tubewell waters varies from 0.1 to as much as 12.9 mg/L (Shah 2014). Few As-contaminated areas are also located at the north of Myil, north of Ara, and east of Ballia. All are confined within the Newer Alluvium older floodplains of the Ganga and Ghaghara rivers system. Low levels of As concentrations are also observed along the left bank of the Ganga river in the Manupur–Hajipur–Myil areas due to the oxidized Pleistocene Older Alluvium upland surfaces (Shah 2014). Tubewells in Ballia, Ara, Chhapra, Patna and Hazipur towns are located on the Older Alluvium upland surfaces and are As-safe in groundwater (Fig. 3.3).

The density of As-contaminated tubewells from Patna to Ballia areas is higher compared to that of Buxar to Mirzapur areas. Table 3.2 shows 66 % of tubewells have As >10 g/L from Buxar to Mirzapur areas, whereas 89 % of tubewells have As >10 g/L from Patna to Ballia areas. Moreover, 36 % of tubewells have As above 50 µg/L from Buxar to Mirzapur areas, whereas 50 % of tubewells have As above 50 µg/L from Patna to Ballia areas. The correlation of As and Fe of 224 tubewells in the Middle Gangetic Basin shows a scatter diagram, where lower value of As corresponds with the higher value of Fe (Fig. 3.5). Therefore, arsenic is always available with iron but high iron with high arsenic is not possible always. Depth information of 224 tubewells in the Middle Gangetic Basin shows that 77 % of tubewells are located in shallow depth (21–40 m). About 40 % of tubewells have As >50 µg/L within the depth of 17–50 m (Shah 2014). However, most of the As contaminated tubewells are located within the depth of 20–50 m in the Newer Alluvium aquifers (Fig. 3.6). Maximum value of As (1,300 µg/L) corresponds to a depth of 33 m at Semariya Ojjhapatti village (25° 36.97′N : 84° 25.71′E).

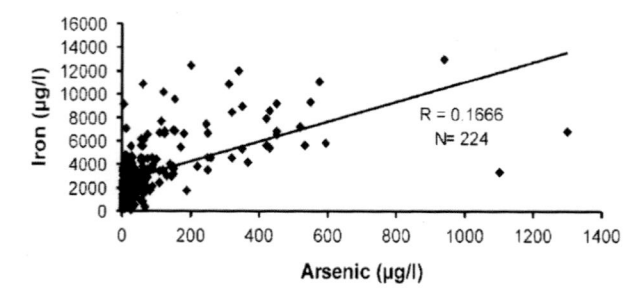

Fig. 3.5 Correlation between As and Fe concentrations of tubewell water in the Middle Gangetic Basin

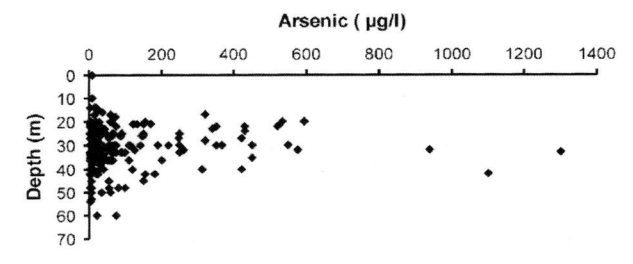

Fig. 3.6 Relation between As concentration and depth of tubewell in the Middle Gangetic Basin

Table 3.3 Tubewell As (µg/L) distribution in percentage in the Ghaghara Basin

Study area	>10	>50	>100	>200	>300	Max. As
Faizabad district (N=101)	38	15	9	5	2	350
Gonda district (N=88)	61	45	33	25	18	510
Basti district (N=42)	42	26	2	–	–	150

3.5.2 Distribution of Groundwater Arsenic in Ghaghara Basin

A result of As concentrations of 231 tubewells in the Ghaghara Basin (Fig. 3.4) are shown in Table 3.3. About 38 %, 61 % and 42 % of tubewells in Faizabad, Gonda and Basti districts, respectively have As > 10 µg/L (WHO guideline). Moreover, 15 %, 45 % and 26 % of tubewells in Faizabad, Gonda and Basti districts, respectively have As above 50 µg/L. Maximum concentrations of As in tubewells from Faizabad, Gonda and Basti districts are 350 µg/L, 510 µg/L and 150 µg/L, respectively. The subsurface information of tubewell has been collected from owners of the tubewells and accordingly a profile section (A-B) is plotted (Fig. 3.7).

The profile section shows that tubewells located in Holocene Newer Alluvium aquifers in the Faizabad side have As concentrations of 25, 23 and 72 µg/L corresponding to the depth of 30, 37 and 12 m, respectively. Moreover, tubewells located in Holocene Newer Alluvium aquifers in the Durgaganj areas have As concentrations of 140, 125 and 410 µg/L corresponding to the depth of 33, 33 and 10 m, respectively. However, As-safe tubewells (<10 µg/L) are mainly located in the Pleistocene Older Alluvium aquifers at the both ends of Faizabad and Durgaganj areas (Fig. 3.4). Durgaganj Village (26° 49.786′N : 82° 11.873′E) and Jayedpur Majha (26° 49.786′N : 82° 11.873′E) are located on the north bank of the Ghaghara River, where As concentrations in tubewells are 410 µg/L and 455 µg/L, respectively. These areas are in entrenched channels and floodplains of the Ghaghara river. However, opposite side of Durgaganj village is Ayodhya town, and tubewells in Ayodhya town are As-safe in groundwater (<10 µg/L) because of their positions on Older Alluvium upland surfaces. As-contaminated tubewells (As $\leq$ 200 µg/L) are also located on the older floodplains (zone 2) to the southeast of Ayodhya town near the bank of Ghaghara river (Fig. 3.4). Further, villages in and around Parbati Tal are

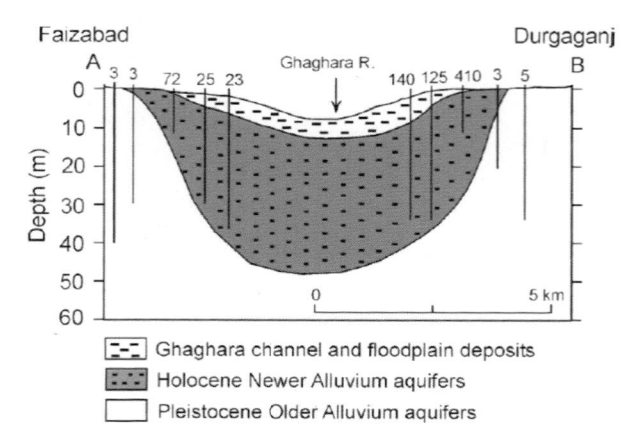

Fig. 3.7 A-B profile section is marked in Fig. 3.1. As value (µg/L) of each tubewell is marked in its *top*

As affected in groundwater, where maximum As 450 µg/L has been reported from Raghuraj Nagar Village (26° 56.101′N : 82° 10.969′E). These inundated areas consist of palaeo-channels and ox-bow lakes of the Ghaghara river and its tributaries (Fig. 3.4). Faizabad and Ayodhya twin towns are historically famous places and are located on the yellow-brown coloured oxidized Pleistocene Older Alluvium upland surfaces with calcareous and ferruginous concretions. Tubewells in twin towns are As-safe (<10 µg/L) in groundwater. However, tubewells in outskirts of Faizabad town near Pathan tolia (26° 47.136′N : 82° 09.990′E) and Kala Majhar (26° 46.999′N : 82° 04.627′E) have maximum As concentrations of 72 µg/L and 350 µg/L, respectively.

About 47 % As-contaminated (As > 10 µg/L) tubewells in these three districts are located within the depth of 10–35 m. About 86 %, 69 % and 35 % of tubewells in Faizabad, Gonda and Basti districts, respectively are within the depth of 21–45 m (Shah 2013). Maximum As concentration 510 µg/L is reported from Bhakir Mahangopur village (25° 51.559′N : 82° 07.669′E) corresponding to a depth of 40 m.

3.5.3 Groundwater Arsenic Contamination from Different Parts of Middle Gangetic Basin

Apart from above studies, groundwater As-contamination has been reported from different parts of the Middle Gangetic Basin. Arsenic contamination in groundwater above 50 µg/L is reported in Shapur, Dhokra, Lilapur, Malikpur, Fathupur villages of Allahabad district in UP (Chakraborti et al. 2009). Moreover, As contamination >50 µg/L has been reported in the Unnao district of UP in the Suklaganj-Kanpur urban areas and Purani Bazar, Misra Colony, Rishi Nagar, Ananda Nagar, Subash Gram, Kanchan Nagar villages. These locations are close to the entrenched channels and floodplains of the Ganga river. Generally the tubewells are of shallow depth ranging from 15 to 35 m (Chakraborti et al. 2009) and most of them are located within the depth of 30 m. However, the temporal variation of As concentrations in Ghazipur district are greater in the pre-monsoon than that of the post-monsoon season (Ahamed et al. 2006; Kumar et al. 2010a; Yano et al. 2012). Lakhimpur Kheri district of UP lies in Terai region along the Himalayan foothills. The Ghaghara and Sharda are the two major rivers draining the area. Arsenic concentrations in tubewell waters above 10 μg/L are found in five blocks (hot spot) viz., Pallia, Nighasan, Ramia Beher, Dhawhara and Isanagar. In these blocks almost all the handpumps were found to be contaminated. About 41.18 % (42 out of 102) of handpumps of these blocks have As above 10 μg/L (Pathak et al. 2013).

Groundwater As in Bihar was first reported in Bhojpur district (Chakraborti et al. 2003). Last several years, Public Health and Engineering Department (PHED), Government of Bihar, initiated the blanket testing of hand pumps and shallow tubewells and reported arsenic contamination in parts of 12 districts on the bank of the Ganga river. The hand pumps and shallow tubewells with As concentration above 10 μg/L being unfit for consumption were coloured red by PHED. Central Ground Water Board (CGWB) initiated exploratory drilling and groundwater sample analysis and identified another three districts with As above 10 μg/L. By now, 57 blocks in 15 districts of Bihar, mainly along the river Ganga are As-affected (10 μg/L). The As-affected districts in Bihar are Buxar, Bhojpur, Patna, Saran, Vaishali, Begusarai, Samastipur, Lakhisarai, Purnea, Kathitar, Khagaria, Darbhanga, Bhagalpur, Kishanganj and Munger. Most of the As-contaminated tubewells are located within the depth of 15–40 m in the Holocene Newer Alluvium sediments (Chakraborti et al. 2003; Nickson et al. 2007; Shah 2008; Ghosh et al. 2009; Kumar et al. 2010a, b; Saha 2010; CGWB 2010). Arsenic contamination in groundwater has been reported in the floodplains of the Kosi River and its confluence with the Ganga River. Contaminated tubewells (As >10 μg/L) are located in the Holocene Newer Alluvium sediments, whereas the upland areas in the Kosi fan are As-safe in groundwater (Nickson et al. 2007; Mukherjee et al. 2012).

The low-lying entrenched channel and floodplain of the Ganga River that flank the northern tip of the Rajmahal Hills and parts of the Sahibganj district in Jharkhand are As-contaminated. Most of the As-contaminated tubewells belong to Sahibganj, Rajmahal and Udhawa blocks. Average 36 % of tubewells have As >10 μg/L in these three blocks (Bhattacharjee et al. 2005; Nayak et al. 2008). In Bhagalpur district, groundwater As-contamination has been reported, where As-contaminated tubewells are located within the depth of 12–32 m. Arsenic concentrations in the groundwater varied from 19.1 to 118 μg/L in the pre-monsoon and from 18.8 to 113 μg/L in the post-monsoon season. The average As-concentration in Bhagalpur district (50 μg/L) was five times than that of Vaishali district (10 μg/L). About 95 % of tubewells (20 samples) were found As-contaminated above the WHO standard for drinking water. Sediment grain-size analysis has indicated the presence of clay at shallow depth with enrichment of As concentrations (Kumar et al. 2010b; Singh et al. 2014). Arsenic and iron concentrations in Older and Newer Alluvium sediments were found in significant amount in the Middle Gangetic Plain (Shah 2014).

Maximum As in the Older Alluvium sediments is 13.73 mg/kg near Saidpur village, UP (25° 23.20′N : 81° 47.58′E), whereas maximum As in the Newer Alluvium flood top clay is 30.91 mg/kg on the bank of Ganga river near Loktola village (25° 18.69′N : 82° 04.59′E), UP. Maximum Fe in the Older Alluvium sediments at Charri village (25° 20.20′N : 81° 55.69′E), and the Newer Alluvium sediments at Pancharukhia village (25° 45.80′N : 84° 21.80′E) are 6.15 g/kg and 6.31 g/kg, respectively (Shah 2014). Moreover, As and Fe concentrations in borehole sediments ranges from 1.1 mg/kg to 11.9 mg/kg, and 0.2 g/kg to 10.97 g/kg, respectively from Varanasi areas in the Middle Ganga Plain (Raju 2012). Arsenic and Fe concentrations in the Newer Alluvium sediments are 6.19 mg/kg and 7.12 g/kg, respectively in the Ghaghara Basin (Shah 2013). From previous published reports, it was known that average As content in clayey sand is 12 mg/kg in Murshidabad district, West Bengal (Acharyya et al. 2000). Arsenic and Fe concentrations in core sample are 7.7 mg/kg and 4.26 g/kg, respectively in Manikganj, Bangladesh (Shamsudduha et al. 2008).

The mineralogical assemblage in the Newer Alluvium sediments in the Middle Gangetic Basin is quartz, muscovite, chlorite, montmorillonite, kaolinite, feldspar and goethite (Shah 2010). However, the same mineralogical assemblage is also reported from borehole sediments in Jessore district of Bangladesh and Nadia district of West Bengal (Akai et al. 2004; Acharyya and Shah 2007). Arsenic and iron concentrations in suspended river sediments were found in significant amount in the suspended river sediments of the Yamuna, Ganga, Gomati, Ghaghara, Sone, Gondak, Buri Gandak and Kosi rivers (Shah 2014). Maximum As in suspended river sediments of the Kosi river is 10.59 mg/kg near Saharsa town (25° 53.07′N : 86° 26.21′E), and minimum As is 5.61 mg/kg in the Yamuna river near Sihonda town (25° 21.27′N : 81° 38.59′E). The Ganges is one of the major rivers of the Indian subcontinent and maximum As in suspended river sediments is 9.52 mg/kg near Buxar town (25° 34.58′N : 83° 58.25′E). The Sone river is the largest of the Ganges' southern tributaries, have As 6.60 mg/kg in suspended sediments near Koelwar town (25° 34.11′N : 84° 48.31′E). Arsenic in suspended river sediments of the Gomati, Ghaghara, Gandak and Buri Gandak rivers ranges from 7.01 to 9.04 mg/kg (Shah 2014).

During high water discharge, the contribution to suspended sediments in rivers most probably originates from erosion and weathering processes of bed sediments and bank scour which results in fluvial transport and sedimentation of As-enriched metal hydroxides especially Fe oxy-hydroxides (Berg et al. 2001). Iron content in suspended river sediments of the Yamuna, Ganga, Gomati, Ghaghara, Sone, Gandak, Buri Gandak and Kosi rivers varies from 5.65 to 1.17 g/kg (Shah 2014). Arsenic concentration in suspended river sediments of the Jamuna, Padma and Meghna rivers during post-flood period varies from 4.07 to 5.47 mg/kg. The As load carried annually with suspended sediments of the Jamuna, Padma and Meghna rivers have been estimated to be 3,054, 4,121 and 4,584 tons, respectively (Chowdhury et al. 2009). Thus, the Himalayan rivers have high content of As and Fe in suspended river sediments that has been deposited in the Ganga-Meghna-Brahmaputra plains.

Quaternary morphostratigraphy play a role on groundwater arsenic contamination. The Ganga is the axial river in the Himalayan foreland basin. The sub-surface extension of the Gangetic Basin is limited to the west by the Aravalli–Delhi ridge and to the east by the Monghyr–Saharsa ridge (Kumar et al. 1996). The Ganga plain has developed a network of small drainages and entrenchment of major rivers during ~20–13 ka BP. The network became dense during 13–8 ka BP (early Holocene) and moved large amounts of sediment-water to the Ganga Delta. During 8–6 ka BP small drainages changed to large lakes and extensive deposition of sediment took place on the upland interfluve surfaces. A large number of small drainages, dense networks of channel and numerous lakes, and swamps have developed (8 ka BP–present) in the Gangetic Basin (Singh 2004). The major parts of the Gangetic plain consist of inter-fluve upland surfaces of Older Alluvium. These Pleistocene Older Alluvium upland surfaces are characterized by yellow-brown coloured sediments with profuse calcareous and ferruginous concretions, and are either exposed or occur under shallow cover of Holocene sediments. Most of the towns in the Gangetic Basin are located on the Pleistocene Older Alluvium upland surfaces. Moreover, Holocene Newer Alluvium sediments are characterised by grey to black coloured organic-rich argillaceous sediments, and occur in the entrenched channels and floodplains of the Gangetic rivers (Figs. 3.2, 3.3, and 3.4).

Geomorphologically, the Gangetic plains are sub-divided into three zones viz. (1) the Pleistocene Older Alluvium upland surfaces, (2) the Older Alluvium flood plains covered with the Holocene sediments on top, and (3) the Holocene Newer Alluvium entrenched channels and flood plains. Tubewells located in the Pleistocene Older Alluvium aquifers are As-safe in groundwater, whereas tubewells located in the Holocene Newer Alluvium aquifers are As contaminated in groundwater in the Ganga and Ghaghara Rivers system. Moreover, the Older Alluvium covered with the Holocene sediments on top are locally As contaminated in groundwater. Arsenic affected villages are located in proximity to meander or abandoned channels (Figs. 3.2, 3.3, and 3.4) in the Ganga and Ghaghara Rivers system. In the Gangetic plains, the Pleistocene Older Alluvium surfaces were dissected by channels and floodplains and buried under the younger Holocene Newer Alluvium sediments. This yellow-brown coloured oxidizing Pleistocene Older Alluvium surfaces were well flushed by groundwater flow due to high-hydraulic head and devoid of organic matter. The environment of the oxidized Pleistocene Older Alluvium aquifer is not favourable to release sorbed As to groundwater (Shah 2008, 2010). However, distribution of high content of groundwater As in alluvial aquifers in Bangladesh and West Bengal is also controlled by regional geologic–geomorphic units of Older Alluvium and Younger Alluvium (Acharyya et al. 2000; von Brömssen et al. 2007; Shamsudduha et al. 2008). The groundwater As status in the Gangetic Basin is shown in the schematic 3D-diagram (Fig. 3.8).

Tubewells located in entrenched channels and flood plains of the Ganga and Ghaghara rivers are mostly As contaminated (zone 3). Moreover, tubewells located in Pleistocene Older Alluvium aquifers with thin cover of Holocene sediments on top are As-safe in groundwater (zone 1). Arsenic affected tubewells are also locally observed in older floodplains (zone 2) covered with Holocene Newer Alluvium sediments on top (Shah 2008, 2010). Regional geologic–geomorphic units of the Older and Younger Alluvium sediments control distribution of high content of groundwater As in alluvial aquifers in fluvial deltaic setting in south and southeast

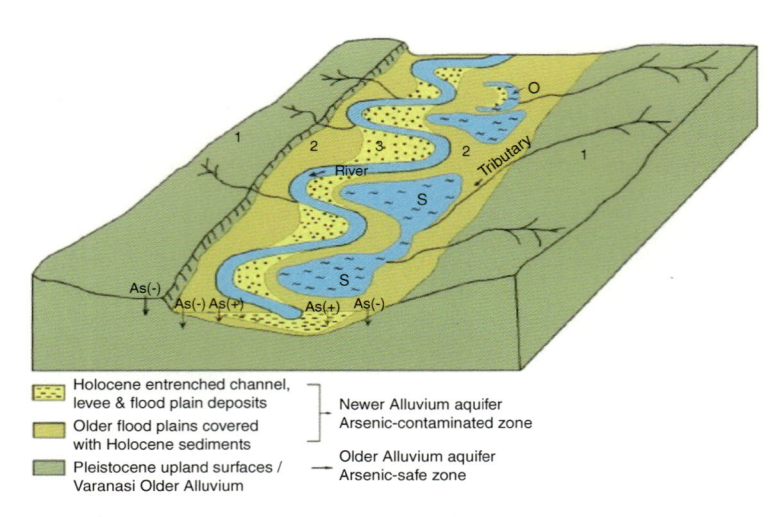

Fig. 3.8 Schematic 3D diagram showing groundwater As status in Holocene and Pleistocene aquifers in the Middle Gangetic Basin. Abbreviations: *O* Ox-bow lake, *S* Back swamp, *As* (−) As-safe aquifers, *As* (+) As-contaminated aquifers

Asia. Groundwater As is mostly found in the delta plains, modern floodplains, marshes, and depressed lowland areas. High groundwater As concentration is found in the Holocene aquifers, while the Pleistocene aquifers have a low As level. Occurrences of high groundwater As are significantly linked with the Pleistocene and Holocene aquifers in the Mekong Delta, Red River Delta and Bengal Basin (Acharyya et al. 2000; Kinniburgh and Smedley 2001; Berg et al. 2001; Acharyya and Shah 2007; von Brömssen et al. 2007; Shamsudduha et al. 2008; Polizzotto et al. 2008; Winkel et al. 2008).

3.5.4 Source and Release of Arsenic in Groundwater

There are several potential sources of As which have been identified both in the Himalayan belt, as well as in the Peninsular India. The Gangetic plains were formed due to accumulation of bulk sediments from the Himalayan hill range, whereas the input of the Peninsular India is minor. There are several As-bearing minerals deposits in the Himalayan hill range including hydrothermal pyrite-chalcopyrite-arsenopyrite-galena mineralization associated with quartz veins in Buniyal, Doda, Almora Garhwal, J and K Hills (Tewari and Gaur 1977). The Indus-Tsangpo suture in north India is marked by ophiolitic rocks, including olivine serpentinites. These ophiolites are composed of serpentinized peridotite, layered mafic to ultramafic rock, volcanic and oceanic sediments that contain high As (Guillot and Charlet 2007). The Peninsular India is also accounted for source of As where pyrite-bearing shale in Amjhore mine has As 2.6 g/kg (Das 1977). In the gold mineralization belt of Sone valley, As concentrations in bedrock locally reach 28 g/kg to 1 g/kg (Mishra et al. 1996). Arsenic contamination of groundwater also affects Terai belt of Nepal (Shrestha et al. 2003; Gurung et al. 2005). Major rivers in Terai belt originate from the Higher Himalaya and minor rivers emanate from the nearby Siwalik hills (Fig. 3.1). Fine sediments and organic material are deposited in inter-fan lowlands in wetlands and swamps. It was observed that sediments carried from the Siwalik hills by the minor rivers release more As than those carried by major rivers from the Higher Himalaya (Shrestha et al. 2003; Gurung et al. 2005).

Most of the rivers in the northern part of the Gangetic Plain viz., Yamuna, Ganga, Gomati, Ghaghara, Sone, Gandak, Buri Gandak and Kosi rivers originated from the Himalayas. The possibility of erosion, oxidation and transportation of As-bearing products in suspension and solution in the Gangetic plains are high. The southern belt of the Himalayas is subjected to high erosion and intense rainfall during the Holocene time (Williams and Clarke 1984). Most of the As-bearing sediments are enriched with Fe oxides. Oxides of Fe and Mn are potentially the most important source/sink for As in aquifer. Hydrous Fe oxide has a very high specific surface area, and thus a very high adsorption capacity to absorb heavy metals. Arsenic adsorbed on Fe/Mn oxides/hydroxides is released into the groundwater due to a decrease of the redox state in the aquifer (Davis and Kent 1990). Reduction of hydrated iron oxide (HFO) is common, and high concentration of dissolved Fe in groundwater indicates strong reducing condition. Mobilisation of As in the Gangetic plains is expected to follow a similar course as observed in the Bengal Delta of West Bengal and Bangladesh. Arsenic sorbed on discrete phases of hydrated Fe-Mn oxide coated sediments grains were preferentially entrapped in grey to black coloured organic-rich argillaceous sediments in entrenched channels and floodplains of the Gangetic rivers. Biomediated reductive dissolution of hydrated iron oxide (HFO) by anaerobic heterotypic Fe^{3+} reducing bacteria (IRB) play an important role to release sorbed As to groundwater (Nickson et al. 1998; Kinniburgh and Smedley 2001; Islam et al. 2004; Polizzotto et al. 2008).

3.6 Conclusions

A survey on arsenic content in groundwater in Mirzapur, Varanasi, Ghazipur, Ballia, Buxar, Bhojpur, Patna, Vaishali, Faizabad, Gonda and Basti districts of Uttar Pradesh and Bihar shows that tubewells located in entrenched channels and floodplains of the Ganga and Ghaghara rivers are prone to As contamination in groundwater. The Holocene Newer Alluvium aquifers are characterised by organic-rich grey to black coloured argillaceous sediments and hence the tubewells located in these aquifers are As contaminated. Most of the As-affected villages are preferentially located close to abandoned or present meander channels of the Ganga and Ghaghara rivers. Tubewells in Mirzapur, Chunar, Varanasi, Saidpur, Ghazipur, Muhammadabad, Ballia, Buxar, Ara, Chhapra, Patna, Hazipur, Faizabad, Ayodhya and Nawabganj towns are located on the Pleistocene Older Alluvium upland surfaces and are As-safe in groundwater. The major parts of the Gangetic plain consist of inter-fluve upland surfaces of the Pleistocene Older Alluvium. These upland oxidizing Pleistocene yellow-brown coloured sediments were well flushed by groundwater flow due to high hydraulic head and devoid of organic matter. The environment of the Pleistocene Older Alluvium aquifers is not favourable to release sorbed As in groundwater and aquifers are generally As-safe. However, in As-contaminated areas, deeper tubewells in the Pleistocene Older Alluvium aquifers would be better option for As-safe drinking water. The potential sources of As-bearing sediments are mainly from the Himalayas. The Older Alluvium and Newer Alluvium sediments have high content of As. The Yamuna, Ganga, Gomati, Ghaghara, Gandak, Buri Gandak and Kosi rivers originated from the Himalayas and have high content of As in suspended river sediments, and most of the As-bearing sediments are deposited in the Gangetic Basin.

Acknowledgements The author thanks School of Environmental Studies, Jadavpur University for As and Fe analyses. The author would like to thank local field assistants from Bihar and UP for helping to collect tubewell waters, riverbank sediments, and turbid river waters. The financial support for this study came from Department of Science & Technology Fast Track Young Scientist Scheme and Council of Scientific and Industrial Research Scientists' Pool Scheme, which is gratefully acknowledged.

References

Acharyya SK, Shah BA (2004) Risk of arsenic contamination in groundwater, affecting Ganga alluvial plain, India. Environ Health Perspect 112:A19–A20

Acharyya SK, Shah BA (2007) Groundwater arsenic contamination affecting different geologic domains in India—a review: influence of geological setting, fluvial geomorphology and Quaternary stratigraphy. J Environ Sci Health A 42:1795–1805

Acharyya SK, Lahiri S, Raymahashay BC, Bhowmik A (2000) Arsenic toxicity of groundwater of the Bengal Basin in India and Bangladesh: the role of Quaternary stratigraphy and Holocene sea level fluctuation. Environ Geol 39:1127–1137

Ahamed S, Sengupta MK, Mukherjee A, Hossain A, Das B, Nayak B, Pal A, Mukherjee SC, Pati S, Dutta RN, Chatterjee G, Mukherjee A, Srivastava R, Chakraborti D (2006) Arsenic groundwater contamination and its health effects in the state of Uttar Pradesh (UP) in upper and middle Ganga plain, India: a severe danger. Sci Total Environ 370:310–322

Akai J, Izumia K, Fukuhara H, Masuda H, Nakano S, Yoshimura T, Ohfuji H, Anawar HM, Akai K (2004) Mineralogical and geomicrobiological investigations on groundwater arsenic enrichment in Bangladesh. Appl Geochem 19:215–230

Berg M, Tran HC, Nguyen TC, Schertenleib R, Giger W (2001) Arsenic contamination of groundwater and drinking water in Vietnam: a human health threat. Environ Sci Technol 35:2621–2626

Bhattacharjee S, Chakravarty S, Maity S, Dureja V, Gupta KK (2005) Metal contents in the groundwater of Sahibganj district, Jharkhand, India, with special reference to As. Chemosphere 58:1203–1217

Bhattacharya P, Chatterjee D, Jacks G (1997) Occurrence of arsenic-contaminated groundwater in alluvium aquifers from delta plains, Eastern India: options for safe water supply. Water Resour Dev 3:79–92

Bureau of Indian Standards (2003) Indian standard: drinking water. Specification (first revision), Amendment no. 2, New Delhi

Central Ground Water Board (2010) Ground water quality in shallow aquifers of India. Central Ground Water Board, Ministry of Water Resources, Government of India, Faridabad

Chakraborti D, Mukherjee SC, Pati S, Sengupta MK, Rahman MM, Chowdhury UK, Lodh D, Chanda CR, Chakraborty AK (2003) Arsenic groundwater contamination in Middle Ganga Plain, Bihar, India: a future danger? Environ Health Perspect 111:1194–1200

Chakraborti D, Ghorai S, Das B, Pal A, Nayak B, Shah BA (2009) Arsenic exposure through groundwater to the rural and urban population in the Allahabad-Kanpur track in the Upper Ganga plain. J Environ Monit 11:1455–1459

Chowdhury MAI, Ahmed MF, Ali MA (2009) Arsenic in water and sediments of major rivers in Bangladesh. J Civ Eng 37:31–41

Das S (1977) A note on prospecting of Amjhore pyrite, Rohtas district, Bihar with discussion on the origin of the deposits. Indian Miner 31:8–22

Davis JA, Kent DB (1990) Surface complexation modeling in aqueous geochemistry. Rev Mineral 23:177–260

Dwivedi CN, Sharma SK (1992) Quaternary geology and geomorphology of a part of Ghaghara Basin, Azamgarh, Ballia and Mau districts, Uttar Pradesh. Rec Geol Surv India 125:51–53

Ghosh AK, Singh SK, Bose N, Singh SK (2009) Arsenic hot spots detected in the state of Bihar (India): a serious health hazard for estimated human population of 5.5 lakh. In: Ramanathan AL, Bhattacharya P, Keshari AK et al (eds) Assessment of ground water resources and management. I.K. International Publishing House, New Delhi

Guillot S, Charlet L (2007) Bengal arsenic, an archive of paleohydrology and Himalayan erosion. Environ Sci Health A 42:1785–1794

Gurung JK, Ishiga H, Khadka MS (2005) Geological and geochemical examination of arsenic contamination in groundwater in the Holocene Terai Basin, Nepal. Environ Geol 49:98–113

Islam FS, Gault AG, Boothman C, Polya DA, Charnock JM, Chatterjee D, Lloyd JR (2004) Role of metal-reducing bacteria in arsenic release from Bengal delta sediments. Nature 430:68–71

Khan AU, Bhartiya SP, Kumar G (1996) Cross faults in Ganga Basin and their surface manifestations. Geol Surv India Spl Pub 21:215–220

Kinniburgh DG, Smedley PL (2001) Arsenic contamination of groundwater in Bangladesh, Report, WC/00/19. British Geological Survey, Dhaka

Kumar G, Khanna PC, Prasad S (1996) Sequence stratigraphy of the foredeep and evolution of the Indo-Gangetic plain, Uttar Pradesh. Geol Surv India Spl Pub 21:173–207

Kumar M, Kumar P, Ramanathan AL, Bhattacharya P, Thunvik R, Singh UK, Tsujimura M, Sracek O (2010a) Arsenic enrichment in groundwater in the Middle Gangetic Plain of Ghazipur district in Uttar Pradesh, India. J Geochem Explor 105:83–94

Kumar P, Kumar M, Ramanathan AL, Tsujimura M (2010b) Tracing the factors responsible for arsenic enrichment in groundwater of the Middle Gangetic Plain, India: a source identification perspective. Environ Geochem Health 32:129–146

Mandal BK, Suzuki KT (2002) Arsenic round the world: a review. Talanta 58:201–235

Mishra SP, Sinha VP, Tripathi AK, Sharma DP, Dwivedi GN, Khan MA, Yadav ML, Mehrotra RD (1996) Arsenic incidence in Son valley gold belt. In: Shanker R, Dayal HM, Shome SK, Jangi BL (eds) Symposium on earth sciences in environmental assessment and management. Geological Survey of India, Lucknow

Mukherjee A, Sengupta MK, Hossain MA, Ahmed S, Das B, Nayak B, Lodh D, Rahman MM, Chakraborti D (2006) Arsenic contamination in groundwater: a global perspective with emphasis on the Asian scenario. J Health Popul Nutr 24:142–163

Mukherjee A, Scanlon BR, Fryar AE, Saha D, Ghosh A, Chowdhuri S, Mishra R (2012) Solute chemistry and arsenic fate in aquifers between the Himalayan foothills and Indian craton (including central Gangetic plain): influence of geology and geomorphology. Geochim Cosmochim Acta 90:283–302

Nayak B, Das B, Mukherjee SC, Pal A, Ahamed S, Hossain MA, Maity P, Dutta RN, Dutta S, Chakraborti D (2008) Groundwater arsenic contamination in the Sahibganj district of Jharkhand state, India in the middle Ganga plain and adverse health effects. Toxicol Environ Chem 90:673–694. doi:10.1080/02772240701655486

Nickson R, McArthur JM, Burgess W, Ahmed KM, Ravenscroft P, Rahman M (1998) Arsenic poisoning of Bangladesh groundwater. Nature 395:338

Nickson R, Sengupta C, Mitra P, Dave SN, Banerjee AK, Bhattacharya A, Basu S, Kakoti N, Moorthy NS, Wasuja M, Kumar M, Mishra DS, Ghosh A, Vaish DP, Srivastava AK, Tripathi RM, Singh SN, Prasad R, Bhattacharya S, Deverill P (2007) Current knowledge on the distribution of arsenic in groundwater in five states of India. Environ Sci Health A 42:1707–1718

Pascoe EH (1964) A manual of the geology of India and Burma, vol III. Geological Survey of India, Calcutta

Pathak BD, Karanth KR, Kidwai AL, Rao AP, Bose BB (1978) Geology and groundwater resources in parts of Jaunpur, Azamgarh, Ballia, Allahabad, Sultanpur and Faizabad districts, Uttar Pradesh. Bull Geol Surv India Sr B 44:1–77

Pathak VK, Agnihotri N, Khatoon N, Khan AH, Rahman M (2013) Hydrochemistry of groundwater with special reference to arsenic in Lakhimpur Kheri district, Uttar Pradesh, India. IOSR J Appl Chem 6:61–68

Polizzotto ML, Kocar BD, Benner SB, Sampson M, Fendorf S (2008) Near-surface wetland sediments as a source of arsenic release to ground water in Asia. Nature 454:505–508

Raju NJ (2012) Arsenic exposure through groundwater in the middle Ganga plain in the Varanasi environs, India: a future threat. J Geol Soc India 79:302–314

Ravenscroft P, Brammer H, Richards K (2009) Arsenic pollution: a global synthesis. Wiley Blackwell, Chichester

Saha D, Sreehari SS, Shailendra ND, Kuldeep GB (2010) Evaluation of hydrogeochemical processes in arsenic contaminated alluvial aquifers in parts of mid-Ganga basin, Bihar, eastern India. Environ Earth Sci 61:799–811

Samanta G, Roy Chowdhury T, Mandal B, Biswas B, Chowdhury U, Basu G, Chanda C, Lodh D, Chakraborti D (1999) Flow injection hydride generation atomic absorption spectrometry for determination of arsenic in water and biological samples from arsenic-affected districts of West Bengal, India, and Bangladesh. Microchem J 62:174–191

Shah BA (2008) Role of Quaternary stratigraphy on arsenic-contaminated groundwater from parts of middle Ganga plain, UP-Bihar, India. Environ Geol 53:1553–1561

Shah BA (2010) Arsenic-contaminated groundwater in holocene sediments from parts of middle Ganga plain, Uttar Pradesh, India. Curr Sci 98:1359–1365

Shah BA (2013) Status of groundwater arsenic pollution in holocene aquifers from parts of the Ghaghara Basin, India: its relation to geomorphology and hydrogeological setting. Phys Chem Earth 58–60:68–76

Shah BA (2014) Arsenic in groundwater, quaternary sediments, and suspended river sediments from the Middle Gangetic Plain, India: distribution, field relations, and geomorphological setting. Arab J Geosci 7:3525–3536. doi:10.1007/s12517-013-1012-4

Shamsudduha M, Uddin A, Saunders JA, Lee MK (2008) Quaternary stratigraphy, sediment characteristics and geochemistry of arsenic-contaminated alluvial aquifers in the Ganges-Brahmaputra floodplain in central Bangladesh. Contam Hydrol 99:112–136

Shrestha RR, Shrestha MP, Upadhyay NP, Pradhan R, Khadka R, Maskey A, Maharjan M, Tuladhar S, Dahal BM, Shrestha K (2003) Groundwater arsenic contamination, its health impact and mitigation program in Nepal. J Environ Sci Health A Tox Hazard Subst Environ Eng 38(1):185–200

Singh IB (2004) Late Quaternary history of the Ganga Plain. J Geol Soc India 64:431–454

Singh SK, Ghosh AK, Kumar A, Kislay K, Kumar C, Tiwari RR, Parwez R, Kumar N, Imam MD (2014) Groundwater arsenic contamination and associated health risks in Bihar, India. Int J Environ Res 8:49–60

Smith AH, Hopenhayn-Rich C, Bates MN, Goeden HM, Hertz-Picciotto I, Duggan HM, Wood R, Kosnett MJ, Smith MT (1992) Cancer risks from arsenic in drinking water. Environ Health Perspect 97:259–267

Smith AH, Goycolea M, Haque R, Biggs ML (1998) Marked increase in bladder and lung cancer mortality in a region of Northern Chile due to arsenic in drinking water. Am J Epidemiol 147:660–669

Srivastava P, Singh IB, Sharma M, Singhvi AK (2003) Luminescence chronometry and late quaternary geomorphic history of the Ganga Plain, India. Palaeogeogr Palaeoclimatol Palaeoecol 197:15–41

Tewari AP, Gaur RK (1977) Geological conditions of formation of pyrite-polymetallic deposits of the Himalaya and the Great Caucasus – a comparison. Himal Geol 7:235–245

von Brömssen M, Jakariya M, Bhattacharya P, Ahmed KM, Hasan MA, Sracek O, Jonsson L, Lundell L, Jacks G (2007) Targeting low-arsenic aquifers in Matlab Upazila, Southeastern Bangladesh. Sci Total Environ 379:121–132

WHO (1993) Guideline for drinking- water quality, vol 1, 2nd edn, Recommendations. WHO Library Cataloguing, Geneva. ISBN 9241544600

Williams MAJ, Clarke MF (1984) Late quaternary environments in north-central India. Nature 308:633–635

Winkel L, Berg M, Amini M, Hug SJ, Johnson CA (2008) Predicting groundwater arsenic contamination in Southeast Asia from surface parameters. Nat GeoSci 1:536–542

Yano Y, Ito K, Kodama A, Shiomori K, Tomomatsu S, Sezaki M, Yokota H (2012) Arsenic polluted groundwater and its counter measures in the Middle Basin of the Ganges, Uttar Pradesh State, India. J Environ Protect 3:856–862

Section II
Groundwater Arsenic Characterisations and Risk Assessments

Chapter 4
Preliminary Assessment of Arsenic Distribution in Brahmaputra River Basin of India Based on Examination of 56,180 Public Groundwater Wells

Chandan Mahanta, Runti Choudhury, Somnath Basu, Rushabh Hemani, Abhijit Dutta, Partha Pratim Barua, Pronob Jyoti Borah, Milanjit Bhattacharya, Krisaloy Bhattacharya, Wazir Alam, Lalit Saikia, Abhijit Mukherjee, and Prosun Bhattacharya

4.1 Introduction

Arsenic (As) rich groundwater in alluvial aquifers is a worldwide problem (Nriagu et al. 2007). Elevated arsenic concentrations have long been detected in Southeast Asia (e.g. Thailand, Myanmar, Vietnam, Cambodia and Lao), India, Bangladesh, China, Mongolia, Nepal and Pakistan (Smedley and Kinniburgh 2002). Recent reports of discovery of arsenic (As) enrichment in groundwater of the Brahmaputra river basin (Bhattacharya et al. 2011) has exposed a significantly large population inhabiting in the river valley to serious health threats, although the actual distribution and extent of the As affected groundwater in the aquifers are yet to be established. Because of its vicinity to the highly As rich groundwater regions of Bengal basin (Bangladesh and West Bengal state of India), the extent of the polluted areas within the Brahmaputra basin may be much wider than what is initially understood.

C. Mahanta • R. Choudhury (✉) • W. Alam • L. Saikia
Department of Civil Engineering, Indian Institute of Technology,
Guwahati 781039, Assam, India
e-mail: runti.choudhury@gmail.com

S. Basu • R. Hemani
UNICEF, Guwahati 781028, Assam, India

A. Dutta • P.P. Barua • P.J. Borah • M. Bhattacharya • K. Bhattacharya
Public Health Engineering Department Assam, Guwahati 781036, Assam, India

A. Mukherjee
Department of Geology and Geophysics, Indian Institute of Technology,
Kharagpur, West Bengal, India

P. Bhattacharya
KTH International Groundwater Arsenic Research Group, Land and Water Resources
Engineering, Royal Institute of Technology, Stockholm, Sweden

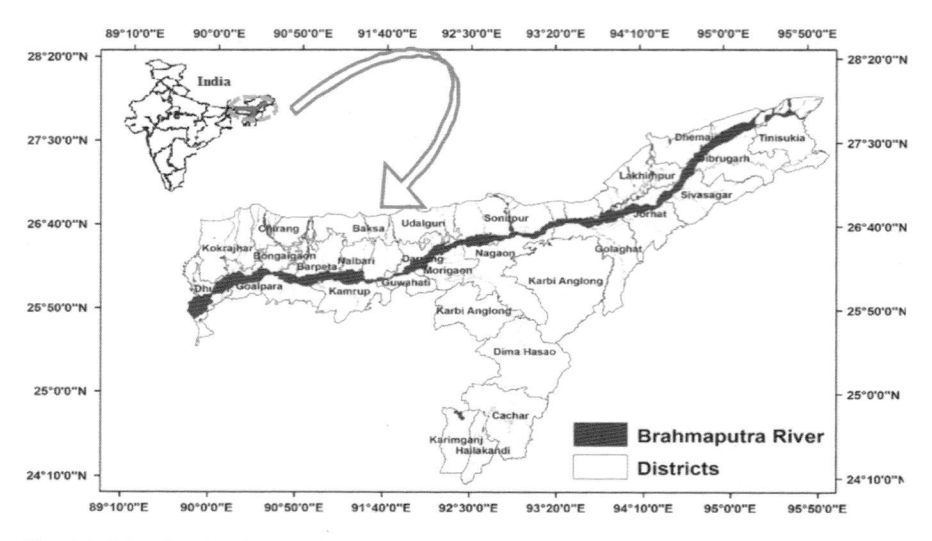

Fig. 4.1 Map showing Assam along with its administrative districts

Groundwater arsenic contamination in the Brahmaputra basin aquifers in Assam, a state in the northeastern part of India, has started gaining attention relatively recently. Singh (2004) reported maximum groundwater arsenic concentrations in Jorhat district (Fig. 4.1), located in the southern bank of the Brahmaputra river in Assam (maximum groundwater As concentration ranges between 194 and 657 µg/L), with relatively lower concentrations in the northern bank like Lakhimpur district (50–550 µg/L). Based on studies conducted in Darrang and Bongaigaon districts located in the northern bank (Fig. 4.1) of the Brahmaputra river in Assam, Enmark and Nordborg (2007) reported the concentration of arsenic in the two districts between 5 and 606 µg/L. In a study conducted in 2010 (Mahanta et al. 2008), concentrations beyond 50 µg/L have been confirmed in 72 blocks out of 214 blocks in 22 districts of Assam. A study by Chetia et al. (2011) in the Golaghat district reported As concentration ranging between 1 and 128 µg/L in six blocks of the district. These studies so far have remained spatially limited and a comprehensive picture is yet to emerge. To comprehensively evaluate the extent of As contamination in the region, a blanket rapid assessment study was undertaken in large parts of the Brahmaputra basin in Assam. This paper reports the preliminary assessment of arsenic distribution in the Brahmaputra basin in Assam based upon results from 56,180 public groundwater wells, tested during the rapid assessment programme.

4.2 Study Area

Brahmaputra river basin is bound by the Eastern Himalayas in its north and east, the Naga and Patkai range in the northeastern end, with the Shillong Plateau on its south (Sharma 2005). The basin is characterized by rocks and sediments of both the

Paleozoic and Cenozoic era, comprising gneisses, high and low grade schists, ultrabasic rocks, shales, sandstones, mottled clays and conglomerates (Sharma 2005). Recent alluvial deposits in the basin are represented by alluvial fans and floodplain sediments carried by the Brahmaputra river and its tributaries (Sharma 2005).

The northern tributaries of the river are braided, constituting the aquifers with recent sedimentary alluvial plains. Some of these sedimentary aquifers are potential key locations of arsenic enrichment in groundwater. The basin is covered by a 200–300 m thick recent alluvium comprising silt, clay, sand and gravels (CGWB 1995), which can be divided into young alluvial and older alluvial sediments (Talukdar et al. 2004). The recent alluvial sediments are light grey to dark grey in colour confined to the flood plain areas adjacent to the Brahmaputra river and its tributaries, while the older alluvial sediments are sandy loam to silty and clay characterized by light yellowish brown to light brown colour (CGWB 2008).

4.3 Methodology

Groundwater was sampled within this blanket rapid assessment programme from community hand pump wells following standard groundwater sampling procedure (APHA 2005).

Groundwater samples were collected from the wells after purging the wells for several times. Samples were collected in 200 mL LDPE sampling bottles acidified with 1 mL dilute hydrochloric acid in the ratio 1:1 to maintain the pH below 2 and to prevent precipitation of iron and possible co-precipitation of arsenic. A stepwise analytical procedure, involving three different instruments—Arsenator, UV Spectrophotometer and Atomic Absorption Spectrophotometer—was adopted that yielded reasonably consistent agreement, showing good analytical performances and helping in establishing contamination levels conclusively.

To check the relative precision and accuracy of the three instruments viz. Arsenator, UV-1 Spectrophotometer and AAS, simultaneous calibration and comparison between the different test results were carried out regularly using calibration standards. The test results were found to be consistent and a 95 % order of confidence was assigned for analytical performance of all the three instruments (Fig. 4.2).

4.4 Results

Based on the three-tier analytical procedure followed, a total of 56,180 samples spread across 76 blocks in the Brahmaputra floodplain in Assam were analysed. 29.75 % of all sampled wells have As concentrations above the WHO guideline of 10 µg/L and 7.95 % sources were found to be with arsenic concentrations above the Bureau of Indian Standard permissible limit of 50 µg/L. An estimated 7,00,000 people located in 1970 affected population centres spread across 76 blocks in 18 districts were found to be exposed to the risk of As contamination (Table 4.1, Figs. 4.3a and 4.3b). The findings revealed that occurrence of elevated arsenic concentrations were localized at several discreet patches of the floodplain (Fig. 4.4).

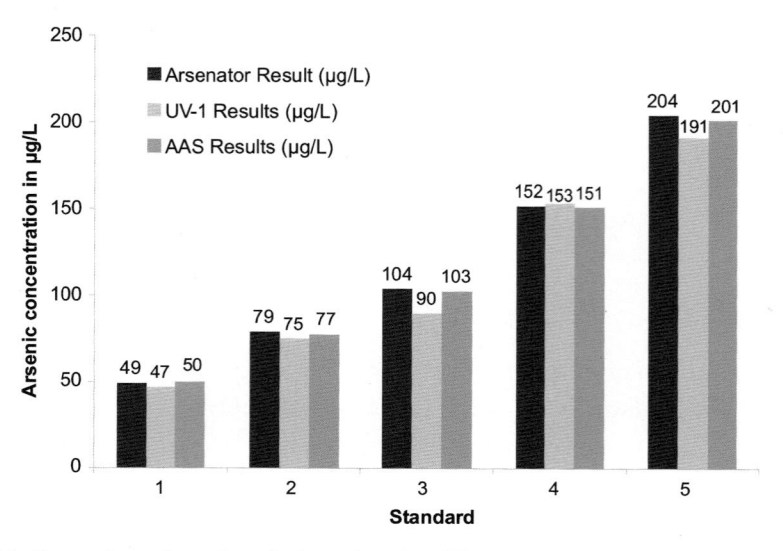

Fig. 4.2 Comparison of arsenic analysis results using different calibration standards for Arsenator, UV-1 and AAS

Table 4.1 Affected habitations spread across 76 blocks based upon findings from the present arsenic screening and surveillance programme

District	Total no of blocks[a]	Surveyed blocks	Total population in each district	Total population in surveyed blocks	Population at risk in the surveyed blocks	% of population at risk in the surveyed blocks
Sonitpur	14	3	1,755,000	269,000	19,981	7.4
Bongaigaon	5	3	533,000	290,000	10,299	3.6
Dhubri	15	7	1,531,000	760,000	47,716	6.3
Goalpara	8	2	850,000	243,000	6,527	2.7
Golaghat	7	4	779,000	415,000	56,773	13.7
Hailakandi	5	5	571,000	571,000	11,659	2.0
Jorhat	8	6	941,000	769,000	104,033	13.5
Bokakhat	1	1	199,000	199,000	6,408	3.2
Karimganj	7	6	1,092,000	991,000	40,146	4.1
Sibsagar	9	5	1,102,000	672,000	29,760	4.4
Cachar	15	8	1,609,000	752,000	132,971	17.7
Nagaon	20	1	2,298,000	112,000	541	0.5
Barpeta	11	6	1,354,000	926,000	51,230	5.5
Morigaon	6	1	846,000	343,000	16,852	4.9
Lakhimpur	9	5	951,000	476,000	37,350	7.8
Darrang	7	4	820,000	678,000	35,413	5.2
Nalbari	9	7	716,000	299,000	99,219	33.2
BTAD	11	2	1,053,000	86,000	15,725	18.3
Total	167	76	19,000,000	8,851,000	722,603	8.2

[a]Block is an administrative unit within the district

4 Preliminary Assessment of Arsenic Distribution in Brahmaputra River... 61

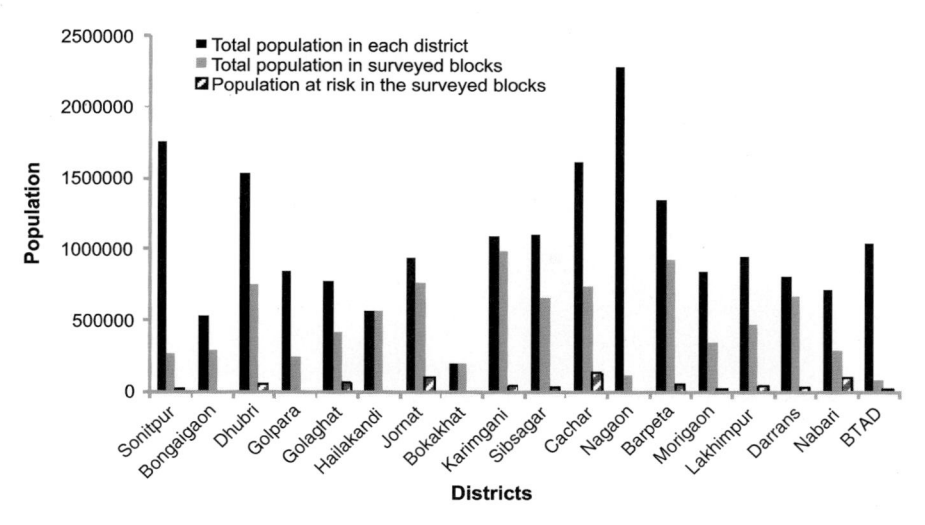

Fig. 4.3a Graph showing total number of blocks surveyed per district during the arsenic screening and surveillance programme

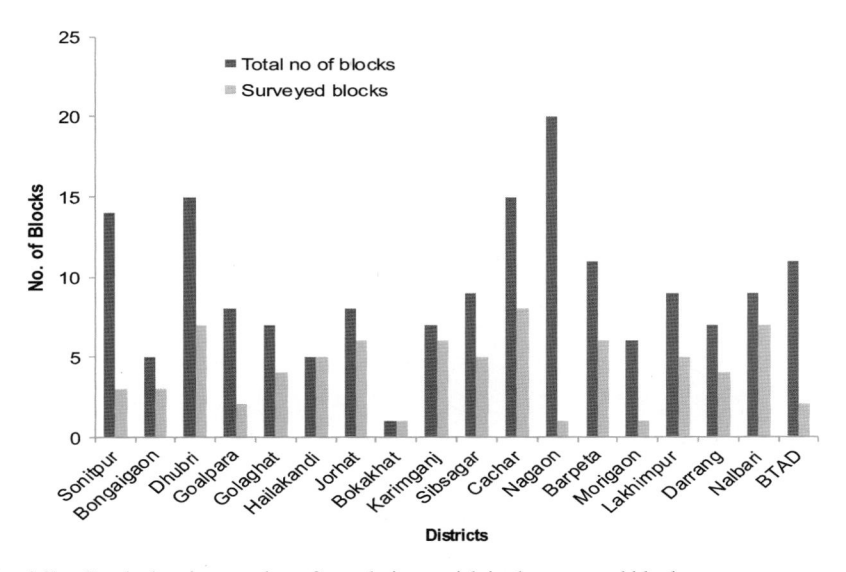

Fig. 4.3b Graph showing number of population at risk in the surveyed blocks

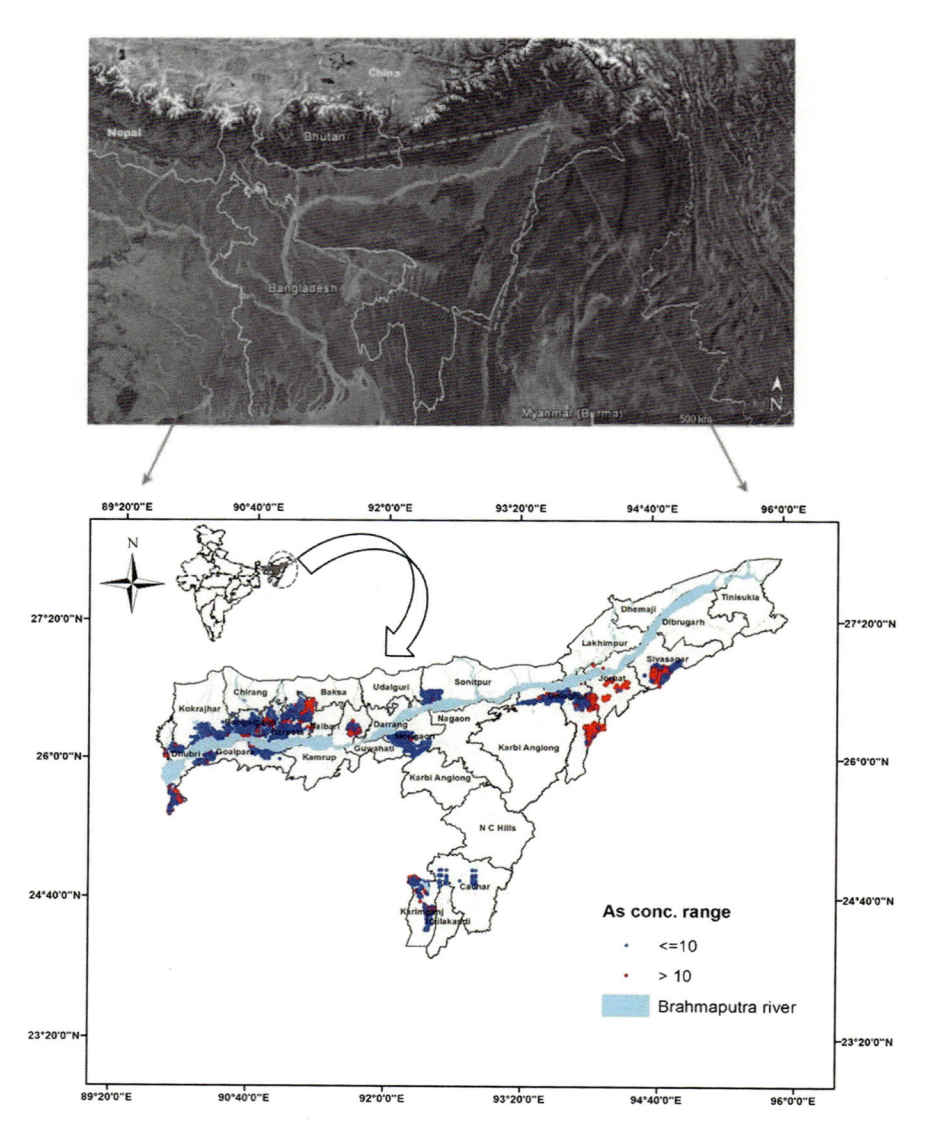

Fig. 4.4 Groundwater arsenic distribution map of the Brahmaputra basin in Assam, India with 10 μg/l as the limit, as identified from the samples collected from the present study

4.5 Discussion and Conclusion

A systematic three-phase study to carry out a primary assessment of groundwater As contamination in the Brahmaputra floodplain revealed groundwater As distributions to be localized at discrete pockets within the river basin, with variable concentration ranges. With little evidence of industrial sources, sources and mobilization mechanisms for arsenic release in the Brahmaputra floodplains in India are most likely dependant on geological, geomorphological and geochemical factors (e.g. sediment source, sediment grain size, density, fluvial dynamics of deposition) which may have played a collective role for such localised nature of arsenic contamination in the region. However, to ascertain the source and causes a detailed scientific study needs to be undertaken. In terms of the health impacts, reported cases of arsenicosis in the region is yet to be verified, indicating the need for a health mapping in the study area relating clinical symptoms to arsenic consumption through drinking water or other modes.

Acknowledgement This work is an outcome of the arsenic screening and surveillance programme in Assam, a joint initiative of Public Health Engineering Department, Assam with technical support from Indian Institute of Technology, Guwahati and financial assistance from UNICEF. The authors are thankful to PHED Assam and UNICEF for all the support provided for this study.

References

American Public Health Association (APHA) (2005) American Water Works Association and Water Pollution Control Federation. Standard methods for the examination of water and wastewater, 13th edn. American Public Health Association (APHA), Washington, DC

Bhattacharya P, Mukherjee A, Mukherjee AB (2011) Arsenic contaminated groundwater of India. In: Nriagu J (ed) Encyclopedia of environmental health. Elsevier B.V, Amsterdam

CGWB (1995) Hydrogeological atlas of Assam. Central Groundwater Board, Ministry of Water Resources, Government of India, Faridabad

CGWB (2008) Information booklet. Jorhat District. Central Ground Water Board, Ministry of Water Resources, Government of India. Retrieved from the url: http://cgwb.gov.in/District_Profile/Assam/JORHAT%20DISTRICT.pdf

Chetia M, Chatterjee S, Banerjee S, Nath MJ, Singh D, Srivastava RB, Sarma HP (2011) Groundwater arsenic contamination in Brahmaputra river basin: a water quality assessment in Golaghat (Assam), India. Environ Monit Assess 173:1393–1398

Enmark G, Nordborg D (2007) Arsenic in the groundwater of the Brahmaputra floodplains, Assam, India—Source, distribution and release, mechanisms. Retrieved from the url: www2.lwr.kth.se/Publikationer/PDF_Files/MFS_2007_131.pdf

Mahanta C, Pathak N, Bhattacharya P, Enmark G, Nordborg D (2008) Source, distribution and release mechanisms of arsenic in the groundwater of Assam floodplains of Northeast India. In: Proceedings of the World Environmental and Water Resources Congress sponsored by Environmental and Water Resources Institute (EWRI) of the American Society of Civil Engineers (ASCE). doi:10.1061/40976(316)78

Nriagu JO, Bhattacharya P, Mukherjee AB, Bundschuh J, Zevenhoven R, Loeppert RH (2007) Arsenic in soil and groundwater: an introduction. In: Bhattacharya P, Mukherjee AB, Bundschuh J, Zevenhoven R, Loeppert RH (eds) Arsenic in soil and groundwater environment: biogeochemical interactions, health effects and remediation. Trace metals and other contaminants in the environment, vol 9 (Ser Ed Nriagu JO). Elsevier, Amsterdam

Sharma JN (2005) Fluvial process and morphology of the Brahmaputra River in Assam, India. Geomorphology 70:226–256

Singh AK (2004) Arsenic contamination in groundwater of North Eastern India. In: Proceedings of 11th national symposium on hydrology with focal theme on water quality. National Institute of Hydrology, Roorkee

Smedley PL, Kinniburgh DG (2002) A review of the source, behaviour and distribution of arsenic in natural waters. Appl Geochem 17:517–568

Talukdar NC, Bhattacharyya D, Hazarika S (2004) Soils and agriculture. In: Singh VP, Sharma N, Shekhar C, Ojha P (eds) The Brahmaputra basin water resources. Kluwer Academic Publishers, Dordrecht

Chapter 5
Problem, Perspective and Challenges of Arsenic Contamination in the Groundwater of Brahmaputra Flood Plains and Barak Valley Regions of Assam, India

Nilotpal Das, Latu Khanikar, Rajesh Shah, Aparna Das, Ritusmita Goswami, Manish Kumar, and Kali Prasad Sarma

5.1 Introduction

The problem of naturally occurring As pollution in groundwater is a burning issue which has now been recognised as one of the greatest environmental hazards, threatening the lives of the millions across the globe (Nickson et al. 1998, 2000; Smith et al. 2000; Berg et al. 2001; Anawar et al. 2002; Smedley and Kinniburgh 2002; Guo et al. 2003; Ravenscroft et al. 2003; Smedley et al. 2003; Li et al. 2005; Polya et al. 2005; Anawar et al. 2006; Enmark and Nordborg 2001; Nriagu et al. 2007; Kumar et al. 2010a, b; Bundschuh et al. 2012). Long-term ingestion of drinking water having As concentration beyond the permissible limit of 50 µg/L leads to detrimental effects on human health. Epidemiological studies have shown that inorganic As is a serious toxicant and can cause a variety of adverse health effects, such as dermal changes, respiratory, pulmonary, cardiovascular, gastrointestinal, haematological, hepatic, renal, neurological, developmental, reproductive, immunologic lead to cancer and other degenerative effects of the circulatory and nervous system (Golub et al. 1998; Lin et al. 1998; NRC 2001; Ahamed et al. 2006). In view of the above perspective WHO in 1993 has lowered its earlier permissible limit of 50 µg/L in drinking water to 10 µg/L. The BIS has also endorsed 10 µg/L as the permissible limit for As in drinking water.

The most widespread As enrichment in groundwater has been documented from parts of Asia, including Bangladesh, India, Nepal, Cambodia, Thailand, Vietnam, China and Taiwan; this region is known as south and southeast Asian As belt (SSAAB) (Bhattacharya et al. 1997, 2002a, 2006, 2009; Acharyya et al. 1999; Smedley and Kinniburgh 2002; Bhattacharyya et al. 2003; Ahmed et al. 2004; Nickson et al. 2005). In terms of the clinical manifestation due to As induced

---

N. Das • L. Khanikar • R. Shah • A. Das • R. Goswami • M. Kumar • K.P. Sarma (✉)
Department of Environmental Science, Tezpur University,
Napaam Sonitpur, Assam 784 028, India
e-mail: sarmakp@tezu.ernet.in

diseases and percentage of tube wells with high As concentration, Bangladesh is the worst affected nation in the world (Nickson et al. 1998; Smith et al. 2000; Kinniburgh and Smedley 2001; Ravenscroft et al. 2005, 2009; Shamsudduha et al. 2008). In the recent decades, occurrence of elevated concentration of As in groundwater has been reported extensively from different parts of India namely the middle Gangetic plains (Bihar, UP, Jharkhand), lower Ganga (West Bengal) (Chakraborti et al. 2003; Ahamed et al. 2006; Nayaka et al. 2008; Shah 2008, 2010; Kumar et al. 2010a, b) and several states of northeast India (Chakraborti et al. 2004, 2008; Singh 2004; Shah 2007, 2012). However, in India the worst cases of As related health problem have been reported from the state of West Bengal where As problem is endemic and the first case of chronic As poisoning due to drinking water was diagnosed from the state way back in 1983 (Bundschuh et al. 2000, 2004; Bhattacharya et al. 2001, 2002a; Bhattacharyya et al. 2003; Smedley et al. 2002, 2003; Smedley and Kinniburgh 2002; Ben et al. 2003; Ahmed et al. 2004; Rahman et al. 2005; Samanta et al. 2004; Kumar et al. 2010a, b).

Occurrence of geogenic As in excess of the permissible limit of 10 µg/L in potable groundwater resources of north-east India was first reported in the year 2004 (Chakraborti et al. 2004; Singh 2004). However, the scale of the problem is potentially large in the state of Assam where 21 of the total 27 districts are severely affected by groundwater As contamination (Singh 2004). The maximum level of As was found in the flood plain of Jorhat, Lakhimpur, Nalbari and Nagaon districts (Hazarika et al. 2003; Singh 2004; Borah et al. 2009; CGWB 2009; Linthoingambi Devi et al. 2009; Baviskar et al. 2011; Chetia et al. 2011). Reductive dissolution of ferric hydroxides is believed to be the release mechanism for controlling the mobility of As in Assam (Mahanta et al. 2008). The reducing condition by microbial activities controls the dissociation of Fe from its compounds (iron hydr (oxides), ferromanganese compounds), which is responsible for the leaching of Fe into the groundwater from the aquifer sediments (Bhattacharya et al. 2002b; Roden 2006). As the As is adsorbed on these iron oxides or oxyhydroxides, the leaching of Fe facilitates As mobilization.

Almost all the As affected groundwaters in Assam and Manipur are located in the floodplains of Brahmaputra and Barak respectively (Singh et al. 1996; Chakraborti et al. 2004; Goswami et al. 2014). All the As affected districts of Assam (18 from BFP and three in the BVP) are overlain by successions of alluvial Quaternary sediments. The source of the As enriched sediments is widely believed to be the mountainous rocks. Himalayas towards the north appear to be the source of As in the BFP; and the Barail in the south appears to be the source of the As in the BVP (Chakraborti et al. 2004; Enmark and Nordborg 2001; Baviskar et al. 2011; Shah 2012). However, the probable As release medium in these two flood plains appear to be reductive dissolution of iron hydr (oxides) (Bhattacharya et al. 1997; Nickson et al. 1998, 2000; McArthur et al. 2001; Harvey et al. 2002; Ahmed et al. 2004; Kumar et al. 2010a, b). Under such an anoxic condition reduction of iron hydr(oxides) coatings on sand, silt and clay particles is a common phenomenon, which ultimately leads to the release of the adsorbed As (Nickson et al. 2000; Harvey et al. 2002;

Smedley and Kinniburgh 2002; Meharg and Rahman 2003; MacArthur et al. 2004; Zheng et al. 2004).

Very similar conditions are also present in the state of Assam, and in recent times the state has witnessed the problem of As contamination in its groundwater. Assam is drained by a number of important rivers, the entire northern and middle half of the state is drained by the river Brahmaputra, while towards the south the Barak river is a very important river system. The economy of Assam is agrarian and the people here depend on groundwater for a variety of utilities, most importantly drinking. Due to a strong monsoon season and the presence of two well developed drainage systems i.e. BFP and BVP, the dependence of the people on groundwater for cropping and irrigation is limited. The present work is a review of the state of As contamination and its release mechanism in the groundwater of the Brahmaputra Flood Plain (BFP) and Barak Valley Plain (BVP). Therefore, an attempt has been made to bring to light the current scenario of As contamination in the state of Assam. In this study we have mainly used secondary data as our source material.

5.2 Study Area Characterisation

The vast alluvial plains of Brahmaputra valley occupy most of the North Assam and it extends from Sadiya in the east to Dhubri in the west. The greater part of Assam lies within the Brahmaputra valley and its length is nearly 640 km and has the width of about 100 km. The Brahmaputra plain is formed of the sediments carried by the river Brahmaputra and its tributaries from the Himalaya and surrounding hills. It is bounded by Arunachal Himalaya in the north and northeast, Patkai-Naga-Lushai range of Nagaland and the Shillong Plateau in the south and southeast. The southern part of Assam (Barak Valley) consists of three districts viz. Cachar, Karimganj and Hailakandi (24°8′ and 25°8′ N latitudes and 92°15′ and 93°15′ E longitudes). The valley is bounded by the North Cachar Hills District of Assam and the state of Meghalaya to the north, Manipur to the east, Mizoram to the south, and Tripura and Bangladesh to the west (Fig. 5.1). The climate of the region is described as 'Tropical Monsoon Climate', which is characteristically warm and humid, and experiences a high rainfall under the influence of the south-east monsoon from June to September. Mean annual rainfall is 320 cm with a variability of 10–15 %. Rainfall varies along the stretch of the plains, from as low as 175 cm in the Kapili basin in central Assam to as high as 410 cm in Jidhal basin in the north-eastern part (Jain et al. 2007). Towards the south the rains are scantier in the rain shadow zones of Nagaon and Karbi Anglong. Minimum temperature during winter in the plains and the valley areas of the basin varies from 9 °C in the western part to 4 °C in the north-eastern part, summer temperatures are around an average of 35 °C, maximum temperature can go higher than 40 °C, especially in regions like Silchar in the extreme south (Jain et al. 2007).

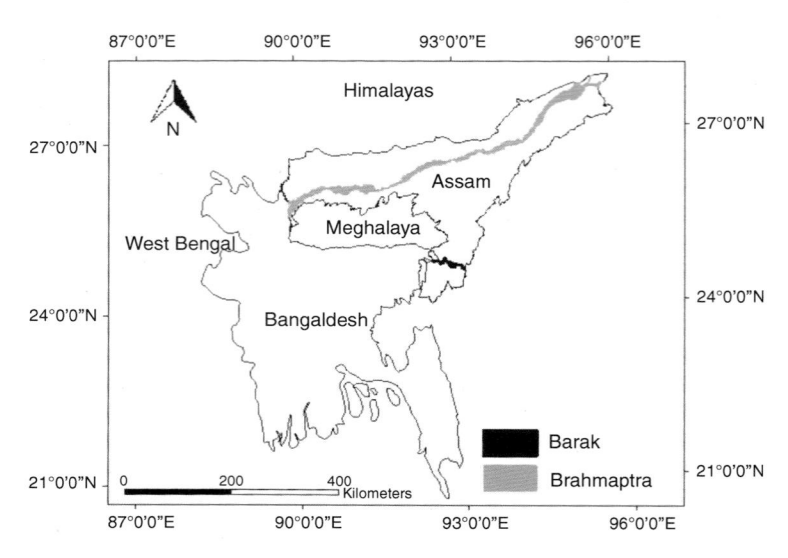

Fig. 5.1 Map of the study area, showing Brahmaputra Flood Plains (BFP) and Barak Valley Region (BVR)

5.2.1 Geology

In this region rocks representing diverse geological ages from Archaean to Recent are found to occur. The Brahmaputra and the Barak valleys (Fig. 5.1) of Assam were formed by the deposition of older and newer alluvium and more than 75 % of the state is underlain by unconsolidated formation comprising clay, silt, sand, gravel and pebble where ground water is available at low to moderate depth. The state of Assam has a unique geology, the northern plains (BFP) are dominated by deposits of new Quaternary Holocene alluvium, but tertiary sandstones along the fractures and solution cavities in limestones also occur (Jain et al. 2007). The BFP which is underlain by recent alluvial deposits are about 200–300 m thick and consist of clay silt, sand and gravels. In BFP groundwater occurrence in hard rock inselberg areas are controlled by foliations, fractures/joints and weathered joints. Both confined and unconfined aquifers are found in the BFP: groundwater in the piedmont and the alluvial fan deposits occurs generally under water table conditions. In relation to the surface topography, the water table has a roughly southerly slope.

Unconfined aquifers occur mainly in the older alluvium, while both confined as well as unconfined aquifers occur in the newer alluvium. Shallow and deeper aquifers also have different distribution, the former occur in the flood plain and inselberg zones, while the latter are more profound in the Chapar formation. The present configuration of BFP is known to have evolved during the last two million years of Pleistocene and Recent era. The valley is situated in an extremely unstable seismic zone. The morphology of the BFP in Assam is determined by two geological factors. First of all the Himalayas are uplifting at a rate of the order of one metre per century, and secondly as explained earlier the whole region is an extremely unstable seismic zone. Some of the most devastating earthquakes in the recent times have occurred in this region, like the earthquakes of 1987 and 1950 both of which had a magnitude of 8.7 on the Richter scale (Jain et al. 2007). Towards the south ortho and paragneisses, schist, migmatite, granulites, quartzite, phylite, quartz-sericite schist, conglomerate and porphyritic coarse granite constitute majority of the Karbi Anglong hills. Much of the North Cachar Hills are underlined by sandstones, shale, sandy shale, carbonaceous shale and coal seam, while much of Cachar is underlined by alluvium, sandstone, clay, shale, conglomerate, siltstones, mudstones and sandy shale.

The Surma basin of south Assam (Barak valley) forms a part of the greater Bengal basin (Sarkar and Nandy 1977). Barak valley consists of hilly terrain surrounded by bowl shaped synclinal valley elongated towards east to west. In Assam, Surma group is exposed in Surma valley and north Cachar hills. Geologically the Surma (Lower to Middle Miocene) are the oldest rock and has been subdivided into a lower arenaceous facies (Bhuban) formation and an upper argillaceous facies (Bokabil Formations). A major part of the Barak valley consists of Tertiary upland surfaces with the presence of shale, sandstone, ferruginous sandstone, mottled clay, pebble and boulder beds or lowland valley areas with thin cover of Holocene Newer Alluvium sediments on top. Both newer and older alluviums are found in the area. The semi-consolidated older rocks consists of shale, ferruginous sandstone, mottled clay, pebble bed and boulder bed etc., belonging to Bhuban and Barail groups of rocks formed under marine condition (Mathur and Evans 1964). Newer Alluvium Holocene deposits along Barak valley in the central parts mainly consist of sand, silt, carbonaceous shale and clay with gravel (Dasgupta and Biswas 2000).

5.3 Distribution of As in the State of Assam

In Northeastern India, As has been detected in 21 of the total 27 districts of Assam. Maximum As content was observed in Jorhat, Dhemaji, Golaghat and Lakhimpur districts (Singh 2004). In a study conducted by NERIWALM (Singh 2004) 1,500 water samples were collected from tube wells and dug wells during post-monsoon in the year 2003. The NERIWALM study shows that 21 out of Assam's then 24 districts have groundwater with an As content exceeding 50 µg/L and As concentration was found to be higher adjacent to Himalayan foothills. The concentration of As was relatively high in shallow tube wells (15–40 m deep) as compared to deep tube wells and rings wells. No report of *Arsenocosis* from any area of the region has been known till date. Chakraborti et al. (2004) also reported high groundwater As in the plains of upper Brahmaputra. Analysis of 137 tubewells samples from the area revealed that 43 % of the samples contained As above 10 µg/L and 26 % above 50 µg/L; the maximum concentration detected was 490 µg/L.

The PHED (Public Health Engineering Department) Assam conducted an assessment of presence of arsenic in groundwater of Assam in collaboration with UNICEF covering an area of 192 blocks spread across 22 districts and observed that 76 blocks in 18 districts were indeed at risk, with As level greater than 50 µg/L (JOPA 2005). Moreover, the PHED, UNICEF and IIT Guwahati jointly implemented the As screening and surveillance programme in which the surveillance of an estimated 56,180 public water sources was carried out. The result of the analysis was used for preparation of district level As concentration maps.

Available literature shows that groundwater of Assam valleys is highly ferruginous (Aowal 1981; Singh 2004). The problem of As contamination in groundwater of Assam has been investigated by various researchers (Baruah et al. 2003; Borah et al. 2009; Bhuyan 2010; Chetia et al. 2011; Buragohain and Sarma 2012).

5.3.1 Distribution in the Brahmaputra Flood Plain

Ministry of Drinking Water & Sanitation, Govt. of India report (2014) suggests that in many states of India, such as in Assam, Bihar, Uttar Pradesh, West Bengal and Karnataka, As contamination co-exists with Fe, salinity and nitrate (Table 5.1). The figure suggests that co-contamination percentage is highest in Uttar Pradesh in comparison to other states. Arsenic occurrence in India have been predominantly

Table 5.1 State-wise representation in the percentage of total groundwater wells found affected with arsenic contamination

State	No of wells analyzed	As	Fe	Salinity	NO_3
Andhra Pradesh	3,875	0.00	3.54	25.68	8.67
Assam	14,645	**3.86**	96.06	0.00	0.00
Bihar	14,068	**7.76**	77.82	0.00	0.01
Chhattisgarh	7,191	0.00	92.45	2.78	0.00
Gujarat	366	0.00	0.00	0.27	53.28
Haryana	15	0	0	0	0
Jharkhand	100	0	79	0	0
Karnataka	5,123	**0.43**	16.49	10.68	24.50
Kerala	928	0.00	62.82	18.10	7.22
Madhya Pradesh	2,888	0.00	6.82	2.74	0.00
Maharashtra	1,659	0.00	19.11	17.18	32.31
Orissa	10,264	0.00	87.97	7.81	0.11
Punjab	262	0.00	87.40	2.67	0.00
Rajasthan	24,437	0.00	0.05	64.91	5.85
Tamil Nadu	670	0.00	83.13	11.94	2.69
Uttar Pradesh	946	**16.38**	8.14	42.39	0.11
Uttarakhand	49	0.00	85.71	0.00	8.16
West Bengal	3,190	**18.93**	77.46	0.09	0.00

Source: Ministry of Drinking Water and Sanitation, 2014. Database for Delhi is not available but both As and F contamination are reported in this region

reported in aquifers with variable thicknesses but boreholes with depth ranging from around 10–80 m from the youngest, Holocene, alluvial deposits (CGWB 1999). Interestingly, high As concentrations have also been found in groundwater from some areas of crystalline basement rocks in northern India. Acharyya (2002) found locally high concentrations in groundwater from weathered granitic rocks from the Dongargarh rift belt of Chhattisgarh. These occurrences are under wetter conditions than those in the central region of India as the climatic regime is humid subtropical over here. Therefore due to presence of similar climatic pattern and Precambrian crystalline basement of granitic rock in Assam, we suspected the susceptibility of contamination of As in the groundwater.

The Majuli river island (in Jorhat district of Assam) is the largest inhabited riverine island in the world (Sarma and Phukan 2004) and is located in the Brahmaputra river. It has two administrative blocks, Majuli and Ujani Majuli. The island has seen many episodes of deposition and erosion in the recent years. The alluvial deposits of this island harbour aquifers are especially very high in As (Goswami et al. 2014). A preliminary survey of As content in ground water from shallow tube wells/hand pumps within 50 m depth of Majuli Island, was first conducted by Saha et al. (2009). Results of the chemical analysis of the 24 groundwater samples collected from shallow groundwater abstraction structures, mainly tubewells, have shown that the As ranged from 6 to 90 µg/L in the area. The concentration of As has been found above permissible limit of 10 µg/L at all locations except two.

A recent study on the As in groundwater Majuli (Goswami et al. 2014) revealed that As in excess of 50 µg/L occur in the groundwater of many of the villages of the Majuli sub-division. The result of analysis of 380 tube-well samples collected from all 20 blocks of Majuli sub-division for As concentration showed that 37 % and 16 % of the water samples have As concentration more than 10 µg/L (WHO guideline value of As in drinking water) and 50 µg/L (Indian standard of As in drinking water) respectively and 1.8 % samples had As concentrations >300 µg/L. A maximum concentration of 468 µg/L was reported from the Sriluhit gram Panchayat where 58.3 % of the tubewell contained As level more than 10 µg/L. The study also reported elevated level of Fe (36–22,000 µg/L) in the groundwaters of Majuli. Arsenic concentration in groundwater of Majuli was found to decrease gradually beyond 25 m depth of tube-wells and no correlation was observed between the concentration of As in water and depth. It was observed that As levels in tubewells deeper than 45 m are found to be less than 10 µg/L and is considered as safe. Similar results were also reported by Kumar et al. (2010a, b) from the middle Gangetic plains where elevated As levels were mostly observed in the tube-wells with a depth range of 10–20 m and As level decreased beyond a depth of 40 m. The range of As concentration in bore-hole sediment was 0.29–1.44 mg/kg.

Hydrogeological studies are required to understand the source and mobilization process of As in groundwater of Majuli. Early mitigation measures are urgently needed to save the inhabitants of Majuli from As exposure and possible health effects. It has been established by research that the dissolved As in the groundwater of GMB plain mostly originates from the sediments deposited by rivers. However, the Majuli research on As in groundwater (Goswami et al. 2014) has revealed that levels of As in the aquifer sediment is not rich in As and concentration is lower than background levels (5–10 mg/kg) suggesting that the mobilization of As in groundwater could be derived locally or proximally (Goswami et al. 2014). Besides, the As/Fe (identical values) in sediments from consecutive layers of a single borehole or different boreholes indicate that both Fe and As are coming from a definite source. Similar result was also observed in the earlier study of Chakraborti et al. (2001) in the Bengal delta region.

Statistically a poor correlation was found between As and Fe levels in sediment and the possible explanation of such poor relationship could be other processes by which Fe may be removed from groundwater and that some of these do not involve As (Harvey et al. 2002). Similarly poor correlation ($r=0.23$) between As and Fe was observed in the groundwater of Majuli. Researchers have already reported such poor relation between As level and Fe level in both water and sediments from GMB delta plain (Nickson et al. 2000; Ravenscroft et al. 2001; Ahmed et al. 2004; Mukherjee and Bhattacharya 2001). One of the possible reasons of this behaviour could be difference in the conservation status of the two species (Fe and As) in groundwater; another reason could be spatial difference in the ratio of As and iron hydr(oxides) based on the chemical characteristics, abundance and concentrations of the mineral phases (Ravenscroft et al. 2001; Goswami et al. 2014). Several other processes like mineral weathering, oxygen consumption and nitrate reduction may influence As release in groundwater, which can explain the poor relation observed between As and Fe in the groundwaters of the middle Gangetic plains of Ghazipur, Uttar Pradesh, India (Kumar et al. 2010a).

The level of As in biological samples (hair, nail and urine) of the inhabitant residing in the high As affected areas of the Majuli river island was investigated by Goswami et al. (2014) to evaluate the As body burden. The result of analysis of biological samples revealed that about 90 % urine samples, 100 % hair samples and 97 % nail samples contained As level above the maximum normal level. The maximum normal level of As in urine, hair and nail were 50 µg/kg, 200 µg/kg and 500 µg/kg respectively. It is noteworthy that the As in the hairs of the residents of Majuli was found out to be higher than those recorded in previous studies from Vietnam and Cambodia (Agusa et al. 2006; Berg et al. 2007; Nguyen et al. 2009).

But mean As concentrations in hair, nail and urine of Majuli residents were lower than those in other contaminated areas of the GMB flood plain (Table 5.2) (Goswami et al. 2014). Significant positive correlations were observed between As in drinking water and As concentrations in hair, nail and urine samples ($r=0.71$–0.78) (Goswami et al. 2014). Arsenic concentration in bore-hole sediment was 0.2–1.44 mg/kg. Early mitigation measures are urgently needed to save the inhabitants of Majuli from As exposure and possible health effects (Goswami et al. 2014).

Mahanta et al. (2008) had established that reductive dissolution of iron hydr(oxides) was the cause of As release in the flood plains of Assam. To investigate the possible release and mobilization mechanism of As in the aquifer, Baviskar et al. (2011) studied the borehole sediments from As enriched areas of Jorhat, Assam. It was observed that the groundwater in the study area was under reducing conditions and Fe(II) and As(III) were the dominant species. High concentrations of

Table 5.2 Comparison of arsenic in urine, hair and nail of the study group from arsenic affected locations of West Bengal and Bangladesh with that of Majuli island

Parameter	Urine (μg/L)			Hair (μg/kg)			Nail (μg/kg)		
	West Bengal	Bangladesh	Majuli	West Bengal	Bangladesh	Majuli	West Bengal	Bangladesh	Majuli
Sample no.	9,295	1,043	41	7,135	2,942	40	7,381	2,940	33
Mean	181	495	157	1,480	4,050	1,223	4,560	9,250	2,507
Max	3,147	3,086	697.5	20,340	28,060	5,461	44,890	79,490	11,725
Min	10	24	20.8	180	280	224	380	260	426
Median	116	302	101.8	1,320	2,490	943	3,870	6,740	1,797
SD	269	493	154.9	1,550	4,040	1039.8	3,980	8,730	2327.4
%[a]	92	99	90	57	97	100	83	96	97

[a]% of samples having arsenic above normal/toxic level

As ranging between 1 mg/kg and 18.6 mg/kg in the sediment samples was also observed as contrary to earlier report from Majuli where As concentration in sediments was very low (Goswami et al. 2014).

The chemical characteristics and nature of As release into the groundwater from borehole sediments were investigated by Baviskar et al. (2011). The objective of the study was to gain an understanding about the process of Fe leaching and As mobilization in Assam by investigating the sediments and groundwater samples collected from As enriched areas of Jorhat, Assam. Combined Eh-pH stability diagram of Fe and As indicated the presence of Fe(II) and As(III) species in the groundwater, indicating the presence of a reducing condition. Under reducing conditions arsenic generally exists in a trivalent form such as H_2AsO^{3-} and $HAsO_3^{2-}$ and the qualitative relationships between arsenic species can be understood with the help of Eh-pH diagram. However in oxidising environment As is present in the arsenate form (Ravenscroft et al. 2009). Concentrations of elevated As in the sediment samples collected at variable depths of the three boreholes ranged between 1 mg/kg and 18.6 mg/kg, with maximum concentrations found in the fine sand fractions. The silt and the fine sand sediment fractions seem to be the key repositories of high Fe and As concentrations in the study area, which is supported by the SEM/EDX and XRD analysis. Similar results are also reported from the fluvio deltaic sediments of West Bengal and Bangladesh (Nickson 1997; Perrin 1998; DPHE 1999; Ahmed et al. 2004).

The groundwater and sediment chemistry of Bongaigaon and Darrang districts of Assam were investigated to find out the distribution, origin and release mechanisms of arsenic (Enmark and Nordborg 2001). Fifty groundwater samples from domestic wells and public water supply schemes, and five sediment samples were collected. Fifteen groundwater samples had As concentration above the Indian national drinking standard (BIS) of 10 µg/L, and 33 of them had As concentration above the WHO guideline of 10 µg/L, besides DOC levels were moderately high ranging 0.3–4.2 mg/L. The study could not identify distinct zones or depths with specific sedimentological features producing arsenic-free water. However, the sediments with a green-olive colour are more likely to bear As contaminated water than white sediments and probably that reductive dissolution of ferric hydroxides is thought to be the release mechanism controlling the mobility of As. No relation was found between dissolved As and SO_4^{2-}. According to McArthur et al. (2004), aquifers having grey coloured sediments have elevated concentration of As as compared to brownish due to reduction of iron hydr (oxides).

In the extreme north-eastern part of the state the district of Dhemaji which lies in the northern bank of the river Brahmaputra experiences annual episodes of mass flooding and deposition. The district is also known to be one of the most As contaminated areas in Assam and some of the earliest incidences of As detection in groundwater come from this district. Arsenic level as high as 100–200 µg/L has been detected in Dhemaji district; the locations where high As levels in groundwater has been detected are Sissiborgaon, Dhemaji town and Jonai (Buragohain and Sarma 2012). Spatial distribution maps for arsenic in different seasons were prepared using curve fitting method in Arc View GIS software. Chetia et al. (2011)

investigated the As content in the 222 groundwater samples collected from shallow and deep tubewells of six blocks of Golaghat district (Assam). The results of the analysis showed that 67 %, 76.4 %, and 28.5 % were contaminated with higher As, Fe and Mn, respectively. A strong significant correlation was observed between As and Fe (0.697 at $p<0.01$), suggesting a possible reductive dissolution of As-Fe-bearing minerals for the mobilization of As in the groundwater of the region. Borah et al. (2009) studied the As, F$^-$ and Fe contamination in the tea gardens of Darrang district. Bhuyan (2010) found high levels of As in three developmental blocks of Lakhimpur district, Assam.

5.3.2 Arsenic Distribution in the Barak Valley Plain

In Cachar and Karimganj districts of Barak valley, Assam the presence of naturally occurring As in shallow tube wells (14–40 m) have been reported where 66 % and 26 % tubewells have As concentration above the 10 µg/L and 50 µg/L respectively. The earlier study carried by Shah (2007) also detected elevated As in groundwater from parts of Barak valley, Cachar and Karimganj districts. Most of the As affected areas of these two districts are covered by newer alluvial sediments of the Barak river basin, and are characterized by grey to black coloured fine grained organic rich argillaceous sediments. The As affected villages of these two districts are situated in entrenched channels and flood plains of Barak-Surma-Langai rivers system. In contrast, Plio-Pleistocene older alluvium aquifers, which are located at higher altitude or have thin cover of Newer Alluvium sediments, are safe from As (Shah 2012). In Barak valley As is getting released from the As bearing sediments which are brought down by the rivers from the surrounding Tertiary Barail hill range. The study also revealed that Fe content in the tube wells were especially high; 48 and 40 mg/L in Cachar and Karimganj districts respectively were the maximum recorded Fe levels. Even wells which were As safe had very high Fe (arsenic 3 µg/L and iron 23 mg/L). It was observed from the study that Fe and As didn't show any proper relation; in fact a reverse relation was observed between the two. Like in many other reported studies it was found that As decreased with an increase in depth. The study lays importance of Quaternary stratigraphy in As mobilization; most of the high As level groundwater was detected in aquifers which belonged to Holocene, while Plio-Pleistocene aquifers were As safe.

5.4 Conclusion

A uniform characteristic observed throughout the As affected areas in the state of Assam is that, these are mostly located in the alluvial plains of the rivers Brahmaputra and Barak. The process of reductive hydrolysis of iron hydr(oxides) appears to be the principal process of As release in the flood plains of the Brahmaputra and Barak.

Bacterial oxidation of organic matter in the newer sediments creates a reducing environment which is conducive for As release. Although reductive dissolution is the process of As mobilization in BFP and BVP, the actual source of As appears to be the sediments which are deposited by the rivers. Both the Brahmaputra as well as the Barak are mountainous rivers, the former originating from the Himalayas while the latter originating from the Barail hills in the border areas of Assam, Nagaland and Manipur. The ultimate source of As in BFP could be the Himalayan mountains while the Barail hill range could be the source of As in the BVP. Also it is important to note that not a single case of arsenicosis has been reported in the state of Assam due to the widespread use of sand filters among the common people. These filters, although crude, are quite effective in removing As from the water due to the presence of iron oxides and hydroxides which effectively adsorbs the As from the water.

Acknowledgements The authors are thankful to Mrs. Sanghita Dutta and Miss Jinu Devi, research scholars from the department of Environmental Science, Tezpur University, for their help in the preparation of the manuscript.

References

Acharya SK (2002) Arsenic contamination in groundwater affecting major parts of southern West Bengal and parts of western Chhattisgarh: source and mobilization processes. Curr Sci 82:740–744

Acharyya SK, Chakraborty P, Lahiri S, Raymahasay BC, Guha S, Bhowmik A (1999) Arsenic poisoning in Ganges Delta. Nature 401(6753):545–546

Agusa T, Kunito T, Fujihara J, Kubota R, Minh TB, Trang PTK, Iwata H, Subramanian A, Viet PH, Tanabe S (2006) Contamination by arsenic and other trace elements in tube-well water and its risk assessment to humans in Hanoi, Vietnam. Environ Pollut 139:95–106

Ahamed S, Kumar Sengupta M, Mukherjee A, Amir HM, Das B, Nayak B, Pal A, Mukherjee CS, Pati S, Nath DR, Chatterjee G, Mukherjee A, Srivastava R, Chakraborti D (2006) Arsenic groundwater contamination and its health effects in the state of Uttar Pradesh (UP) in upper and middle Ganga plain, India: a severe danger. Sci Total Environ 370(2–3):310–322

Ahmed KM, Bhattacharya P, Hasan MA, Akhter SH, Alam SMM, Bhuyan MAH, Imam MB, Khan AA, Sracek O (2004) Arsenic enrichment in groundwater of the alluvial aquifers in Bangladesh: an overview. Appl Geochem 19:181–200

Anawar HM, Akai J, Mostofa KMG, Safiullah S, Tareq SM (2002) Arsenic poisoning in groundwater: health risk and geochemical sources in Bangladesh. Environ Int 27:597–604

Anawar HM, Akai J, Yoshioka T, Konohira E, Lee JY, Fukuhara H, Tari Kul Alam M, Garcia Sanchez A (2006) Mobilization of arsenic in groundwater of Bangladesh: evidence from an incubation study. Environ Geochem Health 28:553–565

Aowal AFSA (1981) Design of an iron eliminator for hand tube wells. J IWWA XIII:63–64

Baruah MK, Kotoky P, Baruah J, Borah GC, Bora PK (2003) Arsenic association and distribution in carbonaceous materials in northeastern India. Curr Sci 85:204–208

Baviskar SM, Mahanta C, Choudhary R (2011) Leaching of iron, controlling factors and implication to arsenic mobilization in an aquifer of the Brahmaputra Floodplain. In: Annual international conference 2011 programme RGS-IBM, London, UK, 28 August 2011

Ben DS, Berner Z, Chandrasekharam D, Karmakar J (2003) Arsenic enrichment in groundwater of West Bengal, India: geochemical evidence for mobilization of As under reducing conditions. Appl Geochem 18:1417–1434

Berg M, Tran HC, Nguyen TC, Pham HV, Schertenlrib R, Giger W (2001) Arsenic contamination of groundwater and drinking water in Vietnam: a human health threat. Environ Sci Technol 35:2621–2626

Berg M, Stengel C, Trang PTK, Viet PH, Sampson ML, Leng M, Samreth S, Fredericks D (2007) Magnitude of arsenic pollution in the Mekong and Red River deltas – Cambodia and Vietnam. Sci Total Environ 372:413–425

Bhattacharya P, Chatterjee D, Jacks G (1997) Occurrence of arsenic contamination of groundwater in alluvial aquifers from Delta Plain, Eastern India: option for safe drinking supply. Int J Water Resour 13:79–92

Bhattacharya P, Jacks G, Jana J, Sracek A, Gustafsson JP, Chatterjee D (2001) Geochemistry of the Holocene Alluvial sediments of Bengal Delta Plain from West Bengal, India: implications on arsenic contamination in groundwater. In: Jacks G, Bhattacharya P, Khan AA (eds) Groundwater arsenic contamination in the Bengal delta plain of Bangladesh, TRITA-AMI report 3084. KTH Special Publication. Department of Geology, University of Dhaka, Dhaka, Bangladesh

Bhattacharya P, Frisbie SH, Smith E, Naidu R, Jacks G, Sarkar B (2002a) Arsenic in the environment: a global perspective. In: Sarkar B (ed) Handbook of heavy metals in the environment. Marcell Dekker Inc., New York

Bhattacharya P, Jacks G, Ahmed KM, Khan AA, Routh J (2002b) Arsenic in groundwater of the Bengal Delta Plain aquifers in Bangladesh. Bull Environ Contam Toxicol 69(4):538–545

Bhattacharya P, Ahmed KM, Hasan MA, Broms S, Fogelström J, Jacks G, Sracek O, von Brömssen M, Routh J (2006) Mobility of arsenic in groundwater in a part of Brahmanbaria district, NE Bangladesh. In: Naidu R, Smith E, Owens G, Bhattacharya P, Nadebaum P (eds) Managing arsenic in the environment: from soils to human health. CSIRO Publishing, Collingwood

Bhattacharya P, Hasan MA, Sracek O, Smith E, Ahmed KM, von Brömssen M, Huq SMI, Naidu R (2009) Groundwater chemistry and arsenic mobilization in the Holocene flood plains in south-central Bangladesh. Environ Geochem Health 31:23–44

Bhattacharyya R, Jana J, Nath B, Sahu S, Chatterjee D, Jacks G (2003) Groundwater arsenic mobilization in the Bengal Delta Plain: the use of ferralite as a possible remedial measure—a case study. Appl Geochem 18:1435–1451

Bhuyan B (2010) A study on arsenic and iron contamination of groundwater in three development blocks of Lakhimpur District, Assam, India. Report Opin 2(6):82–87. http://www.sciencepub.net/report

Borah K, Bhuyan B, Sarma HP (2009) Lead, arsenic, fluoride and iron contamination of drinking water in the tea garden belt of Darrang district, Assam, India. Environ Monit Assess 169:347–352

Bundschuh J, Bonorino G, Viero AP, Albouy R, Fuertes A (2000) Arsenic and other trace elements in sedimentary aquifers in the Chaco-Pampean Plain, Argentina: origin, distribution, speciation, social and economic consequences. In: Bhattacharya P, Welch AH (eds) Arsenic in groundwater of sedimentary aquifers, pre-congress workshop. 31st International Geological Congress. Rio de Janeiro, Brazil

Bundschuh J, Farias B, Martin R, Storniolo A, Bhattacharya P, Cortes J et al (2004) Groundwater arsenic in the Chaco-Pampean Plain, Argentina: case study from Robles Country, Santiago del Estero Province. Appl Geochem 19:231–243

Bundschuh J, Litter MI, Parvez F, Román-Ross G, Nicolli HB, Jean J-S, Liu C-W, López D, Armienta MA, Guilherme LRG, Cuevas AG, Cornejo L, Cumbal L, Toujaguez R (2012) One century of arsenic exposure in Latin America: a review of history and occurrence from 14 countries. Sci Tot Environ 429:2–35

Buragohain M, Sarma HP (2012) A study on spatial distribution of Arsenic in groundwater samples of Dhemaji district of Assam, India by using Arc View GIS software. Sci Revs Chem Commun 2(1):7–11, ISSN 2277–2669

CGWB (1999) Hydrogeological framework for urban development of Bhopal City, Madhya Pradesh. Central Ground Water Board, North Central Region, Bhopal

CGWB (2009) Central Ground Water Board, Ministry of Water Resources, Government of India. Retrieved from url: http://cgwb.gov.in/

Chakraborti D, Basu GK, Biswas GK, Chowdhury UK, Rahman MM, Paul K, Chowdhury TR, Ray SL (2001) Characterization of arsenic bearing sediments in Gangetic Delta of West Bengal, India. In: Chappell WR, Abernathy CO, Calderon RL (eds) Arsenic exposure and health effects. Elsevier Science, New York, pp 27–52

Chakraborti D, Mukherjee SC, Pati S, Sengupta MK, Rahman MM, Chowdhury UK, Lodh D, Chanda CR, Chakraborti AK, Basu GK (2003) Arsenic groundwater contamination in middle Ganga plain, Bihar, India: a future danger. Environ Health Perspect 111:1194–1201

Chakraborti D, Sengupta MK, Rahman MM, Ahamed S, Chowdhury UK, Hossain MA et al (2004) Groundwater arsenic contamination and its health effects in the Ganga-Meghna-Brahmaputra plain. J Environ Monit 6:75N–83N

Chakraborti D, Singh E, Das B, Shah BA, Hossain MA, Nayak B, Ahamed S, Singh NR (2008) Groundwater arsenic contamination in Manipur, one of the seven North-Eastern Hill states of India: a future danger. Environ Geol 56:381–390

Chetia M, Chatterjee S, Banerjee S, Nath MJ, Singh L, Srivastava RB, Sarma HP (2011) Groundwater arsenic contamination in Brahmaputra river basin: a water quality assessment in Golaghat (Assam), India. Environ Monit Assess 173(1–4):371–385. doi:10.1007/s10661-010-1393-8

Dasgupta AB, Biswas AK (2000) Geology of Assam. Geological Society of India, Bangalore

DPHE (1999) Groundwater studies for arsenic contamination in Bangladesh. Rapid investigation phase. Final report, Mott MacDonald Ltd and British Geological Survey. Report for the Department for Public Health Engineering and the Department for International Development

Enmark G, Nordborg D (2001) Arsenic in the groundwater of the Brahmaputra floodplains, Assam, India – Source, distribution and release mechanisms. Committee of Tropical Ecology ISSN 1653–5634 minor field study 131. Uppsala University, Sweden

Golub MS, Macintosh MS, Baumrind N (1998) Developmental and reproductive toxicity of inorganic arsenic: animal studies and human concerns. J Toxicol Environ Health Part B 1(3):199–241

Goswami R, Rahman MM, Murril M, Sarma KP, Thakur R, Chakraborti D (2014) Arsenic in the groundwater of Majuli – the largest river island of the Brahmaputra: magnitude of occurrence and human exposure. J Hydrol 5:354–362

Guo HM, Wang YX, Shpeizer GM, Yan SL (2003) Natural occurrence of arsenic in shallow groundwater, Shanyin, Datong basin, China. J Toxicol Environ Health Part A, Environ Sci Eng Toxic Hazard Subst Control 38:2565–2580

Harvey CF, Swartz CH, Badruzzaman BM, Keon-Blute N, Yu W, Ali MA et al (2002) Arsenic mobility and groundwater extraction in Bangladesh. Science 298(5598):1602–1606

Hazarika S, Baruah MK, Kotoky P, Baruah J, Borah GC, Bora PK (2003) Arsenic association and distribution in carbonaceous materials in northeastern India. Curr Sci 85:204–208

Jain KS, Agarwal PK, Singh VP (2007) Hydrology and water resources of India. Water Sci Technol Libr 57:419–472. Springer

JOPA (2005) Joint plan of action for arsenic and fluoride mitigation in Assam. Public Health Engineering Department, Assam/New Delhi

Kinniburgh DG, Smedley PL (2001) Arsenic contamination of groundwater in Bangladesh, vol 3. Hydrochemical atlas. British geological survey technical report, Keyworth

Kumar M, Kumar P, Ramanathan AL, Bhattacharya P, Thunvik R, Singh UK, Tsujimura M, Sracek O (2010a) Arsenic enrichment in groundwater in the middle Gangetic Plain of Ghazipur District in Uttar Pradesh, India. J Geochem Explor 105:83–94

Kumar P, Kumar M, Ramanathan AL, Tsujimura M (2010b) Tracing the factors responsible for arsenic enrichment in groundwater of the middle Gangetic Plain India: a source identification perspective. Environ Geochem Health 32:129–146

Li J, Wang Z, Cheng X, Wang S, Jia Q, Han L et al (2005) Investigation of the epidemiology of endemic arsenism in Ying County of Shanxi Province and the content relationship between water fluoride and water arsenic in aquatic environment. Chin J Endem 24:183–185

Lin T-H, Huang Y-L, Wang M-Y (1998) Arsenic species in drinking water, hair, fingernails and urine of patients with Blackfoot disease. J Toxicol Environ Health 53:85–93

Linthoingambi Devi N, Yadav Ishwar Chandra, QI Shihua (2009) Recent status of arsenic contamination in groundwater of Northeastern India – a review. Report Opin 1(3):22–32

Mahanta C, Pathak N, Bhattacharya P, Enmark G, Nordborg D (2008) Source, distribution and release mechanisms of arsenic in the groundwater of Assam floodplains of Northeast India. In: Proceedings of the World Environmental and Water Resources Congress sponsored by Environmental and Water Resources Institute (EWRI) of the American Society of Civil Engineers, pp 1–19

Mathur LP, Evans P (1964) Oil in India. 22nd session International Geological Congress proceedings, New Delhi

McArthur JM, Ravenscroft P, Safiulla S, Thirlwall MF (2001) Arsenic in groundwater: testing pollution mechanisms for sedimentary aquifers in Bangladesh. Water Resour Res 37(1):109–117

McArthur JM, Banerjee DM, Hudson-Edwards KA et al (2004) Natural organic matter in sedimentary basins and its relation to arsenic in anoxic groundwater: the example of West Bengal and its worldwide implications. Appl Geochem 19:1255–1293

Meharg AA, Rahman M (2003) Arsenic contamination of Bangladesh paddy fields soils: implications for rice contribution to arsenic consumption. Environ Sci Technol 37:229–234

Mukherjee AB, Bhattacharya P (2001) Arsenic in groundwater in the Bengal Delta Plain: slow poisoning in Bangladesh. Environ Rev 9:189–220

National Research Council (2001) Arsenic in drinking water 2001 update. National Academy Press, Washington, DC

Nayaka B, Dasa B, Mukherjee SC, Pala A, Ahamed S, Amir Hossain M, Maitya P, Dutta RN, Dutta S, Chakraborti D (2008) Groundwater arsenic contamination in the Sahibganj district of Jharkhand state, India in the middle Ganga plain and adverse health effects. Toxicol Environ Chem 90(4):673–694

Nguyen VA, Bang S, Viet PH, Kim KW (2009) Contamination of groundwater and risk assessment for arsenic exposure in Ha Nam province, Vietnam. Environ Int 35:466–472

Nickson RT (1997) Arsenic in groundwater, Central Bangladesh. Unpublished M.Sc. thesis, University College London, London

Nickson R, McArthur JM, Burgess W, Ahmed KM, Ravenscroft P, Rahman M (1998) Arsenic poisoning of Bangladesh groundwater. Nature 395:338

Nickson RT, McArthur JM, Ravenscroft P, Burgess WG, Ahmed KM (2000) Mechanism of arsenic release to groundwater, Bangladesh and West Bengal. Appl Geochem 15(4):403–413

Nickson RT, McArthur JM, Sengupta B, Kyaw-Myint TO, Lowry D (2005) Arsenic and other drinking water quality issues, Muzaffargarh District, Pakistan. Appl Geochem 20:55–68

Nriagu JO, Bhattacharya P, Mukherjee AB, Bundschuh J, Zevenhoven R, Loeppert RH (2007) Arsenic in soil and groundwater: an introduction. In: Bhattacharya P, Mukherjee AB, Bundschuh J, Zevenhoven R, Loeppert RH (eds) (Series Editor Nriagu JO) Arsenic in soil and groundwater environment: biogeochemical interactions, health effects and remediation. Trace metals and other contaminants in the environment, vol 9. Elsevier, Amsterdam, pp 1–58

Perrin J (1998) Arsenic in groundwater at Meherpur, Bangladesh: a vertical porewater profile and rock/water interactions. M.Sc. thesis (unpub), University College London

Polya DA, Gault AG, Diebe N, Feldman P, Rosenboom JW, Gilligan E et al (2005) Arsenic hazard in shallow Cambodian groundwaters. Mineral Mag 69:807–823. doi:10.1180/0026461056950290

Rahman MM, Sengupta MK, Ahamed S, Chowdhury UK, Das B, Hossain MA, Lodh D, Saha KC, Palit SK, Chakraborti D (2005) A detailed study of the arsenic contamination of groundwater and its impact on residents in Rajapur village of the Domkal block, district Murshidabad, West Bengal, India. Bull World Health Organ 83(1):49–57

Ravenscroft P, McArthur JM, Hoque BA (2001) Geochemical and palaeohydrological controls on pollution of groundwater by arsenic. In: Chapell WR, Abernathy CO, Calderon R (eds) Arsenic exposure and health effects, vol IV. Elsevier Science Ltd, Oxford

Ravenscroft P, Burgess GW, Ahmed KM, Burren M, Perrin J (2003) Arsenic in groundwater of the Bengal Basin, Bangladesh: distribution, field relations and hydrogeological setting. Hydrogeol J 13:727–751

Ravenscroft P, Burgess WG, Ahmed KM, Burren M, Perrin J (2005) Arsenic in groundwater of the Bengal Basin, Bangladesh: distribution, field relations and hydrological setting. Hydrogeol J 13:727–751

Ravenscroft P, Brammer H, Richards K (2009) Arsenic pollution: a global synthesis. Wiley, Chichester. ISBN 978-1-405-18602-5

Roden EE (2006) Geochemical and microbiological controls on dissimilatory iron reduction. Compt Rendus Geosci 338:456–467

Saha GC, Samanta SK, Kumar S (2009) Arsenic contamination in ground water in Majuli Island, Jorhat District, Assam. Bhujal News, Quarterly Journal of Central Ground Water Board, pp 107–113

Samanta G, Sharma R, Roychowdhury T, Chakraborti D (2004) Arsenic and other elements in hair, nails and skin-scales of arsenic victims in West Bengal, India. Sci Total Environ 326(1–3):33–47

Sarkar K, Nandy DR (1977) Structures and tectonics of Tripura-Mizoram area, India. Geol Surv India Misc Publ 34(1):141–148

Sarma JN, Phukan MK (2004) Origin and some geomorphological changes of Majuli Island of Brahmaputra River in Assam, India. Geomorphology 60:1–19

Shah BA (2007) Arsenic in groundwater from parts of Barak Valley, Chachar and Karimganj districts, Assam. Indian J Geol 79:59–62

Shah BA (2008) Role of Quaternary stratigraphy on arsenic contaminated groundwater from parts of Middle Ganga Plain. Environ Geol 53:1553–1561

Shah BA (2010) Arsenic contaminated groundwater in Holocene sediments from parts of Middle Ganga Plain, Uttar Pradesh. Curr Sci 98:10

Shah B (2012) Role of Quaternary stratigraphy on arsenic-contaminated groundwater from parts of Barak Valley, Assam, North-East India. Environ Earth Sci 66(8):2491–2501

Shamsudduha M, Uddin A, Saunders JA, Lee MK (2008) Quaternary stratigraphy, sediment characteristics and geochemistry of arsenic-contaminated alluvial aquifers in the Ganges-Brahmaputra floodplain in central Bangladesh. J Contam Hydrol 99:112–136

Singh AK (2004) Arsenic contamination in groundwater of North Eastern India. In: Proceedings of National seminar on Hydrology with focal theme on "Water Quality" held at National Institute of Hydrology, Roorkee during Nov 22–23

Singh DB, Prasad G, Rupainwar DC (1996) Adsorption technique for the treatment of As(V) rich effluents. Colloids Surf A Physiochem Eng Asp 111:49–56

Smedley PL, Nicolli HB, Macdonald DMJ, Barros AJ, Tullio JO (2002) Hydrogeochemistry of arsenic and other inorganic constituents in groundwaters from La Pampa, Argentina. Appl Geochem 17:259–284

Smedley PL, Kinniburgh DG (2002) A review of the source, behaviour and distribution of arsenic in natural waters. Appl Geochem 17(5):517–568

Smedley PL, Zhang M, Zhang G, Luo Z (2003) Mobilisation of arsenic and other trace elements in fluvio-lacustrine aquifers of the Huhhot Basin, Inner Mongolia. Appl Geochem 18(9):1453–1477

Smith AH, Lingas EO, Rahman M (2000) Contamination of drinking-water by arsenic in Bangladesh: a public health emergency. Bull World Health Organ 78(9):1093–1103

Zheng Y, Stute M, van Geen A, Gavrieli I, Dhar R, Simpson HJ, Schlosser P, Ahmed KM (2004) Redox control of arsenic mobilization in Bangladesh groundwater. Appl Geochem 19(2):201–214

Chapter 6
Arsenic Contamination of Groundwater in Barak Valley, Assam, India: Topography-Based Analysis and Risk Assessment

Abhik Gupta, Dibyajyoti Bhattacharjee, Pronob Borah, Tushar Debkanungo, and Chandan Paulchoudhury

6.1 Introduction

Arsenic (As) contamination of groundwater is a major environmental and public health issue in the Ganga-Brahmaputra-Meghna (GBM) plains, including almost all states in the Ganga basin, large areas of Bangladesh (GBM basin), and some districts of Assam (Brahmaputra and Meghna sub-basins of the GBM basin) in North East India (Chowdhury et al. 2000; Chakraborti et al. 2003; Singh 2004; Ahamed et al. 2006; Nickson et al. 2007). Further, the detection of As in the groundwater of several districts of Manipur, which are part of the Chindwin-Irrawaddy basin (Chakraborti et al. 2008; Oinam et al. 2011, 2012; Singh et al. 2013), suggests possible As contamination of groundwater in the river valleys of Myanmar as well. Arsenic in the groundwater of this area is derived from the microbial reductive dissolution of iron (Fe) oxyhydroxide and subsequent release of the sorbed As (Nickson et al. 2000; Winkel et al. 2008). The presence of arsenic in this region is, therefore, a natural phenomenon that warrants realistic assessment of the risks involved followed by proper, often locale-specific management in order to reduce public misery while achieving safe and sustainable utilization of surface- as well as ground-water resources. The problem of arsenic contamination of groundwater in the lower Ganga basin state of West Bengal and in Bangladesh has been amply highlighted (Nickson

---

A. Gupta (✉) • P. Borah
Department of Ecology & Environmental Science, Assam University, Silchar, Assam, India
e-mail: abhik.eco@gmail.com

D. Bhattacharjee
Department of Business Administration, Assam University, Silchar, Assam, India

T. Debkanungo
Department of Chemistry, G.C. College, Silchar, Assam, India

C. Paulchoudhury
Department of Geology, G.C. College, Silchar, Assam, India

et al. 1998, 2000; Chowdhury et al. 2000; Smith et al. 2000; Rahman et al. 2001; Fazal et al. 2001; Chakraborti et al. 2009) since long, while that in the middle Ganga plain is relatively less widely known, although the situation may be no less alarming in this region as well.

Arsenic in groundwater in this area was first detected in the state of Bihar a decade earlier (Chakraborti et al. 2003). Subsequently, arsenic was also detected in Uttar Pradesh in the upper and the middle Gangetic plain. In the three Uttar Pradesh districts of Ballia, Varanasi and Ghazipur, 46.5 % of samples tested exceeded 10 µg L^{-1}, 26.7 % 50 µg L^{-1}, and 10 % 300 µg L^{-1}, with the highest concentration recorded being 3,192 µg L^{-1}. However, in the worst-affected *gaon panchayats* (village cluster), 91.8 % samples were above 10 µg L^{-1}, 75.4 % above 50 µg L^{-1}, and 41 % above 300 µg L^{-1} (Ahamed et al. 2006). Nickson et al. (2007) detected arsenic at levels greater than 50 µg L^{-1} in 10.8 % of samples tested in Bihar, 2.4 % of samples from Uttar Pradesh and 3.7 % from Jharkhand. Arsenic concentrations of up to 468 µg L^{-1} was found in the groundwater of Ballia district, Uttar Pradesh, at a depth of 30–33 m, while it was 12–20 µg L^{-1} at a depth of 66–75 m (Chauhan et al. 2009). Fluvial input was found to be the main source of arsenic in the middle Gangetic plain with 47 % of the samples exceeding the Indian permissible limit of 50 µg L^{-1}, and with a high percentage contribution (66 %) of toxic As III to total arsenic concentration. High As concentrations were associated with low NO_3^- concentrations, while there was poor correlation between As and Fe (Kumar et al. 2010a, b). In the Ganga and Sone river floodplains in Bihar, shallow aquifers contained arsenic up to 178 µg L^{-1}. Flood-prone areas with fluvial swamps were more arsenic-prone. The newer alluvial tracts of Patna and Bhojpur districts had a two-tier aquifer system of which the shallow (<50 m depth) aquifer was contaminated with arsenic, while the deeper (>130 m depth) aquifer had extremely low As load (Saha 2009; Saha et al. 2011).

It is noteworthy that Bangladesh as well as the Indian states of Uttar Pradesh, Bihar, West Bengal, Assam and Manipur in this arsenic-affected stretch of the Indian subcontinent are rich in surface water ecosystems like rivers and wetlands, and receive adequate average rainfall of *c.a.* 1,000–3,000 mm annually. Traditionally, the people of this area met their drinking and other domestic water needs from surface water bodies, especially ponds and tanks, and shallow groundwater in open wells. Poor sanitation resulting in large-scale coliform contamination of surface water ecosystems as well as open wells led to widespread morbidity and mortality, especially of infants. Switching over to groundwater as the main source of domestic water was seen as a deliverance from the scourge of enteric diseases. Use of shallow groundwater in irrigation was also instrumental in augmenting agricultural production in this region. Detection of arsenic in the groundwater of this area, therefore, imposes constraints on its continued use for both domestic and agricultural use. Widespread presence of arsenic also thwarts the recent trend of trying to achieve self-sufficiency in water supply among many urban as well as rural consumers by investing in a hand- or electrically-operated shallow (<100 m) tubewell. Such self-sufficiency is increasingly sought because of the uncertainty as well as insufficiency of domestic water supply operated by government or quasi-government service providers to the burgeoning consumers. The proliferation of multistorey apartments in urban centres that has been at least partly able to address the housing problem in this region has been made possible, among other factors, by the exploitation of groundwater. Groundwater management is, therefore, an important issue in the entire GBM basin, as well as in the other As-affected areas of south and Southeast Asia.

Most of the existing studies on arsenic contamination in different states or regions of India have taken the approach of assessment at the administrative district or block level (Chowdhury et al. 2000; Ahamed et al. 2006; Nickson et al. 2007; Chakraborti et al. 2008). While such mapping has been useful in bringing to light the status and extent of arsenic contamination in a given area, it is not able to show the local, micro-level differences in the risk involved in consumption of groundwater in that area. Furthermore, the boundaries between administrative units are often more political or linguistic rather than physical or topographical. Consequently, adjacent blocks or districts may share highlands or low-lying floodplain areas that may differ considerably in the arsenic concentration of groundwater or in the frequency of occurrence of arsenic-contaminated wells or both. Thus low arsenic pockets are likely to exist in blocks or districts labeled as highly or mildly affected and vice versa. Block or district level assessments, therefore, do not provide to the potential users a realistic appraisal of the risk involved in investing in a groundwater well in their respective localities.

Keeping this in mind, the present study has classified the groundwater collection sites in Barak valley, Assam, India, which represents the extreme northeastern corner of the Meghna sub-basin of the GBM basin, into clusters on the basis of their geographical location and broad topographical features, such as the upstream or downstream sections of a river valley, low-lying floodplains, and relatively high ground between river valleys and amidst floodplains. Similar topographic variations were observed to influence groundwater As distribution in Bangladesh and Cambodia (Buschmann et al. 2007; Shamsudduha et al. 2009; Hoque et al. 2011). Subsequently, the potential risk of consuming groundwater in different areas of Barak valley was estimated by Bayesian Statistics and Monte Carlo Simulation. Such an approach could help the existing and potential groundwater users in the GBM basin and other areas affected through similar mechanisms of arsenic contamination to objectively assess the risk of finding arsenic in their tubewell water and adopt appropriate management strategies.

6.2 Material and Methods

6.2.1 The Study Area

The state of Assam in the northeastern region of India comprises two valleys: the larger Brahmaputra valley in the north and the smaller Barak valley, which is a part of the Meghna basin of India-Bangladesh, in the south. Barak valley includes the three south Assam districts of Cachar, Hailakandi and Karimganj, situated at 24°8′–25°8′ N and 92°15′–93°15′ E with a geographical area of 6,921 km². Barak valley is flanked by the North Cachar Hills district of Assam and the state of Meghalaya to the north; Manipur to the east; Mizoram to the south; and Tripura and Bangladesh to the west. This area is drained by R. Barak and its tributaries such as Jiri, Chiri, Dolu-Jatinga, Madhura, Larang, Harang and Kalainchera on the north bank and Sonai-Rukni, Ghagra, Katakhal and Dhaleswari on the south bank. The Barak bifurcates into two branches – Surma and Kusiara – before entering the Sylhet district of Bangladesh. Singla is among the major tributaries of Kusiara, while Longai, another important river that originates in the hills of Mizoram, flows through Karimganj district to flow into a wetland in Bangladesh. All these rivers, especially the Barak, which adopt a meandering path through the alluvial plains, frequently shift their courses and form ox-bow lakes.

The topography of this area is heterogeneous comprising small hillocks strewn within plain areas and low lying river floodplains that are locally called as *beels* or *haors*. The climate is humid sub-tropical with temperatures ranging from *c.a.* 9–38 °C and an annual average rainfall of >2,500 mm. Geologically the region comprises unconsolidated alluvial deposits of sub-recent to recent age and semi-consolidated Bhuban, Bokabil, Tipam, Dupitila and Dihing formations of Miocene to Pliocene age. The alluvial deposits are in the central parts of the valleys and made up of sand, silt and clay. The semi-consolidated rocks in the hills are composed of shale, sandstone, pebble and boulder beds. Groundwater occurs in phreatic condition in shallow aquifer and semi-confined condition in deeper aquifer. Flow of groundwater is north to south in the northern and south to north in the southern parts of the valley.

The 644 tubewells sampled in this study were distributed in the three districts in 15 clusters, of which 12 were in the floodplains of River Barak and its major tributaries, while three were in low hills and in terrain elevated from the floodplain (Fig. 6.1). A brief description of the 15 sites is given in Table 6.1.

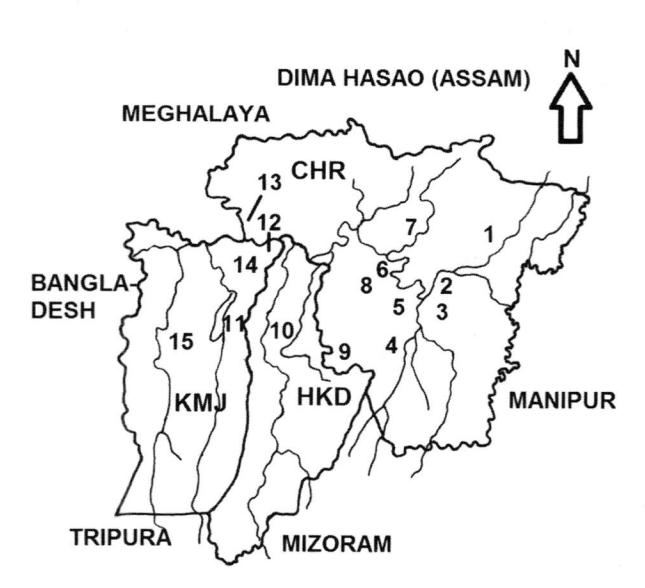

Fig. 6.1 Map of Barak valley, Assam, India, showing the groundwater collection sites. The number assigned to a given site indicates the area from which the groundwater samples were collected. The Site Code is the same as in Table 6.1. *CHR* Cachar, *HKD* Hailakandi, *KMJ* Karimganj

6 Arsenic Contamination of Groundwater in Barak Valley, Assam, India

Table 6.1 Groundwater collection sites in Barak valley, Assam, India

Site code	Site name	Nature and location
1.	Chiri	This area is in the floodplain of R. Chiri, which after originating in North Cachar Hills, flows into R. Barak. Its lower course runs almost parallel to that of R. Barak, and the catchment comprises tea gardens, small villages and townships
2.	Barak 1	This is the upstream stetch of R. Barak extending from its confluence with Chiri to a place called Kashipur. Most tubewells are on the north bank of Barak in cultivated areas
3.	Sonai	The combined flow of Rukni and Sonai joins Barak. The area is dotted with floodplain lakes called *beels* and ox-bows
4.	Rukni	R. Rukni, a south bank tributary of Barak, originates in Mizoram and flow through southern Cachar before joining R. Sonai, another tributary of Barak
5.	Barak 2	This section of Barak is after its confluence with Sonai till Silchar, the largest urban centre in this region. The river flows through a number of sharp meanders
6.	Silchar LL	This site covers the urban sprawl of Silchar, which has grown along the banks of R. Barak and by reclaiming low-lying wetlands on its eastern fringes. The urban area also has a few low hills and relatively elevated ground strewn amidst low-lying areas. Shallow groundwater is increasingly being exploited in this area for domestic supply
7.	Madhura	This is an important north-bank tributary of R. Barak flowing down from the North Cachar Hills through rural settlements in a meandering course
8.	Silchar HL	This area in the southern parts of Silchar has urban development on relatively elevated ground above the high flood level
9.	Borjalinga	This is a hilly area fringing the Chatla floodplain with altitude ranging *c.a.* 25–50 msl. Many tubewells in this area go dry during the dry months of December-February
10.	Dhaleswari-Katakhal	These two rivers flow down from the hills of Mizoram through Hailakandi district through a relatively narrow valley
11.	RK Nagar	This is in a hilly area that forms the water divide between Katakhal and Singla rivers. Several artesian wells are present in this area
12.	Barak 3	This is the downstream section of R. Barak between Badarpur, an industrial town and Bhanga, where the river bifurcates into two branches
13.	Surma	R. Surma is the right arm of R. Barak after bifurcation and flows through a low-lying area dotted with numerous floodplain wetlands
14.	Kusiara	R. Kusiara is the left arm of R. Barak and flows through Karimganj district with a few wetlands in tis floodplain
15.	Longai	R. Longai flows from Mizoram through a relatively narrow valley and crosses the international border with Bangladesh to flow into a floodplain lake in the Sylhet district

6.2.2 Collection and Analysis of Water Samples

Water samples were collected in polythene bottles, which had been kept overnight in dilute laboratory grade nitric acid (1:1) and finally washed with distilled water. The samples were preserved by adding a drop of dilute nitric acid (1:1) GR Grade immediately after collection. Arsenic concentrations were determined in these samples by flow-injection hydride generation atomic absorption spectrometry (FIHG-AAS) at the School of Environmental Studies Laboratory, Jadavpur University, Kolkata, India. The accuracy of estimation was verified by analyzing standard reference materials.

6.3 Statistical Analysis

The data were normalized by log transformation and then subjected to one-way ANOVA. Least square difference (LSD) test was used for multiple comparisons. The software used was SPSS 20 for Windows.

6.3.1 Risk Assessment

The potential risk of boring a shallow hand-operated tubewell in Barak valley was estimated by Bayesian Statistics and Monte Carlo Simulation as follows:

Let θ_I be the actual concentration of arsenic (µg l^{-1}) that was available in the ith locality. The actual value of θ_I was unknown.

Let the observed rate of occurrence of As be

$$t_i = \frac{Z_i}{n_i} \quad (6.1)$$

where n_i was the number of water samples considered from the ith locality; and Z_i was the concentration of arsenic (µg l^{-1}) that was found in all the n_i samples taken together.

Thus, considering θ_I to be a random variable we can have $E\,(\theta_I) = \mu_i$ and Var $(\theta_I) = \sigma_i^2$. Under this assumption it can be demonstrated that the best Bayes estimate of θ_I is a linear combination of the observed rate t_i and μ_i (Guttman et al. 1971; Carvalho and Carvalho 2004).

The estimated value of

$$\hat{\theta}_i = w_i t_i + (1 - w_i)\mu_i \quad (6.2)$$

where $0 \le w_i \le 1$.

This estimator is useful for improving the estimation by reducing the mean squared error towards zero. The value of w_i is given by,

$$w_i = \frac{\sigma_i^2}{\sigma_i^2 + \frac{\mu_i}{n_i}} \quad (6.3)$$

The Bayes estimate of θ_I given in (6.2) is also called the shrinkage estimator. Shrinkage is implicit in Bayesian inference. This estimate shrinks or moves the individual estimate ti toward the pooled estimate μ_i where the parameter $0 < w_i < 1$ determines the size of the shrinkage.

Now, if we assume that μ_i and σ_i^2 were the same in all the i localities then we could estimate them as follows:

$$\hat{\mu} = \hat{\mu}_i = \frac{\sum Z_i}{\sum n_i} \text{(Pooled estimate)} \quad (6.4)$$

$$\hat{\sigma}^2 = \hat{\sigma}_i^2 \frac{\sum n_i (t_i - \hat{\mu})^2}{\sum n_i} - \frac{\hat{\mu}}{n} \quad (6.5)$$

However, if μ_i and σ_i^2 were not considered to be the same in all the i localities, Monte Carlo simulation could be employed to determine the values of μ_i and σ_i^2 for all the localities individually.

Based on the samples collected from the ith locality, the distributional pattern (relative frequency) of the arsenic level was arranged. The resultant distribution took the form of a discrete probability distribution. The arsenic content is a measurable quantity and was, therefore, continuous. Yet it could be well approximated by a discrete distribution provided large samples were considered. Even though the discrete distribution was an approximation to reality, it could still give important insights into the actual problem (Albright et al. 2009). On identification of the probability distribution of arsenic content ($\mu g\ l^{-1}$) in the ith locality, Monte Carlo simulation was used to generate several random samples of size 1,000 each. Computation based on such large samples was expected to provide consistent estimates of the parameters with considerably diminished standard error (Lynch 2007).

One hundred such samples, each of size 1,000 were generated. The distribution of sample means were obtained using the Central Limit Theorem (CLT) which in turn could be gainfully applied to estimate the values of μ_i and σ_i^2. The central limit theorem states that for any population distribution with mean μ and standard deviation σ the sampling distribution of the sampling mean X^- is approximately normal with mean μ and standard deviation $\sigma/\sqrt{n}$ (Albright et al. 2009). Notationally, for this example we would have

$$f(\overline{x}|\mu_i, \sigma_i, n) \overset{asy}{\sim} N\left(\mu_i, \frac{\sigma_i}{\sqrt{n}}\right)$$

The values of μ_i and σ_i^2 obtained in this fashion can then be replaced in (6.3) and then in (6.2) to obtain the corresponding estimate of θI. This estimate is obtained under the consideration that the arsenic contents in the different localities are not equal.

It may be noted that when estimation of θI is done using the second method, one part of the Bayes estimate of θ_1 [Eq. (6.2)] is determined from the observed sample and hence is deterministic for a given sample (i.e., t_i). However, the other part is stochastic i.e. μ_i. So we can generate 95 % or 99 % credibility interval for the stochastic part and hence express the estimated value of θI in terms of an interval instead of a point value. The interval estimate of a parameter is a robust approach compared to point estimate, where one can express with a given level of confidence the expected range in which a particular parameter is supposed to lie (Freund 1992).

6.4 Results and Discussion

6.4.1 District-Wise Distribution of Arsenic in Groundwater

The distribution of arsenic in the groundwater of the three districts in the Barak Valley region of south Assam, viz., Cachar, Hailakandi and Karimganj is shown in Table 6.2. When all 644 samples irrespective of sites were taken into consideration, 395 (61.34 %) had arsenic at or below the detection limit (3 µg L^{-1}) of the equipment used in the present study. The remaining 249 (38.66 %) had detectable levels of arsenic, with 39 (6.06 %) samples having arsenic ≤ 10 µg L^{-1}.

Thus a total of 434 (67.4 %) groundwater sources in the study area were within the safe limit of 10 µg L^{-1} recommended by the World Health Organization (WHO). Of the 210 samples (32.6 %) above the WHO standard, 10.86 % also exceeded the Indian Standard of 50 µg L^{-1}. Chakraborti et al. (2008) reviewed the groundwater arsenic concentration status of several Indian states as well as Bangladesh. The percentage of samples in Barak valley exceeding the 10 µg L^{-1} WHO Standard was comparable to that in Bihar and Jharkhand (32.7 and 35.0 %, respectively), but considerably less than that in states like Manipur (only valley area), Uttar Pradesh and West Bengal (63.3, 46.5 and 48.1 %, respectively) as well as Bangladesh (40.3 %). Similarly, the percentage of samples exceeding the Indian standard of 50 µg L^{-1} was also comparatively less in this area. The mean arsenic concentration of 4.39 µg L^{-1} in Hailakandi district was well below the permissible concentration of 10 µg L^{-1} prescribed by the World Health Organization (WHO), while those of 16.55 and 18.01 µg L^{-1}, respectively, in Cachar and Karimganj districts were considerably less than the Indian Standard of 50 µg L^{-1}. Kumar et al. (2010a, b) showed that 47 % of the samples tested from the central Gangetic plain exceeded the Indian permissible limit of 50 µg L^{-1} along with a high percentage contribution (66 %) of toxic AsIII to total arsenic concentration. Hence, compared to the situation in both middle and lower Gangetic plains, the arsenic contamination scenario in Barak valley is relatively less alarming. Further, the percentage of tubewells having >300 µg L^{-1}

6 Arsenic Contamination of Groundwater in Barak Valley, Assam, India

Table 6.2 Distribution of arsenic (%) in the groundwater of Cachar, Hailakandi and Karimganj districts, Barak valley

District	≤3 µg L^{-1}	4–10 µg L^{-1}	11–50 µg L^{-1}	51–100 µg L^{-1}	101–200 µg L^{-1}	201–300 µg L^{-1}	>300 µg L^{-1}	Mean ± SE µg L^{-1} (range)	% Samples ex-ceeding 10 µg L^{-1}	% Samples ex-ceeding 50 µg L^{-1}
Cachar	58.2	5.9	23.5	7.9	4.1	0	0.5	16.55 ± 5.53	35.97	12.5
Hailakandi	78.6	3.0	17.4	1.0	0	0	0	4.39 ± 1.0 (0–58)	18.4	1
Karimganj	58.4	8.4	20.1	6.5	6.5	0	0	18.01 ± 4.06	33.1	13.0
Total Barak Valley	61.3	6.1	21.7	6.5	4.1	0	0.3	17.14 ± 1.47 (0–383)	32.6	10.9

arsenic in their water is relatively less in the Barak valley districts. This was perhaps the reason for no confirmed incidence of *arsenicosis* in this area, as opposed to those in West Bengal, Bangladesh, and parts of the central Gangetic plain.

6.4.2 Topography-Based Distribution of Arsenic in Groundwater

A topography-based categorization of groundwater sources allowed a closer examination of arsenic contamination in the study area. The predominant landforms in Barak valley comprise river floodplains and low hills. The lithology and landscape of this area is typically fluvial, characterized by active channels of river Barak and its tributaries, abandoned channels, natural levees, back swamps and flood plains. The hill areas of Borjalinga-Dwarbond in Cachar and Ramkrishnanagar (RK Nagar) in Karimganj districts comprise Tipam sandstones and Girujan clay formations in certain places. Table 6.3 shows the distribution of arsenic in the groundwater of tubewells located in different river floodplains and hill areas of Barak valley, Assam.

Tubewells sampled in the hill areas of Borjalinga and RK Nagar were arsenic free (<3 µg L^{-1}). Nickson et al. (2007) identified the Ramkrishna Nagar (RK Nagar) administrative block as As-affected. However, as the topography of this block comprises both hills and the low-lying floodplain of river Singla (*shonbeel* wetland), block-level categorization did not distinguish between As-free areas in high-relief topography, where users could safely access groundwater for domestic consumption, and the low-relief floodplain with its As-affected wells. This difference, therefore, calls for different drinking water management strategies for the two topographically different areas in the same administrative unit. Hence, it may be useful to supplement block or district level characterization with topography-based distribution of arsenic. In contrast to these high-relief areas, tubewells in the floodplain of river Longai in Karimganj district had the highest mean arsenic concentration (47.3 ± 7.59) followed by Barak 2 and Surma sites (39.15 ± 19.5 and 36.81 ± 4.78, respectively). However, only two tubewells among those sampled had arsenic concentrations of >300 µg L^{-1}, and were in Barak 2 site, while the maximum percentage of tubewells exceeding the Indian Standard of 50 µg L^{-1} were in Longai, followed by Surma. At the Barak 2 site, which is downstream of the confluence of the combined flow of river Rukni and river Sonai with river Barak, the river follows a meandering course through low-lying areas. Among the upper tributaries of river Barak, river Chiri, Rukni and Sonai floodplains also had moderately high arsenic concentrations (12.85 ± 7, 11.16 ± 3.35 and 19.13 ± 5.32 µg L^{-1}, respectively). Downstream from Barak 2, all tubewells barring one in the Silchar HL site in Silchar urban area were arsenic-free. This area comprises elevated ground (*c.a.* 25–28 msl) in the southern part of Silchar city that is above the high flood level, and was largely As-free. The relatively low-lying areas of Silchar (Silchar LL) had a higher mean arsenic concentration of 15.87 ± 2.81 µg L^{-1}.

Table 6.3 Groundwater quality from Barak valley, Assam, India

Site (n)	Number of groundwater samples (% of total)							Mean ± SE μg L^{-1} (range)	% Samples exceeding 10 μg L^{-1}	% Samples exceeding 50 μg L^{-1}
	≤3 μg L^{-1}	4–10 μg L^{-1}	11–50 μg L^{-1}	51–100 μg L^{-1}	101–200 μg L^{-1}	201–300 μg L^{-1}	>300 μg L^{-1}			
Chiri (13)	7 (53.8)	2 (15.4)	3 (23.1)	1 (7.7)	0	0	0	12.85 ± 7 (≤3–91)	30.8	7.7
Barak 1 (14)	8 (57.1)	1 (7.1)	4 (28.4)	1 (7.1)	0	0	0	9 ± 4.2 (≤3–58)	35.5	7.1
Sonai (48)	23 (47.9)	8 (16.8)	12 (25.2)	3 (6.3)	2 (4.2)	0	0	19.13 ± 5.32 (≤3–186)	35.7	10.5
Rukni (32)	21 (65.6)	0	10 (31.1)	1 (3.1)	0	0	0	11.16 ± 3.35 (≤3–70)	34.2	3.1
Barak 2 (26)	11 (42.3)	4 (15.4)	8 (30.8)	1 (3.8)	0	0	2 (7.6)	39.15 ± 19.5 (≤3–383)	42.2	11.4
Silchar LL (84)	45 (53.6)	4 (4.8)	27 (32.1)	7 (8.3)	1 (1.2)	0	0	15.87 ± 2.81 (≤3–161)	41.6	9.5
Madhura (23)	11 (47.8)	1 (4.3)	8 (34.8)	2 (8.6)	1 (4.3)	0	0	21.13 ± 7.59 (≤3–156)	47.7	12.9
Silchar HL (30)	29 (96.7)	0	1 (3.3)	0	0	0	0	0.73 ± 0.73 (≤3–22)	3.3	0
Borjalinga (37)	37 (100)	0	0	0	0	0	0	0	0	0
Dhaleswari-Katakhal (98)	77 (78.6)	3 (3.0)	17 (17.4)	1 (1.0)	0	0	0	4.39 ± 1.0 (≤3–58)	18.4	1
RK Nagar (22)	22 (100)	0	0	0	0	0	0	0	0	0
Barak 3 (48)	25 (52.1)	6 (12.5)	13 (27.1)	2 (4.2)	2 (4.2)	0	0	17.04 ± 5.31 (≤3–196)	35.5	8.4
Surma (85)	36 (42.4)	3 (3.5)	19 (22.4)	15 (17.6)	12 (14.2)	0	0	36.81 ± 4.78 (≤3–150)	54.2	31.8
Kusiara (44)	36 (81.8)	2 (4.5)	3 (6.8)	2 (4.5)	1 (2.3)	0	0	7.71 ± 3.33 (≤3–110)	13.6	6.8
Longai (40)	7 (17.5)	5 (12.5)	15 (37.5)	6 (15.0)	7 (17.5)	0	0	47.3 ± 7.59 (≤3–154)	70.0	32.5
Total Barak Valley (644)	395 (61.34)	39 (6.06)	140 (21.74)	42 (6.52)	26 (4.04)	0	2 (0.3)	17.14 ± 1.47 (≤3–383)	32.6	10.86

River Madhura, which is an important north bank tributary of river Barak that flows down from North Cachar Hills, also had a fairly high mean arsenic concentration (21.13 ± 7.59 µg L^{-1}) in the tubewells located in its floodplain. Further downstream, the Badarpur-Bhanga stretch (Barak 3) of river Barak had a mean arsenic level of 17.04 ± 5.31 µg L^{-1}. Beyond Bhanga, the Barak bifurcates into two branches, viz., Surma and Kusiara at a place called Haritikar. River Surma flows north through the Katigorah area forming the international boundary with Bangladesh, while the Kusiara flows in a more or less westerly direction along the Bangladesh border. Tubewells in the Surma catchment contained significantly higher arsenic concentrations (36.81 ± 4.78 µg L^{-1}) than those in the Kusiara catchment (7.71 ± 3.33 µg L^{-1}). Because As concentrations in the tubewells were not normally distributed, non-parametric Mann-Whitney tests among sites revealed that As concentrations at Silchar HL site were significantly lower than those in all the other sites; and those at Longai significantly greater than all sites other than Surma. The Katigorah area through which the Surma flows has numerous low-relief floodplain lakes locally called as *beels* or *haors*, while the Kusiara catchment is on slightly higher ground. The Sylhet district of Bangladesh is contiguous to Barak valley, and is in the lower catchment of Surma and Kusiara rivers. Arsenic data from Bangladesh show that the two districts of Sylhet and Sunamganj, which are in the Surma catchment, had higher groundwater arsenic levels than those in Moulavi bazar, which is in the Kusiara basin (Fazal et al. 2001).

Thus the present study shows that the low-lying river floodplains in Barak valley have relatively high As concentrations. High As concentrations were observed to be associated with low-relief topography in the flat land between the Mekong and the Bassac rivers in Cambodia (Buschmann et al. 2007). High arsenic (>50 µg L^{-1}) tubewells in Bangladesh were found to be located in low-lying areas and within extremely low slopes (<0.7°). Low elevation and gentle slopes resulted in the accumulation of fine sediments, As-containing iron oxyhydroxide minerals and organic matter that facilitated As release by microbial activity (Shamsudduha et al. 2009). Hoque et al. (2009) observed that low arsenic aquifers had thinner silt/clay layer or permeable sandy materials at the surface at slightly higher topography. Spatial variations in distribution of As in shallow groundwater were likely to be due to differences in topography on multiple scales, where even slightly elevated areas with sandy soil could be largely free of As (Fendorf et al. 2010). In the present study, areas above the flood level in Silchar HL had As-free groundwater. Hoque et al. (2011) identified several indicators such as a shallow water table (≥ 1- ≤ 7 m depth), ≥ 50 % loamy silt in the surface soil, and <15 m thickness of the uppermost silty-clay layer that characterize low-arsenic shallow groundwater in the holocene floodplains of Bangladesh. Viewed in this perspective, it is important to delineate possible low-arsenic areas of groundwater in the floodplain-predominant topography of the GBM, Mekong and other river basins in south and southeast Asia. In these areas, the depth distribution of wells vary considerably, being up to 350 m deep in India and Bangladesh and mostly within 100 m in Nepal, Cambodia and Vietnam (Fendorf et al. 2010). However, considerable local variations exist within this general pattern. For example, in the Barak valley region of Assam, which is a part of the Meghna sub-basin of the GBM basin, the majority of the tubewells are in the depth range of c.a. 30–40 m. This is because the most commonly available tubewell (*Tara* pump) package in this area is equipped with a set of pipes running to a depth of 100–120 ft (30.5–36.5 m), which can be installed manually with ease, the expenditure being c.a. INR 40,000 (c.a. US $ 700), which an average middle income family can afford to spend to own a tubewell and ensure access to pathogen-free safe water. Tubewells going beyond 100 m are difficult to install as this cannot be done manually, and the drilling rig has to be hired and transported from outside the Valley with the resultant cost becoming prohibitive for most consumers. Therefore, the delineation of areas with a minimal risk of As contamination of shallow groundwater could improve access to microbiologically safe water for a large segment of population not only in Barak Valley but in the entire region, where diarrhea still remains a killer disease. Of course, the vulnerability of these low-As zones to fresh As contamination needs to be better understood before they can be extensively used in a given area. A better alternative could be to install deeper (>100 m) community tubewells in low-As zones (Fendorf et al. 2010). In most countries of this region, installation of such wells would need funding or other forms of subsidies from government or aid organizations, and till such programmes could be planned and executed, shallow groundwater wells in low-As zones could comprise an interim source of relatively safe drinking water to rural and suburban populations.

6.4.3 Risk Assessment

The results of the risk assessment study are presented in Table 6.4, which provides the estimated values of θ_1 obtained through two different methods. For each locality the estimates were very close to each other.

This implies that it is not unreasonable to think that μ_i and σ_i^2 were equal in all the *i* localities. Based on the interval estimates (Table 6.4) we may conclude that Barak 2, Longai and Surma sites had the highest risk in terms of arsenic contamination and investing money in a tubewell is more than likely to be an unwise decision. Nevertheless, the estimated level in all these three sites were less than the Indian Standard for arsenic (50 µg L^{-1}) but more than the international level set by the WHO. The safest sites were Barak 1, Silchar HL, Dhaleswari-Katakhal and Kusiara which were safe even by the international standard, and Borjalinga and RK Nagar, where the groundwater was found to be free from arsenic contamination.

6.5 Conclusion

The results of this study further confirm some previous observations that high groundwater As concentrations were largely associated with low-relief topography, especially in those areas where As mobilization took place through microbial

Table 6.4 Estimated arsenic concentration in the study sites with interval estimates at $p \leq 0.05$ and ≤ 0.01

Site	Mean As concentration (µg L^{-1})	Estimated concentration (µg L^{-1})		Interval Estimates	
		Method 1	Method 2	95 %	99 %
Chiri	12.85	12.89189	12.9788	[12.86, 13.09]	[12.83, 13.13]
Barak 1	9.0	9.069652	9.0087	[8.89, 9.12]	[8.86, 9.16]
Sonai	19.13	19.12446	19.1376	[19.03, 19.24]	[18.99, 19.27]
Rukni	11.16	11.18016	11.15625	[11.05, 11.28]	[11.01, 11.31]
Barak 2	39.15	39.0903	39.0727	[39.01, 39.17]	[38.99, 39.19]
Silchar LL	15.87	15.87259	15.8239	[15.69, 15.96]	[15.65, 16]
Madhura	21.13	21.12066	21.1134	[21.00, 21.23]	[20.96, 21.27]
Silchar HL	0.73	0.793306	0.8048	[0.76, 0.84]	[0.74, 0.86]
Borjalinga	≤3	≤3	≤3	≤3	≤3
Dhaleswari-Katakhal	4.388	4.402447	4.3878	[4.31, 4.48]	[4.28, 4.51]
RK Nagar	≤3	≤3	≤3	≤3	≤3
Barak 3	17.04	17.04544	17.03974	[16.94, 17.12]	[16.92, 17.16]
Surma	36.81	36.79076	36.80845	[36.66, 36.99]	[36.61, 37.04]
Kushiara	7.705	7.729739	7.7074	[7.63, 7.77]	[7.61, 7.79]
Longai	47.3	47.22939	47.3106	[47.12, 47.51]	[47.06, 47.57]

reductive dissolution of organic matter. Such areas were common in most south and Southeast Asian countries, which also have ample surface water resource that is annually replenished by heavy monsoon rainfall. This resource needs to be conserved and measures taken to improve its microbial quality. Nevertheless, as the risk of arsenic is very low in even slightly elevated areas, people living there could safely enjoy the cost-effective option of exploitation of shallow groundwater for achieving water self-sufficiency in individual households, multistorey apartments, educational institutions, government offices and business concerns alike.

Acknowledgements We are grateful to Prof. Dipankar Chakraborti, Director, School of Environmental Studies, Jadavpur University, India, and his associates for helping us in sample collection and conducting arsenic estimation in all the samples free of cost.

References

Ahamed S, Sengupta MK, Mukherjee A, Hossain MA, Das B, Nayak B, Pal A, Mukherjee SC, Pati S, Dutta RN, Chatterjee G, Mukherjee A, Srivastava R, Chakraborty D (2006) Arsenic groundwater contamination and its health effects in the state of Uttar Pradesh (UP) in upper and middle Ganga plain, India: a severe danger. Sci Total Environ 370:310–322

Albright SC, Winston WL, Zappe CJ (2009) Decision making with Microsoft Excel. South Western Cengage Learning, Stamford

Buschmann J, Berg M, Stengel C, Sampson ML (2007) Arsenic and manganese contamination of drinking water resources in Cambodia: coincidence of risk areas with low relief topography. Environ Sci Technol 41:2146–2152

Carvalho G, Carvalho M (2004) A tutorial on spatial analysis of areas. International Post Graduate Course on Geographic Information Technologies. http://edugi.uji.es/Camara/spatial_analysis_areas.pdf. Accessed 25 June 2013

Chakraborti D, Mukherjee SC, Pati S, Sengupta MK, Rahman MM, Chowdhury UK, Lodh D, Chanda CR, Chakraborti AK, Basu GK (2003) Arsenic groundwater contamination in Middle Ganga Plain, Bihar, India: a future danger? Environ Health Perspect 111(9):1194–1201

Chakraborti D, Singh EJK, Das B, Shah BA, Hossain MA, Nayak B, Ahamed S, Singh NR (2008) Groundwater arsenic contamination in Manipur, one of the seven North-Eastern states of India: a future danger. Environ Geol 56:381–390

Chakraborti D, Das B, Rahman MM, Chowdhury UK, Biswas B, Goswami AB, Nayak B, Pal A, Sengupta MK, Ahamed S, Hossain A, Basu G, Roychowdhury T, Das D (2009) Status of groundwater arsenic contamination in the state of West Bengal, India: a 20-year study report. Mol Nutr Food Res 53:542–551

Chauhan VS, Nickson RT, Chauhan D, Iyengar L, Sankararamakrishnan N (2009) Ground water geochemistry of Ballia district, Uttar Pradesh, India, and mechanism of arsenic release. Chemosphere 75(1):83–91

Chowdhury UK, Biswas BK, Chowdhury TR, Samanta G, Mandal BK, Basu GC, Chanda CR, Lodh D, Saha KC, Mukherjee SK, Roy S, Kabir S, Quamruzzaman Q, Chakraborti D (2000) Groundwater arsenic contamination in Bangladesh and West Bengal, India. Environ Health Perspect 108(5):393–397

Fazal MA, Kawachi T, Ichion E (2001) Extent and severity of groundwater arsenic contamination in Bangladesh. Water Int 26(3):370–379

Fendorf S, Michael HA, van Geen A (2010) Spatial and temporal variations of groundwater arsenic in South and Southeast Asia. Science 328:1123–1127

Freund JE (1992) Mathematical statistics, 5th edn. Prentice-Hall, Inc., Englewood Cliffs

Guttman I, Wilks SS, Hunter JS (1971) Introductory engineering statistics, 2nd edn. Wiley, New York

Hoque MA, Khan AA, Shamsudduha M, Hossain MS, Islam T, Chowdhury SH (2009) Near surface lithology and spatial variation of arsenic in the shallow groundwater: southeastern Bangladesh. Environ Geol 56:1687–1695

Hoque MA, Burgess WG, Shamsudduha M, Ahmed KM (2011) Delineating low-arsenic groundwater environments in the Bengal aquifer system, Bangladesh. Appl Geochem 26:614–623

Kumar P, Kumar M, Ramanathan AL, Tsujimura M (2010a) Tracing the factors responsible for arsenic enrichment in groundwater of the middle Gangetic Plain, India: a source identification perspective. Environ Geochem Health 32:129–146

Kumar P, Kumar M, Ramanathan AL, Bhattacharya P, Thunvik R, Singh UK, Tsujimura M, Sracek O (2010b) Arsenic enrichment in groundwater in the Middle Gangetic Plain of Ghazipur district in Uttar Pradesh, India. J Geochem Explor 105(3):83–94

Lynch SM (2007) Introduction to applied Bayesian statistics and estimation for social scientists. Springer, New York

Nickson R, McArthur J, Burgess W, Ahmed KM, Ravenscroft P, Rahmann M (1998) Arsenic poisoning of Bangladesh groundwater. Nature 395:338

Nickson RT, McArthur JM, Ravenscroft P, Burgess WG, Ahmed KM (2000) Mechanism of arsenic release to groundwater, Bangladesh and West Bengal. Appl Geochem 15:403–413

Nickson R, Sengupta C, Mitra P, Dave SN, Banerjee AK, Bhattacharya A, Basu S, Kakoti N, Moorthy NS, Wasuja M, Kumar M, Mishra DS, Ghosh A, Vaish DP, Srivastava AK, Tripathi RM, Singh SN, Prasad R, Bhattacharya R, Deverill P (2007) Current knowledge on the distribution of arsenic in groundwater in five states of India. J Environ Sci Health A 42:1707–1718

Oinam JD, Ramanathan AL, Linda A, Singh G (2011) A study of arsenic, iron and other dissolved ion variations in the groundwater of Bishnupur District, Manipur, India. Environ Earth Sci 62:1183–1195

Oinam JD, Ramanathan AL, Singh G (2012) Geochemical and statistical evaluation of groundwater in Imphal and Thoubal district of Manipur, India. J Asian Earth Sci 48:136–149

Rahman MM, Chowdhury UK, Mukherjee SC, Mondal BK, Paul K, Lodh D, Biswas BK, Chanda CR, Basu GK, Saha KC, Roy S, Das R, Palit SK, Quamruzzaman Q, Chakraborti D (2001) Chronic arsenic toxicity in Bangladesh and West Bengal, India – a review and commentary. Clin Toxicol 39(7):683–700

Saha D (2009) Arsenic groundwater contamination in parts of Middle Ganga Plain, Bihar. Curr Sci 97(6):753–755

Saha D, Sahu S, Chandra PC (2011) Arsenic-safe alternate aquifers and their hydraulic characteristics in contaminated areas of Middle Ganga Plain, Eastern India. Environ Monit Assess 175:331–348

Shamsudduha M, Marzen LJ, Uddin A, Lee MK, Saunders JA (2009) Spatial relationship of groundwater arsenic distribution with regional topography and water-table fluctuations in the shallow aquifers in Bangladesh. Environ Geol 57:1521–1535

Singh AK (2004) Arsenic contamination in groundwater of North Eastern India. In: Proceedings of national seminar on hydrology, National Institute of Hydrology, Roorkee, India. http://users.physics.harvard.edu/~wilson/arsenic/references/singh.pdf. Accessed 25 June 2013

Singh EJK, Gupta A, Singh RM (2013) Groundwater quality in Imphal West district, Manipur, India, with multivariate statistical analysis of data. Environ Sci Pollut Res 20(4):2421–2434

Smith AH, Lingas EO, Rahman M (2000) Contamination of drinking-water by arsenic in Bangladesh: a public health emergency. Bull World Health Organ 78:1093–1103

Winkel L, Berg M, Amini M, Hug SJ, Johnson CA (2008) Predicting groundwater arsenic contamination in Southeast Asia from surface parameters. Nat Geosci 1:536–542

Chapter 7
Hydrogeochemistry and Arsenic Distribution in the Gorakhpur District in the Middle Gangetic Plain, India

Hariom Kumar, Rajesh Kumar Ranjan, Shailesh Yadav, Alok Kumar, and AL. Ramanathan

7.1 Introduction

About one-third of the earth's land surface is facing challenge of water scarcity in both quantitative and qualitative terms (Postel 1997). In the last few decades, the global finding of the presence of elevated concentrations of various toxic solutes from natural and anthropogenic sources has limited the available volume of safe drinking water. Understanding the aquifer hydraulic properties and hydrochemical characteristics of water is crucial for proper planning, management and sustainable utilization of groundwater. Generally, the motion of groundwater along its flow paths below the ground surface increases the concentration of the chemical species; hence the groundwater chemistry could reveal important information on the geological history of the aquifers and the suitability of groundwater for domestic, industrial and agricultural purposes (Kortatsi 2007).

One of the main reasons for the increased concern over dissolved arsenic (As) in middle Gangetic plain is the recognition of its acute carcinogenicity and other related health effects, and its natural, ubiquitous presence in varied geologic, geographic, geomorphic (e.g. fluvial, aeolian, marine and lacustrine deposits), hydrogeochemical and hydrologic systems (Welch et al. 2000; Smedley et al. 2005; Bhattacharya et al. 2006; Mukherjee et al. 2008). Research on As in environmental systems has substantially increased in the last two decades. During the same period, the lowering of the As drinking-water standard from 0.05 to 0.01 mg/L by regulatory agencies and advisory bodies (e.g. WHO 2011) has caused previously "safe" drinking water sources to be labelled as unsafe. Recent studies have revealed the

---

H. Kumar • R.K. Ranjan (✉)
Centre for Environmental Sciences, Central University of Bihar, Patna 800014, Bihar, India
e-mail: rajeshkranjan@gmail.com

S. Yadav • A. Kumar • AL. Ramanathan
School of Environmental Sciences, Jawaharlal Nehru University, New Delhi 110067, India

prevalence of natural As in hydrologic systems in scores of geological settings in India (Bhattacharya et al. 2007; Mukherjee et al. 2008). The impact of natural As toxicity from groundwater has been particularly prevalent in Southeast and East Asia. This region has some of the highest population densities and is drained by some of the largest river systems in the world, which generally originate from the Himalayan or Tibetan highlands. The Indo-Gangetic plain occupied by states of Uttar Pradesh (U.P.), Bihar and West Bengal, is extremely fertile for agriculture and rich in water resources, providing favourable conditions for the growth of rural and urban centres. Thus, groundwater has extensively been exploited for domestic supplies and irrigation in the last few decades (Bhattacharya et al. 2011). The excessive pumping of shallow and deep aquifers caused imbalance in the groundwater dynamics and resulted in induced leakage of arsenic contamination from shallow aquifers to the deep aquifers (Muralidharan 1998). The present study gives an account of the hydrogeochemistry of the Gorakhpur district in U.P. showcasing groundwater As risk in the region.

7.2 Study Area

Gorakhpur is located between 26.5°–27.9°N and 83.4°–84.26°E in the north-east "Tarai" region of Middle Gangetic plain in Uttar Pradesh, India. It is located on the bank of river Rapti and Rohini of Ghaghara tributary originating in Nepal. The Rapti is interconnected through many other small rivers following meandering courses across the interfluve region of the Gangetic plain. The district is spread over an area of 3483.8 sq. km, with a total population of 44,36,275 (Census of India 2011). On the eastern side of the city, a shallow but perennial eutrophic lake, Ramgarh Lake is situated. The changing course of river Rapti has resulted into formation of ponds, lakes and depressions, which act as buffer zones for groundwater recharge. However, rapid urbanization in recent years has led to infilling of these recharge areas. There are four layers of aquifers at different depth from surface viz. (1) 6–8 m, (2) 22–34 m, (3) 40–46 m and (4) >60 respectively. Between these aquifers, an impermeable bed of clay, sandy clay, and hard clay are found which separate the two aquifers (Verma 2009). Groundwater is the only source of drinking water for the city. Due to shallow groundwater table, the groundwater is easily accessible to majority of population.

7.2.1 Geology and Geomorphology

Geology of the district is dominated by older and newer alluvium which is gradually deposited by these rivers. Khadar is the new alluvial and Bhangar is the old alluvial and it is the dominant lithological constituent of the study area. On the basis of the cross section of the tube wells of Jal Nigam and private colonization, lithology of the study area is mostly deposited by clay and sand (medium and fine quality). Sand have different composition and structure while clay is found with or without sand. Therefore in this area fine to moderate sand and clay are found. Thus, the geological structure of the city area is very favourable for infiltration of surface water to the underground because most of the sediments are made of sand and clay through which water can percolate easily to the permeable strata. Likewise the sub-surface water can easily be drafted out for various uses due to soft rock lithology.

7.3 Methodology

7.3.1 Groundwater Sampling

Total 29 groundwater samples were collected from bore wells (hand pumps) at different depths (20–135 ft.) in flood plain of the middle Gangetic plain of Gorakhpur district during May and June 2012 (Fig. 7.1). The locations of the tube wells were determined with the help of a global positioning system (Garmin, GPS map 76CSx).

Groundwater samples were collected following the standard procedures which included: (1) filtered samples using filtration unit with prebaked 0.45 μm Whatman GF/C filters and acidified with HNO_3 (1:1) for the analysis of heavy and trace elements including As and (2) Non-acidified samples for the analysis of major ions. Temperature, pH, conductivity (EC), oxidation-reduction potential (ORP) and total dissolved solids (TDS) were determined in situ by portable corresponding digital instruments (HANNA HI9828) multi-parameter. Collected samples were stored at 4 °C in the laboratory until analyzed.

7.3.2 Analytical Techniques

Groundwater samples were analyzed to understand the hydrochemistry of water using standard methods (APHA 2005). Bicarbonate was measured by acid titration method. Major anions like chloride, nitrate, sulphate and phosphate were measured by standard titrimetric method using UV-spectrophotometer (JENWAY 6505 UV spectrophotometer accuracy of ±0.1 % @10 % T). Fluoride was measured using ion-selective electrode (ISE). Major cations like sodium and potassium were estimated using Flame photometer with Na '4' ppm and K '2' ppm FSD standard accuracy. Calcium, iron and magnesium were measured using atomic absorption spectroscopy (AAS) (Thermo Fischer M series). Concentrations of As were measured on acidified samples using Hydride Generator (HG) AAS on wavelength in absorption mode using chemical standards with detection limit of 2 μg/L.

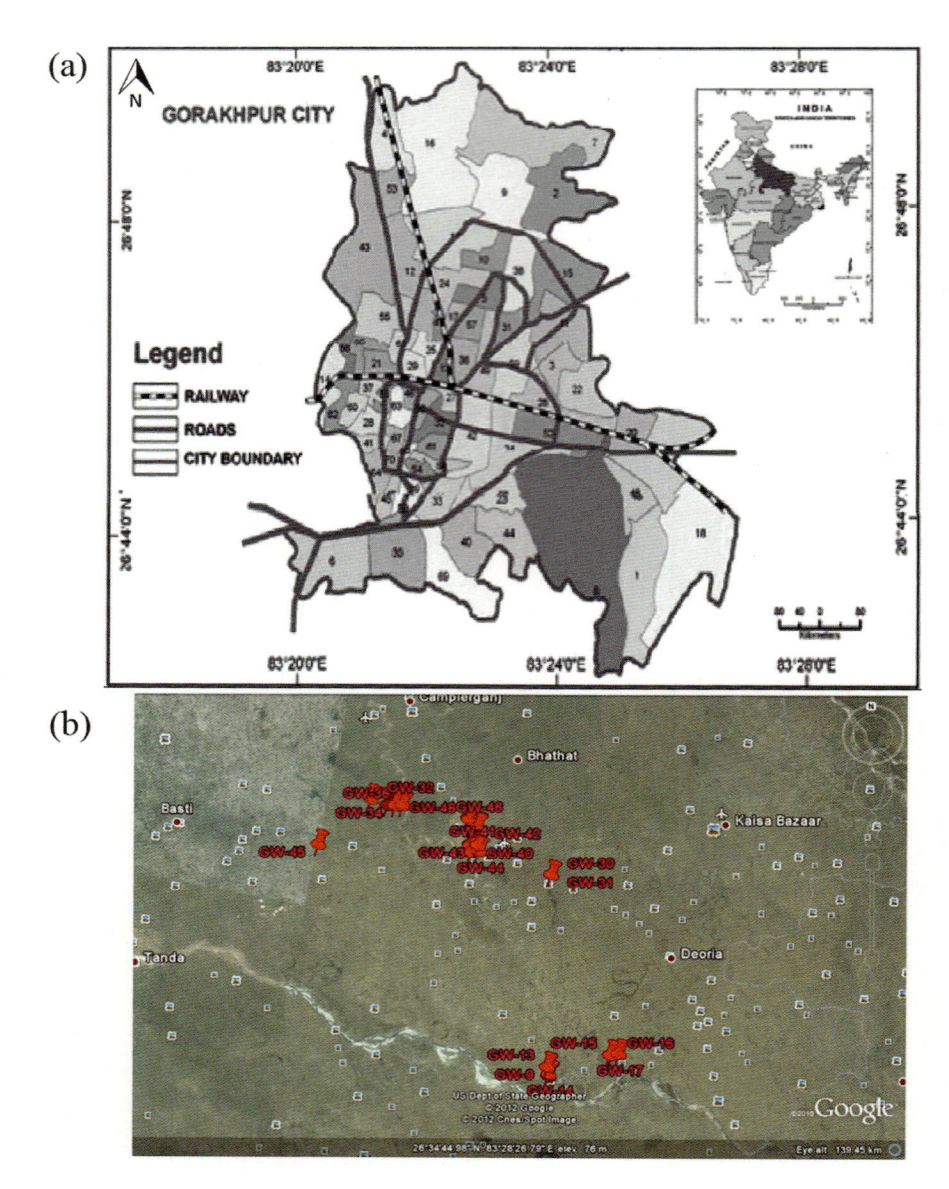

Fig. 7.1 (**a**) Location map of study area; (**b**) Sampling sites in google maps snapshot

7.4 Results and Discussion

7.4.1 Major Ion Geochemistry

Descriptive statistical data of various physico-chemical analyses of the ground water samples is presented in Table 7.1. Groundwater was predominantly near neutral to mildly alkaline (pH value 6.38–7.72) in nature. Electrical conductivity (EC) of groundwater widely varied from 459 to 1,714 µS/cm at different sampling sites. High EC values of some samples (viz. GW 40, GW12, GW14) indicated the presence of alkaline groundwater. Oxidation reduction potential (ORP) data varied from 86.6 to 162.8 mV. In study area, bicarbonate (HCO_3^-) concentration ranged between 156.27 and 783.3 mg/L indicating bicarbonate forms the dominant source of alkalinity in the region. High HCO_3^- concentration is mainly due to presence of carbonaceous sandstones in the aquifers and weathering of carbonate minerals related to the flushing of CO_2 rich water from unsaturated zone, where it is formed by decomposition of organic matter (Chauhan et al. 2009). The concentration of major cations, e.g. Ca^{2+} (18–94.5 mg/L), Mg^{2+} (3.461–7.679 mg/L), Na^+ (10.32–53.75 mg/L) and K^+ (0.235–17.79 mg/L) levels do not change systematically between different sampling locations. The high concentration of Ca^{2+} could be due to weathering of carbonate and plagioclase feldspar minerals, which are abundant in flood plain regions (Bhattacharya et al. 1997). The groundwater of the study area showed a wide variability in the concentration of Cl^- (bdl to 46.98 mg/L), which is consistent with EC values. Sulphate (SO_4^{2-}) concentrations varied from 0.29 to 105 mg/L

Table 7.1 Summary of the physico-chemical parameters of groundwater samples ($n=29$)

	Unit	Min	Max	Average	SD
pH		6.38	7.72	6.89	0.30
TDS	mg/L	279	1,036	547.03	232.24
EC	µS/cm	459	1,714	904.14	386.07
ORP	mV	86.6	162.8	145.35	18.28
Ca^{2+}	mg/L	18	94.5	48.88	19.92
Mg^{2+}	mg/L	3.461	7.679	4.57	1.17
Na^+	mg/L	10.32	53.73	28.82	13.27
K^+	mg/L	0.235	17.79	4.23	4.38
Cl^-	mg/L	bdl[a]	46.98	12.84	13.21
HCO_3^-	mg/L	156.27	783.3	374.68	193.45
SO_4^{2-}	mg/L	0.29	105.86	31.86	26.60
PO_4^{3-}	mg/L	0.001	0.008	0.00	0.00
F^-	mg/L	0.1	0.605	0.28	0.16
NO_3^-	mg/L	bdl[a]	54.22	6.53	13.37
Fe	mg/L	0.121	11.65	2.96	3.25
As	µg/L	10	400	60	0.10

[a]Below detection limit

while NO_3^- ranged from 0.67 to 54.22 mg/L. High concentration of anions such as Cl^-, SO_4^{2-} and NO_3^- are part of major inorganic components and may be responsible in deteriorating the quality of groundwater as drinking water. High concentrations of NO_3^- could be due to leaching of fertilizer residues from agricultural activities and microbial mineralization on the groundwater systems. Piper plot (Piper 1944) showed that most of the samples were clustered at left hand side of the central diamond, thus two hydrochemical facies were identified in cation facies Ca^{2+}-HCO_3^- type (Fig. 7.2). Most of the samples were Ca–HCO_3 type in central diamond suggesting carbonate dissolution and cation exchange may be the dominant processes influencing the major-ion composition.

Overall, the general trend of various ions was found to be $Ca^{2+} > Na^+ > Mg^{2+} > K^+$ and $HCO_3^- > SO_4^{2-} > Cl^- > NO_3^- > F^- > PO_4^{3-}$. Such trend could be explained by attributes such as Ca^{2+} and Mg^{2+} could be possibly derived from both silicate and carbonate weathering whereas Na^+ and K^+ may enter the groundwater from incongruent dissolution of feldspars, micas and pyroxenes (Kumar et al. 2010). Other than carbonate and silicate weathering, HCO_3^- in groundwater can originate from the vadose zone or biogenic CO_2 gas dissolution (Garrels 1967). Mg^{2+} may be introduced from weathering of biotite, and to a lesser extent from pyroxenes, chlorites and garnets (Galy and France-Lanord 1999), to form hydrobiotite, vermiculite or smectite,

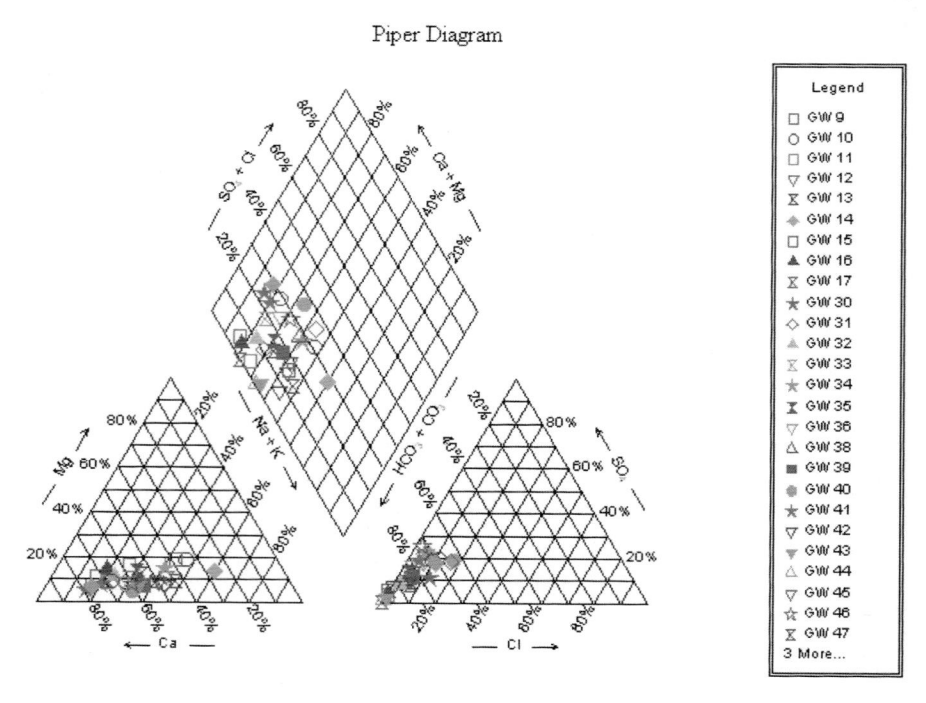

Fig. 7.2 Piper plot showing various hydrochemical facies distribution for the groundwater samples ($n = 29$)

which are found in the Himalaya-derived sediments (Grout 1995). Thus Ca^{2+} in the groundwater of middle Gangetic plains might be introduced from carbonate dissolution, while Mg^{2+} will probably have mixed contributions.

7.4.2 Arsenic Concentration

Groundwater As concentrations ranged from <0.01 to 0.4 mg/L for all samples ($n=29$). About 59 % of all the groundwater samples collected from the study area have dissolved As concentrations ≥0.01 mg/L (WHO 1993; BIS 1991), and 24 % have As ≥0.05 mg/L. The highest concentration of As were observed in GW 48 (0.4 mg/L) and GW 40 (0.5 mg/L). It may be noted that higher concentration of total As were more prevalent in the samples that were in proximity of the Ganges river, suggesting role of fluvial inputs in the groundwater arsenic. High As concentration in the groundwater is mainly because of infiltration of river water through contaminated river bed sediment (Ravenscroft et al. 2005). The weathering of carbonate and silicate minerals along with surface-groundwater interactions, ion exchange, and anthropogenic activities seem to be the processes governing groundwater contamination, including arsenic.

7.4.3 Relationship Between Hydrogeochemical Variable

Multivariate statistical analysis such as Pearson correlation and factor analysis of various hydrogeochemical variables revealed strong pairing and affinity amongst each other. Strong positive correlation exists between pH ~ K^+, the EC and TDS. TDS also showed positive correlation with Cl^-, HCO_3^-, SO_4^{2-}, Na^+, K^+; and $Cl^- \sim SO_4^{2-}$, Na^+, K^+ can be linked to leaching of fertilizers from agricultural areas. $HCO_3^- \sim Mg^{2+}$ and $SO_4^{2-} \sim (Na^+, K^+)$ showed significant correlation between pairs. Ca^{2+} and PO_4^{3-} also exhibited positive correlation ($r=0.57$). Also, As and Fe showed strong ($r=0.86$) positive correlation (Table 7.2), which indicate iron dependent mechanism for the release of As into the groundwater. Significant correlations of As with Fe, therefore, suggest microbially mediated processes that lead to Fe (III) oxide dissolution under anoxic conditions which generally promotes the release of As (Rowland et al. 2008).

Factor analysis of physico-chemical parameters of the groundwater indicates five factors which explained 80.74 % of total variability (Table 7.3). The number of significant factors within the data set was established by considering only those with an eigen value >1.0. Factor 1 explained 33.44 % of variation, which is contributed by strong geochemical associations between TDS, EC, Na^+, K^+, Cl^-, HCO_3^- and SO_4^{2-}, which is mainly resultant of water-rock interaction and weathering process. Factor 2 comprising Fe and As explained 13.27 % of variability among parameter, which indicate that reduction of FeOOH was the dominant mechanism for the

Table 7.2 Correlation graph

	pH	temp	EC	ORP	TDS	HCO₃	PO₄	NO₃	SO₄	Cl	F	Na	K	Ca	Mg	Cu	Mn	Fe	As
pH	1.00																		
temp	0.15	1.00																	
EC	0.20	0.00	1.00																
ORP	−0.03	0.40	0.33	1.00															
TDS	0.20	−0.02	**1.00**	0.31	1.00														
HCO₃	−0.09	−0.23	**0.50**	−0.05	**0.50**	1.00													
PO₄	−0.23	−0.09	0.24	0.01	0.24	0.46	1.00												
NO₃	−0.26	−0.06	0.19	0.22	0.19	0.02	0.16	1.00											
SO₄	0.16	0.05	**0.80**	0.34	**0.80**	0.25	−0.04	0.13	1.00										
Cl	0.23	0.01	**0.89**	0.37	**0.89**	0.32	0.25	0.25	**0.79**	1.00									
F	0.17	−0.01	−0.33	0.18	−0.34	−0.14	−0.22	−0.19	−0.33	−0.42	1.00								
Na	0.18	−0.03	**0.82**	0.38	**0.82**	0.48	0.14	0.11	**0.67**	**0.66**	0.05	1.00							
K	**0.63**	−0.02	**0.66**	0.05	**0.67**	0.23	0.00	−0.11	**0.60**	**0.67**	−0.19	0.46	1.00						
Ca	0.16	−0.01	0.01	0.14	0.00	−0.03	0.22	0.14	−0.03	0.30	−0.26	−0.06	0.10	1.00					
Mg	0.08	−0.19	0.40	−0.25	0.41	**0.51**	−0.08	0.05	0.34	0.24	−0.18	0.30	0.26	−0.36	1.00				
Cu	−0.08	0.19	−0.03	0.12	−0.04	0.10	0.31	−0.16	−0.02	−0.09	−0.14	−0.12	−0.12	−0.10	−0.13	100			
Mn	−0.40	−0.13	0.26	0.18	0.26	0.41	**0.83**	0.16	0.08	0.29	−0.26	0.22	−0.05	0.19	−0.10	0.45	1.00		
Fe	−0.02	0.01	0.40	0.14	0.40	0.09	−0.10	0.13	0.21	0.35	−0.34	0.28	0.29	0.19	−0.12	−0.16	0.03	1.00	
As	0.06	0.02	0.40	0.22	0.40	−0.08	−0.05	0.03	0.17	0.43	−0.26	0.31	0.35	0.18	−0.17	−0.16	0.16	**0.86**	1.00

Temperature in °C, EC in µS/cm, ORP in mV, As in µg/L and remaining in mg/L and confidence level is 0.5

Table 7.3 Factor analysis of hydrogeochemical variables of groundwater ($n=29$)

	Factor 1	Factor 2	Factor 3	Factor 4	Factor 5
pH	0.201	−0.083	0.066	**0.887**	0.104
TDS	**0.948**	0.234	0.002	0.047	−0.096
EC	**0.952**	0.233	−0.002	0.043	−0.076
ORP	0.418	0.055	−0.066	−0.195	**0.791**
Ca	0.005	0.104	**0.824**	0.122	0.233
Mg	0.467	−0.256	−0.269	−0.017	**−0.643**
Na	**0.844**	0.129	−0.219	0.033	0.126
K	**0.624**	0.182	0.174	**0.628**	−0.136
Cl	**0.875**	0.211	0.313	0.091	0.041
HCO$_3$	**0.528**	−0.119	−0.072	−0.160	−0.465
SO$_4$	**0.860**	0.043	0.005	0.060	0.008
PO$_4$	−0.102	0.025	**0.864**	−0.103	−0.171
F	−0.230	−0.350	−0.455	0.206	0.491
NO$_3$	0.269	−0.074	0.397	−0.603	0.146
Fe	0.215	**0.913**	0.073	−0.045	0.019
As	0.208	**0.925**	0.070	0.080	0.125
Eigen values	5.35	2.12	2.06	1.70	1.68
% of Variance	33.44	13.27	12.89	10.64	10.50
% Cumulative	33.44	46.71	59.60	70.24	80.74

release of As into the groundwater. Factor 3 explained 12.89 % of variability with strong association between Ca^{2+} and PO_4^{3-}. Factor 4 showed positive loadings of pH and K^+ and negative loading of NO_3^-. Factor 5 showed positive loading of ORP, while inverse association with Mg and HCO_3^-.

7.5 Conclusions

The spatial and temporal variation of groundwater chemistry in the study site suggests that the hydrogeochemical compositions of groundwater have been mainly controlled by groundwater flow pattern that follows the topography of the study area. The groundwater in the region is above desirable limits for Ca at GW 30, GW 40 and GW 50, for K at GW 40 and for Mg and Na the concentration is well below the prescribed desirable limits by WHO. The anions Cl^-, HCO_3^-, SO_4^-, PO_4^{3-}, F^- and NO_3^- are well below desirable limits, indicating the concentration of various ions well below WHO limits. Piper plot clearly depicts that the groundwater was that of Ca^{2+}-HCO_3^- type. Schoeller diagram (Fig. 7.3) indicates that HCO_3^- is the most dominant constituent of the groundwater in the region followed by $Ca^{2+} > Na^+ + K^+ > Cl^- > Mg^{2+} > SO_4^{2-}$. About 59 % of all the groundwater samples collected from the study area have dissolved As concentrations ≥ 0.01 mg/L and 24 % have

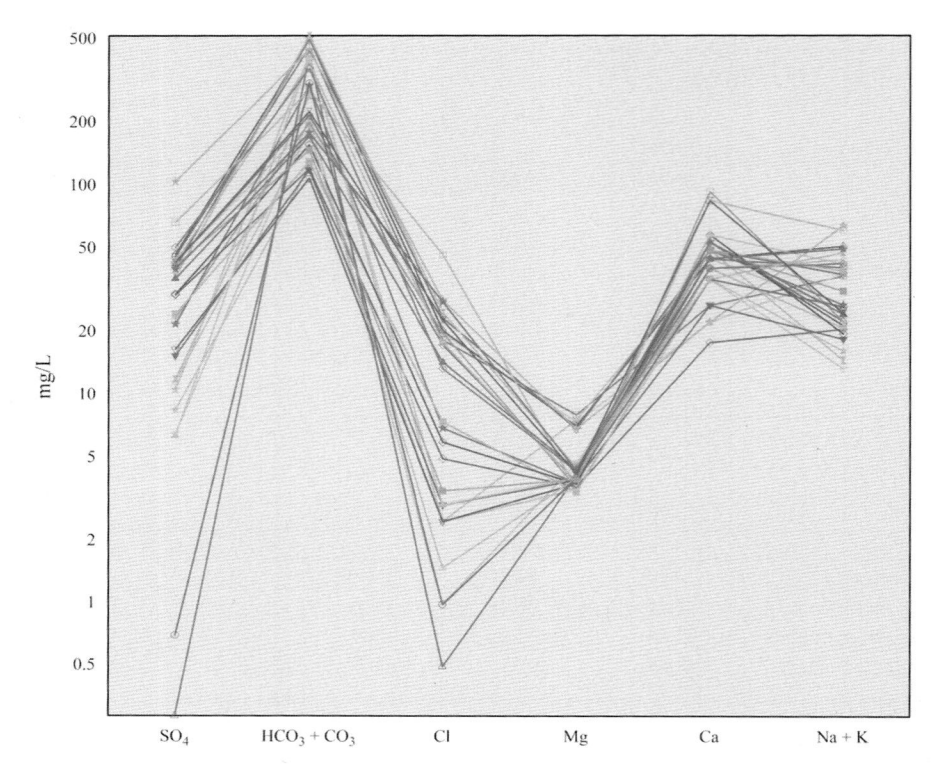

Fig. 7.3 Schoeller diagram showing dominance of ions in the groundwater samples ($n=29$)

As ≥ 0.05 mg/L. Multivariate analysis and Factor analysis indicate that the groundwater of the region is mostly under the influence of natural geochemical processes while at few places it may be affected by anthropogenic practices.

References

APHA (American Public Health Association) (2005) Standard methods for the examination of water and wastewater, 21st edn. American Public Health Association, American Water Works Association, and Water Environment Federation, Washington, DC

Bhattacharya P, Chatterjee D, Jacks G (1997) Occurrence of arsenic contamination of groundwater in alluvial aquifers from Delta Plain, Eastern India: option for safe drinking supply. Int J Water Resour Dev 13:79–92

Bhattacharya P, Claesson M, Bundschuh J, Sracek O, Fagerberg J, Jacks G, Martin RA, Storniolo AR, Thir JM (2006) Distribution and mobility of arsenic in the Río Dulce Alluvial aquifers in Santiago del Estero Province. Argent Sci Total Environ 358(1–3):97–120

Bhattacharya P, Mukherjee AB, Bundschuh J, Zevenhoven R, Loeppert RH (2007) Arsenic in soil and groundwater environment: biogeochemical interactions, health effects and remediation. In: Nriagu JO (ed) Trace metals and other contaminants in the environment. Elsevier, Amsterdam

Bhattacharya P, Mukherjee A, Mukherjee AB (2011) Arsenic contaminated groundwater of India. In: Nriagu J (ed) Encyclopedia of environmental health. Elsevier, Amsterdam

BIS (1991) Specifications for drinking water, IS:10500:1991. Bureau of Indian Standards, New Delhi, India

Chauhan VS, Nickson RT, Chauhan D, Iyengar L, Sankararamakrishnan N (2009) Ground water geochemistry of Ballia district, Uttar Pradesh, India and mechanism of arsenic release. Chemosphere 75:83–91

Galy A, France-Lanord C (1999) Weathering processes in the Ganges-Brahmaputra basin and the riverine alkalinity budget. Chem Geol 159:31–60

Garrels RM (1967) Genesis of some ground waters from igneous rocks. In: Abelson PH (ed) Researches in geochemistry. Wiley, New York

Grout H (1995) Characterisation physique, mineralogique, chimique et signification de la charge particulaire et colloýdale derivieres de la zone subtropicale. Unpublished PhD thesis, Aix-Marseille, France

Kortatsi BK (2007) Hydrochemical framework of groundwater in the Ankobra Basin, Ghana. Aquat Geochem 13(1):41–74

Kumar P, Kumar M, Ramanathan AL, Tsujimura M (2010) Tracing the factors responsible for arsenic enrichment in groundwater of the middle Gangetic Plain, India: a source identification perspective. Environ Geochem Health 32(2):129–146

Mukherjee A, Bhattacharya P, Savage K, Foster A, Bundschuh J (2008) Distribution of geogenic arsenic in hydrologic systems: controls and challenges. J Contam Hydrol 99:1–7

Muralidharan D (1998) Protection of deep aquifers from arsenic contamination in Bengal Basin. Curr Sci 75(4):351–353

Piper AM (1944) A graphic procedure in the geochemical interpretation of water analyses. AGU Trans 25:914–923

Postel SL (1997) Last oasis: facing water scarcity, 2nd edn. W.W. Norton, New York

Ravenscroft P, Burgess WG, Ahmed KM, Burren M, Perrin J (2005) Arsenic in groundwater of the Bengal basin, Bangladesh: distribution, field relations, and hydrological setting. Hydrogeol J 13:727–751

Rowland HAL, Gault AG, Lythgoe P, Polya DA (2008) Geochemistry of aquifer sediments and arsenic-rich groundwaters from Kandal Province. Cambodia Appl Geochem 23:3029–3046

Smedley PL, Kinniburgh DG, Macdonald DMJ, Nicolli HB, Barros AJ, Tullio JO, Pearce JM, Alonso MS (2005) Arsenic associations in sediments from the loess aquifer of La Pampa. Argent Appl Geochem 20:989–1016

Verma SS (2009) Geo-hydrological study of Gorakhpur city. Sectoral study conducted by Department of geography, D.D.U. Gorakhpur University, Gorakhpur

Welch AH, Westjohn DB, Helsel DR, Wanty R (2000) Arsenic in ground water of the United States: occurrence and geochemistry. Ground Water 38:589–604

WHO (1993) Guidelines for drinking water quality, vol 1, 2nd edn, Recommendations. World Health Organization, Geneva, p 188

WHO (2011) Guidelines for drinking water quality: recommendation edn, vol 4. World Health Organization, Geneva

Section III
Arsenic Hydrogeochemistry and Processes

Chapter 8
Arsenic Distribution and Mobilization: A Case Study of Three Districts of Uttar Pradesh and Bihar (India)

Manoj Kumar, Mukesh Kumar, Alok Kumar, Virendra Bahadur Singh, Senthil Kumar, AL. Ramanathan, and Prosun Bhattacharya

8.1 Introduction

Tectonic evolution of Himalayas is related to high erosional potential and substantial sediment transport. Fluvial deposition of clastic material in the Middle Gangetic plain (MGP) is mainly governed by crustal deformation and climatic condition of Himalayas (Singh et al. 2007). Seven large Asian rivers—Ganga, Indus, Brahmaputra, Yangtze, Huang He or Yellow River, Salween and Mekong—are fed by Himalayan glaciers which are supplying ~30 % of the global sediments to the ocean (Milliman and Meade 1983; Singh et al. 2014, 2005). High flux of sediment transported from different terrain of Himalayas is product of geologically young rock formation (Singh et al. 2014). It provides an opportunity to study the fluvial system and post-depositional changes in sediment water interaction depending on the degree of mobility of element under the altered environmental conditions. Arsenic (As) contamination of groundwater is a global problem. Understanding of As mobilization from sediments to As-contaminated aquifers is important for water quality management in areas of MGP of India.

The acute and chronic toxic effects of As are well recognized and As has also been declared as human carcinogen, contributing to a high incidence of skin and

---

M. Kumar • M. Kumar • A. Kumar • V.B. Singh • S. Kumar • AL. Ramanathan (✉)
School of Environmental Sciences, Jawaharlal Nehru University, New Delhi 110067, India
e-mail: alr0400@mail.jnu.ac.in

P. Bhattacharya
KTH-International Groundwater Arsenic Research Group,
Department of Land and Water Resources Engineering,
Royal Institute of Technology (KTH), SE-100 44 Stockholm, Sweden

other cancers in populations exposed to high levels of As in drinking water. Standard limit 10 µg/L of As concentration for safe drinking water has been set by the World Health Organisation (WHO 1993).

The Lithostratigraphic studies of the MGP covering parts of Uttar Pradesh and Bihar show that the terraces of older alluvium (Pleistocene) are free from As contamination, whereas the organic rich recent alluvium deposit (Holocene) are rich in As content (Acharyya and Shah 2007; Shah 2008). Several studies have been conducted on the mobility of As in the ground water of MGP of India (Chakraborti et al. 2003; Acharyya and Shah 2004, 2005, 2007; Shah 2008; Chauhan et al. 2009; Kumar et al. 2010a, b, 2012) but the processes leading to the mobilization of As into the groundwater are not properly understood. Quaternary sediments around the Ganges delta have high proportions of clay and contain relatively large amounts of organic carbon whereas Quaternary sediments in the Ganges alluvial tract in Uttar Pradesh and Bihar contain more sand and are much narrower than sediments in the Bengal basin, which may explain why they retain less As (Acharyya et al. 1999).

A knowledge of fluvial processes involved in the transportation and deposition of geogenic contaminants passing through river sediments is of fundamental importance in the hydrogeochemistry of groundwater to enhance our understanding of the fate and transport of As in the MGP. It will also help to strengthen our understanding to make credible long-term plans for better use and management of the groundwater in the MGP for millions of people. The principal objective of this paper is to emphasize the present As contamination situation in Uttar Pradesh and Bihar based on chemical analysis of groundwater to make immediate remedial actions to avert further worsening of the situation.

8.1.1 Geological Setting and Location of Study Area

Study area (districts: Ballia, Buxar and Bhojpur) located along the active floodplain in the MGP, States of eastern Uttar Pradesh and Bihar (Fig. 8.1). Middle Gangetic floodplains have numerous natural riparian features like small ponds, Oxbow lakes, Point bar complexes, Cut off meanders, Buried channels, and other natural depressions. Ballia district (Uttar Pradesh) is situated in the alluvial plains between the Ganga and Ghaghara river systems. Ballia district covers 329,023 ha and lies on the border of Uttar Pradesh and Bihar. The Ganges flows along its southern border while the Ghaghara flows along its northern border, joining around at some 60–70 km east of Ballia town.

Geological formation of the Bhojpur district forms a part of the Indo-Gangetic plain consisting of newer and older alluvial. The alluvial sediments consist of clay, fine to coarse-grained sand, kankar, gravel and have a maximum thickness along the present course of the Ganga.

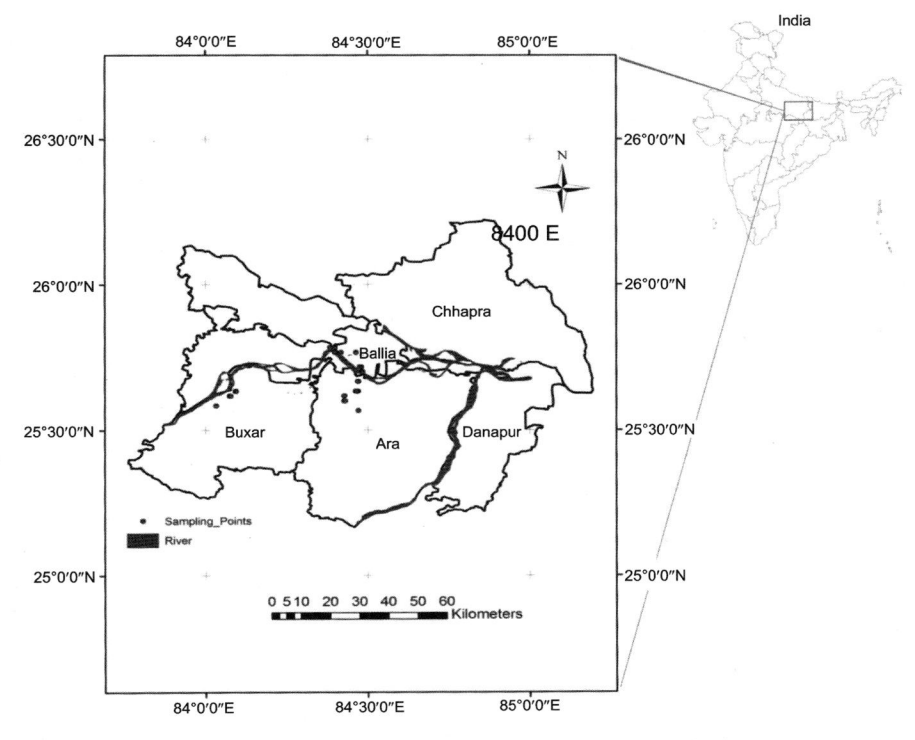

Fig. 8.1 Map showing sampling locations in Ballia, Buxar and Bhojpur districts

8.2 Material and Methods

8.2.1 Sampling and Field Parameter Analysis

A set of 84 water samples were collected in the winter season of 2012 from three districts namely Ballia, Buxar and Bhojpur of MGP. Geo-referencing Lat/Log of each tube well was done with the help of portable Global Positioning System (GPS) device at the time of field survey. A multi-parameter instrument (HANNA HI 9828) was used to measure environmental variable (pH, electrical conductivity, oxidation/reduction potential).

For approximate As concentration an arsenator field kit (Wagtech Arsenator W9000167) was used for selected tube wells. Measurement of bicarbonate (HCO_3^-) was done by acid titration method (APHA 2005) on site. Samples for analysis of major cations and heavy metals were filtered through (0.45 μm cellulose filter) and then acidified with HNO_3 and other samples filtered but not acidified were used for analysis of anions.

8.2.2 Laboratory Analysis of Water Samples

The major ions were determined in the filtered non-acidified water sample. Dissolved chloride (Cl^-), soluble sulfate (SO_4^{2-}), dissolved phosphate (PO_4^{3-}) and nitrate (NO_3^-) were analysed by the Dionex Ion-Chromatography (DX-120). Dissolved silicate (H_4SiO_4) was measured by molybdosilicate blue method (Strickland and Parsons 1968). Atomic Absorption Spectrophotometer (AAS Thermo Scientific M Series) was used to determine metals (Na^+, K^+, Ca^{++}, Mg^{++}, Fe, Mn^{++}, Zn^{++} and Cu^{++}), total As concentration in the field acidified samples was determined by using hydride generation atomic absorption spectro-photometer (HG-AAS, <1 µg/l detection limit). A high purity reagents (Merck) and milli-Q water (Model Milli-Q, Biocel) were used for all the analysis.

8.3 Results and Discussion

8.3.1 Hydrochemistry

Table 8.1 shows the summary of the statistical analysis of the three districts. The pH values vary widely from 7 to 8 with mean value of 7.3 (ranges 7.0–8.0), 7.0 (ranges 6.9–7.9) and 7.4 (ranges 6.7–7.9) for Ballia, Buxar and Bhojpur respectively, indicating that the groundwater is generally neutral to alkaline in nature. The higher values of pH during end of the winter season may be due to high reactive nature of the soil resulting from more elemental dissolution which may alter geochemical conditions resulting in high arsenic concentration dissolution (Kumar et al. 2010a, b; Ramanathan et al. 2012). Electrical conductivity and TDS values were found highest in district Buxar with the mean values for TDS being 793 ± 220 mg/L (ranges 570–1,304 mg/L) and 992 ± 408 µS/cm (ranges 660–2,038 µS/cm) for EC. However the lowest values were observed in Bhojpur with mean values for TDS 667 ± 167 mg/L (ranges 494–1,168 mg/L) and for EC 796 ± 252 µS/cm (ranges 516–1,559 µS/cm). Ballia has moderate values for TDS and EC with compare to both the places. The abundance of cations in the groundwater follows the same trend for Ballia, Buxar and Bhojpur (i.e. $Ca^{++} > Na^+ > Mg^{++} > K^+$ and for anions $HCO_3^- > Cl^- > SO_4^- > NO_3^- > PO_4^-$). The abundance of ions contributes significantly for the water classification in the MGP. The concentration of iron was found higher in Bhojpur with mean values 2.3 ± 1.9 mg/L (ranges 0.1–9.3 mg/L) while for Ballia and Buxar the concentration of Fe were 1.4 ± 1.4 and 1.6 ± 1.9 mg/L respectively. Zinc, Cu and Mn concentration does not vary for the three places. Arsenic concentration was found higher in Ballia with the mean value of 35.1 ± 23.8 µg/L (ranges 16.1–127.9 µg/L) while it was comparatively less 20.3 ± 13.4 and 24.3 ± 27.2 µg/L for Buxar and Bhojpur respectively.

8 Arsenic Distribution and Mobilization

Table 8.1 Summary of the statistical data of Ballia, Buxar and Bhojpur

S.No.	Parameters	Ballia Range	Ballia Mean±SD	Buxar Range	Buxar Mean±SD	Bhojpur Range	Bhojpur Mean±SD
1	Depth	15–46	28±11	15–55	27±14	12–58	28±13
2	pH	7.0–8.0	7.3±0.2	6.9–7.9	7±0.3	6.7–7.9	7.3±0.2
3	ORP	(−252)–(−14)	(−123)±(−67)	(−189)–(−46)	(−117)±(−41)	(−178)–(−49)	(−123)±(−38)
4	TDS	373–1,388	736±191	570–1,304	793±220	494–1,168	667±167
5	Na^+	12–104	30±19	20–136	59±35	17–130	40±28
6	K^+	8–16	10±2	10–62	22±17	9–69	19±13
7	Mg^{++}	13–64	28±11	13–67	28.58±16.96	2–64	22±12
8	Ca^{++}	38–236	89±34	54–214	80±34	54–151	83±20
9	EC	497–1,783	892±267	660–2,038	992±408	516–1,559	796±252
10	HCO_3^-	244–671	468±96	335–671	465±86	293–763	393±85
11	Cl^-	5–245	67±46	20–215	78±67	20–180	65±37
12	PO_4^{3-}	0.02–2.46	0.48±0.62	0.03–2.00	0.20±0.43	0.03–1.03	0.24±0.27
13	SO_4^{2-}	1.9–78.5	18.1±18.7	2.4–80.5	28.4±19.7	1.9–59.2	16.6±17.1
14	NO_3^-	1.1–25.4	3.1±4.3	1.6–35.8	9.8±9.7	1.8–30.7	5.2±5.9
15	H_4SiO_4	14.4–26.3	20.5±2.8	13.0–26.7	19.8±4.0	9.5–29.4	19.4±4.1
16	Fe	0.2–4.4	1.4±1.4	0.2–6.3	1.6±1.9	0.1–9.3	2.3±1.9
17	Zn	0.00–0.35	0.05±0.08	0.00–1.14	0.28±0.35	0.00–0.80	0.06±0.17
18	Cu	0.01–0.33	0.04±0.06	0.00–0.04	0.02±0.01	0.00–0.02	0.01±0.01
19	Mn	0.06–0.98	0.51±0.27	0.15–1.11	0.56±0.22	0.18–1.48	0.56±0.28
20	As(tot)	16.1–127.9	35.1±23.8	0.8–59.1	20.3±13.4	0.5–104.7	24.3±27.2

All values are given in mg/L except depth in metre, pH, EC in µS/cm and As in µg/L

8.3.2 Hydrochemical Facies

Piper trilinear diagram has been used to delineate the hydrochemical facies that exists in the study area. It has been found that major water type that exists is Ca-HCO_3 which is characterized by low conductivity and evolves from water–rock interaction. The Ca-HCO_3 water type has been reported from all three districts of MGP i.e. Ballia, Buxar and Bhojpur (Fig. 8.2a–c). This indicates the presence of carbonaceous sandstone and weathering of carbonate minerals. It is also supported by high log pCO_2 values in its comparison with the carbonate minerals (Fig. 8.4).

8.3.3 Aquifer Vulnerability

In Ballia, all the tested samples have As concentration above 10 µg/L which is WHO and the Bureau of Indian Standards (BIS) notified permissible limit for As. Similarly, for Buxar all the tested samples have As above WHO and BIS permissible limit of 10 µg/L while only 8 % of the samples are contaminated in nature. For Bhojpur, 66 % of the tested samples have As above 10 µg/L while 34 % water samples show As values below 10 µg/L and, thus, are uncontaminated in nature as per WHO and BIS norms. With respect to aquifer depth groundwater As concentrations of ≥ 50 µg/L consistently occurs in the aquifers of depths ranging from 40 to 160 ft and further with depth the As concentration decreases in the groundwater (Fig. 8.3a). As shown in this graph the variation of As with depth shows certain inconsistent trends and it seems to behave in a patched manner with respect to groundwater As contamination. Generally, microbiological reduction of Fe oxyhydroxide (FeOOH) would take place when reduction of free molecular O_2 and NO_3^- has exhausted these more thermodynamically favourable oxygen sources. Also, waters that contain NO_3^- do not contain detectable amounts of dissolved As (Nickson et al. 2000) and the similar scenario is very well represented in the plot between NO_3^- and As (Fig. 8.3b). With respect to HCO_3^- concentration which ranges from 244 to 762 mg/L but most of the samples fall under 300–600 mg/L of HCO_3^- which represents a source of dissolved organic matter and its oxidation in the groundwater (Fig. 8.3c). Dissolved Fe follows a very strong correlation with As in Buxar (≈0.6196) and Bhojpur (≈0.6661) but in Ballia (≈0.2659) it shows a non-conservative behaviour with dissolved Fe (Fig. 8.3d). The occurrence of Mn is low in the groundwater of MGP and does not show any consistent behaviour with groundwater As concentration (Fig. 8.3e). The representation from the present data supports the release and mobilization of As into the groundwater through reduction of arseniferous iron-oxyhydroxides when anoxic conditions develop during sediment. There is no correlation between Fe and NO_3^- like As and NO_3^- (Fig. 8.3f).

8 Arsenic Distribution and Mobilization

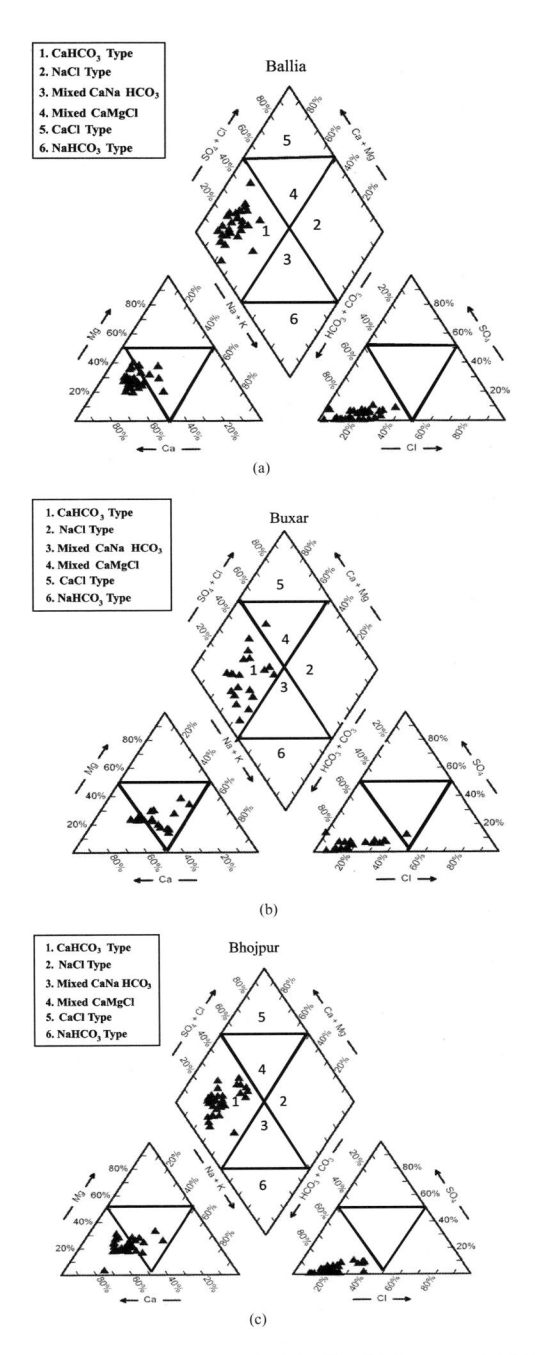

Fig. 8.2 Piper diagrams of the groundwater of (a) Ballia, (b) Buxar and (c) Bhojpur

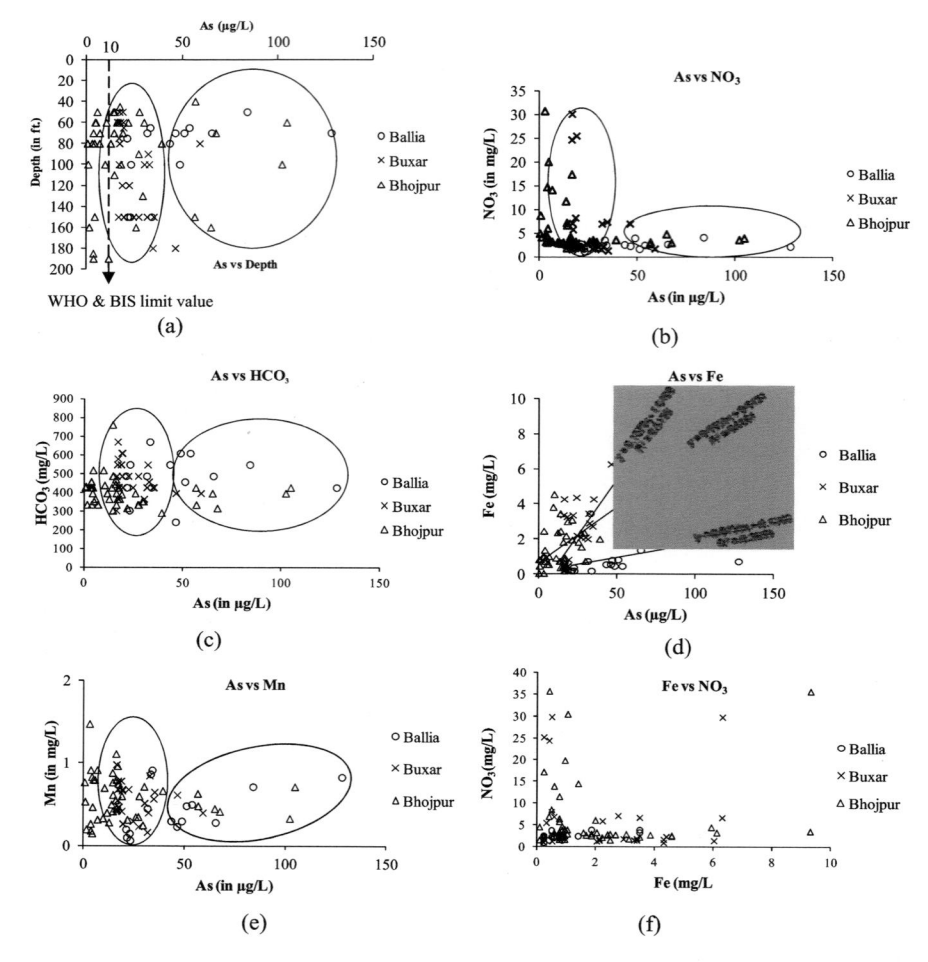

Fig. 8.3 Relation between concentrations of As vs. (**a**) Depth, (**b**) NO_3^-, (**c**) HCO_3^-, (**d**) Fe and (**e**) Mn; and (**f**) Fe vs NO_3^-

8.3.4 Geochemical Modelling for Mineral Phases Identification Using PHREEQC

Saturation Indices (SI) values were calculated to know the different mineral phases present in groundwater of the three districts by using 18 parameters (viz. Temperature, pH, ORP, Na^+, K^+, Mg^{++}, Ca^{++}, HCO_3^-, Cl^-, PO_4^{3-}, SO_4^{2-}, NO_3^-, H_4SiO_4, Fe, Zn, Cu, Mn and $As_{(tot)}$). Computer code WATEQ4F.DAT (The PHREEQC.DAT database extended with many heavy metals) was used to target metal containing minerals (Appelo and Postma 1994). Table 8.2 shows the SI values for some of the important phases known to control the chemistry of the groundwater. Oxygen and carbon dioxide are known as most important reactive gases present in the subsurface (Deutsch 1997). In the subsurface CO_2 gas, normally increased by factors of 100 or more, contrast with atmospheric air values due to result of oxidation of organic matter (Deutsch 1997). Log pCO_2 values (-1.8 ± 0.27, -1.7 ± 0.30 and -1.6 ± 0.24 at Ballia, Buxar and Bhojpur respectively) are greater than 100 times atmospheric value $10^{-3.5}$ atm which indicate that oxidation of organic matter occurs in these aquifers. The groundwater is highly undersaturated with O_2 in all the three locations. The groundwater is undersaturated with respect to the major As phases like arsenolite (As_2O_3) and arsenic (V) oxide (As_2O_5) along with Mn oxide phases like birnessite (MnO_2), nsutite (MnO_2) and pyrolusite (MnO_2) but they show near equilibrium to slightly saturated condition with rhodochrosite ($MnCO_3$) and revealed that these phases will remain in dissolved form once it enters the groundwater (Mukherjee and Fryar 2008). Hence, these phases may be the source of arsenic in the groundwater in these areas.

The groundwater is moderately to highly saturated with iron phases like goethite (FeOOH), hematite (Fe_2O_3), magnetite (Fe_3O_4), greenalite ($Fe_3Si_2O_5(OH)_4$), and siderite ($FeCO_3$) suggesting these phases may be sink for dissolved iron (Kumar et al. 2010a, b). However groundwater is moderately undersaturated with ferrihydrite ($Fe(OH)_3$) and magnetite (Fe_2O_3). The groundwater is nearly equilibrium to slightly supersaturated with respect to carbonate phases like aragonite ($CaCO_3$), calcite ($CaCO_3$) and dolomite ($Ca, Mg(CO_3)_2$) indicating carbonate precipitation in the aquifers. However magnesite ($MgCO_3$) is undersaturated.

8.3.5 Log pCO_2 vs State of Saturation

Log pCO_2 values were compared for all three districts with pH and found that there is negative correlation hence if pH increases log pCO_2 value decreases (Fig. 8.4a). The dissolution of silicate minerals results in consumption of large amount of CO_2. The groundwater is slightly supersaturated with quartz in all three districts. If dissolution of quartz is proceeded under the CO_2 closed system or in the subsurface then the amount of CO_2 required for the dissolution is larger than the amounts generated from biological process (Chidambaram et al. 2011). There is no sharp correlation found between log pCO_2 and As (Fig. 8.4b). A plot of district-wise samples with log pCO_2 was plotted and found that all the ranges between log pCO_2 values -1 to -2.5. In two samples of Bhojpur log pCO_2 value was higher may be due to shallow wells and are in equilibrium with atmospheric CO_2. Saturation indices of the carbonate minerals like aragonite ($CaCO_3$), calcite ($CaCO_3$), dolomite ($CaMg(CO_3)_2$) and magnesite ($MgCO_3$) were calculated and compared with log pCO_2 values (Fig. 8.4d–f) and found that in all three districts decreasing the log pCO_2 increases the SI.

Table 8.2 Summary of calculated values of saturated indices (SI) for selected minerals along with pCO_2 and pO_2 using PHREEQC

Mineral phase	Chemical formula	Ballia, $n=30$		Buxar, $n=20$		Bhojpur, $n=34$	
		Range	Mean ± SD	Range	Mean ± SD	Range	Mean ± SD
Aragonite	$CaCO_3$	−0.37–0.85	0.22 ± 0.28	−0.16–0.85	0.18 ± 0.25	−0.52–0.54	0.17 ± 0.24
Arsenolite	As_2O_3	(−12)–(−5.4)	−11 ± 1.23	(−14)–(−10)	−12 ± 0.79	(−15)–(−10)	−12 ± 1.10
Arsenic(V) oxide	As_2O_5	(−48)–(−28)	−38 ± 4.95	(−43)–(−34)	−38 ± 2.58	(−44)–(−35)	−39 ± 2.06
Birnessite	MnO_2	(−29)–(−20)	−24 ± 2.60	(−26)–(−20)	−23 ± 1.68	(−28)–(−21)	−24 ± 1.63
Calcite	$CaCO_3$	−0.22–1.0	0.37 ± 0.27	−0.01–1.0	0.32 ± 0.26	−0.37–0.68	0.31 ± 0.24
	Log pCO_2	(−2.2)–(−1.2)	−1.6 ± 0.24	(−2.3)–(−1.2)	−1.7 ± 0.30	(−2.4)–(−1.1)	−1.8 ± 0.27
Dolomite	$CaMg(CO_3)_2$	−0.48–1.9	0.58 ± 0.56	−0.2–1.7	0.52 ± 0.46	−1.0–1.3	0.37 ± 0.56
Ferrihydrite	$Fe(OH)_3$	(−5.2)–(−1.0)	−3.3 ± 1.29	(−4.5)–(−1.5)	−3.1 ± 0.80	(−5.2)–(−1.5)	−2.9 ± 0.81
Goethite	$FeOOH$	0.7–4.9	2.6 ± 1.29	1.3–4.4	2.7 ± 0.80	0.65–4.4	3 ± 0.81
Greenalite	$Fe_3Si_2O_5(OH)_4$	−2.9–3.1	0.36 ± 1.87	−2.7–3.5	0.37 ± 2.08	−2.4–4.8	1.5 ± 1.78
Gypsum	$CaSO_4 \cdot 2H_2O$	(−3.4)–(−1.5)	−2.6 ± 0.57	(−3.3)–(−1.5)	−2.3 ± 0.48	(−3.4)–(−1.7)	−2.6 ± 0.56
Hematite	Fe_2O_3	3.4–11	7.2 ± 2.58	4.7–10	7.5 ± 1.59	3.3–10	7.9 ± 1.63
Maghemite	Fe_2O_3	−6.9–1.3	−3.2 ± 2.58	−5.7–0.43	−2.9 ± 1.64	−4.5–0.47	−2 ± 1.30
Magnesite	$MgCO_3$	−0.84–0.27	−0.37 ± 0.29	−0.79–0.16	−0.39 ± 0.25	−1.65–0.13	−0.53 ± 0.35
Magnetite	Fe_3O_4	3.9–14	8.8 ± 2.87	5.7–13	9.1 ± 2.07	4.6–13	10 ± 2.01
Nsutite	MnO_2	(−28)–(−19)	−23 ± 2.60	(−25)–(−19)	−22 ± 1.68	(−27)–(−20)	−22 ± 1.63
	Log pO_2	(−71)–(−54)	−62 ± 4.79	(−67)–(−57)	−61 ± 2.85	(−68)–(−57)	−62 ± 2.79
Pyrolusite	MnO_2	(−26)–(−17)	−21 ± 2.60	(−24)–(−18)	−21 ± 1.68	(−26)–(−19)	−21 ± 1.63
Quartz	SiO_2	0.36–0.62	0.51 ± 0.06	0.32–0.63	0.49 ± 0.09	0.18–0.67	0.48 ± 0.09
Rhodochrosite	$MnCO_3$	−0.36–0.79	0.33 ± 0.37	−0.06–1.0	0.41 ± 0.29	−0.2–0.88	0.42 ± 0.26
Siderite	$FeCO_3$	−0.38–1.2	0.44 ± 0.54	−0.38–1.3	0.46 ± 0.51	−0.6–1.5	0.74 ± 0.44
Vivianite	$Fe_3(PO_4)_2 \cdot 8H_2O$	−5.4–2.1	−1.4 ± 2.22	−4.7–3.0	−1.9 ± 2.07	−5.2–2.4	−0.44 ± 1.51

Note: SI < 0: undersaturation
SI > 0: supersaturation

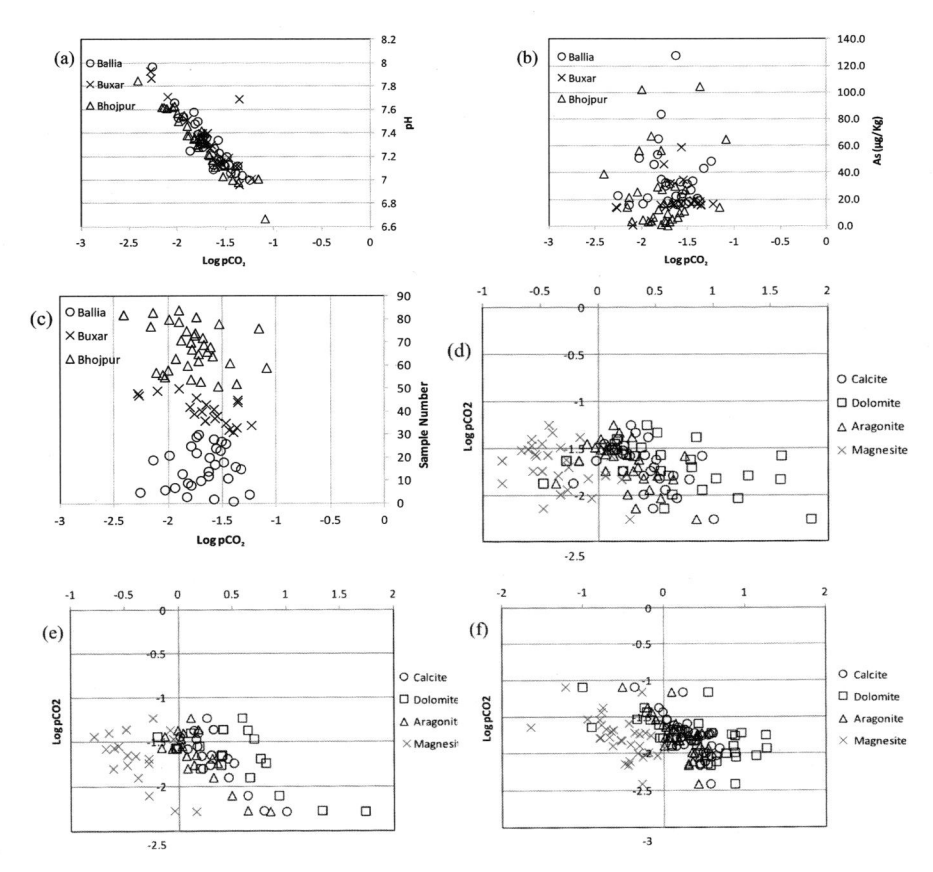

Fig. 8.4 Relationship of log pCO_2 with (**a**) pH, (**b**) As, (**c**) area samples, (**d**) S.I. of Ballia, (**e**) S.I. of Buxar and (**f**) S.I. of Bhojpur

8.4 Conclusion

The Ca-HCO$_3$ water type has been reported from all the three districts of MGP i.e. Ballia, Buxar and Bhojpur. In MGP the aquifers fall under Pleistocene-Recent alluvial aquifers and the concentration of As correlates strongly with concentration of dissolved Fe and consistently with concentrations of HCO_3^-. This strongly suggests that the As in groundwater beneath the MGP is derived by reductive dissolution of Fe oxyhydroxides in the sediments. The groundwater was highly undersaturated with pO_2 whereas pCO_2 is 100 times greater than the atmospheric pCO_2 values in all the three locations. Groundwater was also observed to be undersaturated with respect to major As phases like arsenolite (As_2O_3) and arsenic(V) oxide (As_2O_5) which might have acted as the source of arsenic dissolution in the aquifer. Mn oxide phases like birnessite (MnO_2), nsutite (MnO_2) and pyrolusite (MnO_2) also indicated undersaturation in groundwater. Moderate to high saturation of groundwater was observed with iron phases like goethite (FeOOH), hematite (Fe_2O_3), magnetite (Fe_3O_4), greenalite ($Fe_3Si_2O_5(OH)_4$) and siderite ($FeCO_3$) suggesting that these phases may be a sink for dissolved iron. The groundwater was nearly in equilibrium to slightly supersaturated with respect to carbonate phases like aragonite ($CaCO_3$), calcite ($CaCO_3$) and dolomite ($CaMg(CO_3)_2$) indicating carbonate precipitation in the aquifers. Log pCO_2 showed negative correlation with pH in all the three districts. The saturation indices obtained for all carbonate minerals showed negative correlation with log pCO_2 values which indicated that decrease in log pCO_2 may have enhanced the dissolution of carbonate minerals.

Acknowledgements Authors are thankful to School of Environmental Sciences, JNU for providing lab facilities and SIDA and KTH Sweden for partial financial assistance.

References

Acharyya SK Comment on Nickson et al (1999) Arsenic poisoning of Bangladesh groundwater. Nature 401:545
Acharyya SK, Shah BA (2004) Risk of arsenic contamination in groundwater affecting Ganga Alluvial Plain, India? Environ Health Perspect 112:19–20
Acharyya SK, Shah BA (2005) Genesis of arsenic contamination of ground water in alluvial Gangetic aquifer in India. In: Bundschuh J, Bhattacharya P, Chandrasekharam D (eds) Natural arsenic in groundwater. Balkema/Taylor and Francis, Leiden/London
Acharyya SK, Shah BA (2007) Arsenic contaminated groundwater from parts of Damodar fan delta and west of Bhagirathi River, West Bengal, India: influence of fluvial geomorphology and Quaternary morphostratigraphy. Environ Geol 52:489–501
APHA (2005) Standard methods for the examination of water and wastewater, 21st edn. APHA AWWA WEF, Washington, DC
Appelo CAJ, Postma D (1994) Geochemistry, groundwater and pollution. Balkema, Rotterdam
Chakraborti D, Mukherjee SC, Pati S, Sengupta MK, Rahman MM, Chowdhury UK, Lodh D, Chanda CR, Chakraborti AK, Basu GK (2003) Arsenic groundwater contamination in middle Ganga Plain, Bihar, India: a future danger? Environ Health Perspect 111:1194–1201
Chauhan VS, Nickson RT, Divya C, Iyengar L, Nalini S (2009) Ground water geochemistry of Ballia district, Uttar Pradesh, India. Chemosphere 75:83–91
Chidambaram S, Prasanna MV, Karmegam U, Singaraja C, Pethaperumal S, Manivannan R, Anandhan P, Tirumalesh K (2011) Significance of pCO2 values in determining carbonate chemistry in groundwater of Pondicherry region. India Front Earth Sci 5(2):197–206
Deutsch WJ (1997) Groundwater geochemistry: fundamentals and applications to contamination. Lewis Publishers, Boca Raton/New York
Kumar P, Kumar M, Ramanathan AL, Tsujimura M (2010a) Tracing the factors responsible for arsenic enrichment in groundwater of the middle Gangetic Plain, India: a source identification perspective. Environ Geochem Health 32:129–146
Kumar M, Kumar P, Ramanathan AL, Bhattacharya P, Thunvik R, Singh UK, Tsujimura M, Sracek O (2010b) Arsenic enrichment in groundwater in the middle Gangetic Plain of Ghazipur District in Uttar Pradesh, India. J Geochem Explor 105:83–94
Kumar A, Ramanathan AL, Prabha S, Ranjan RK, Ranjan S, Singh G (2012) Metal speciation studies in the aquifer sediments of Semria Ojhapatti, Bhojpur District, Bihar. Environ Monit Assess 184:3027–3042
Milliman JD, Meade RH (1983) Worldwide delivery of river sediments to ocean. J Geol 9:1–19

Mukherjee A, Fryar AE (2008) Deeper groundwater chemistry and geochemical modeling of the arsenic affected western Bengal basin, West Bengal, India. Appl Geochem 23:863–894

Nickson RT, McArthur JM, Ravenscroft P, Burgess WB, Ahmed KM (2000) Mechanism of As poisoning of groundwater in Bangladesh and West Bengal. Appl Geochem 15:403–413

Ramanathan AL, Tripathi P, Kumar M, Kumar A, Kumar P, Kumar M, Bhattacharya P (2012) Arsenic in groundwaters of the central Gangetic plain regions of India. In: Ng JC (ed) Understanding the geological and medical interface of arsenic. Taylor & Francis Group, London. ISBN 978-0-415-63763-3

Shah BA (2008) Role of Quaternary stratigraphy on arsenic contaminated groundwater from parts of Middle Ganga Plain, UP–Bihar, India. Environ Geol 53:1553–1561

Singh M, Sharma M, Tobschall HJ (2005) Weathering of the Ganga alluvial plain, northern India: implications from fluvial geochemistry of the Gomati River. Appl Geochem 20:1–21

Singh M, Singh IB, Müller G (2007) Sediment characteristics and transportation dynamics of the Ganga River. Geomorphology 86:144–175

Singh VB, Ramanathan AL, Pottakkal JG, Kumar M (2014) Seasonal variation of the solute and suspended sediment load in Gangotri glacier meltwater, central Himalaya, India. J Asian Earth Sci 79:224–234

Stricklan JDH, Parsonas TR (1968) A practical handbook of seawater analysis, Bulletin 167. Fisheries Research Board of Canada, Ottawa

WHO (1993) Guidelines for drinking water quality. Recommendation, vol 1–2. World Health Organization, Geneva

Chapter 9
Understanding Hydrogeochemical Processes Governing Arsenic Contamination and Seasonal Variation in the Groundwater of Buxar District, Bihar, India

Kushagra, Manish Kumar, AL. Ramanathan, and Jyoti Prakash Deka

9.1 Introduction

The presence of high concentrations of arsenic (As) in groundwater is a worldwide well known environmental problem in recent years (Nriagu et al. 2007). Most estimates of As pollution have focused on the predominance of As poisoning in the groundwater of West Bengal (India) and Bangladesh and thought to be limited to the Ganges delta (the lower Ganga plain) (Bhattacharya et al. 1997, 2002a; Ahmed et al. 2004; Ben et al. 2003). High As in the groundwater of the Lower Gangetic plain of Bangladesh and West Bengal was reported by Bhattacharya et al. (1997). Several authors suggested that the reductive dissolution of Fe (III)-oxyhydroxides in strongly reducing conditions of the young alluvial sediments is the cause for groundwater As mobilization (Ahmed et al. 2004; Bhattacharya et al. 1997, 2002b; Harvey et al. 2002; McArthur et al. 2001; Nickson et al. 1998, 2000). The reduction is driven by microbial degradation of sedimentary organic matter, which is a redox-dependent process consuming dissolved O_2 and NO_3^- (Stumm and Morgan 1981; Bhattacharya et al. 1997, 2002a, b; Nickson et al. 2000). Overall, the quality of groundwater depends on the composition of recharging water, the mineralogy and reactivity of the geological formations in aquifers, the impact of human activities and the environmental parameters that may control the geochemical mobility of redox sensitive elements in the groundwater environment (Kumar et al. 2006, 2007; Bhattacharya et al. 2009; Hasan et al. 2009).

Kushagra • AL. Ramanathan (✉)
School of Environmental Sciences, Jawaharlal Nehru University, New Delhi 110067, India
e-mail: alrjnu@gmail.com

M. Kumar • J.P. Deka
Department of Environmental Science, Tezpur University,
Napaam, Sonitpur, Assam 784-028, India

© Capital Publishing Company 2015
AL. Ramanathan et al. (eds.), *Safe and Sustainable Use of Arsenic-Contaminated Aquifers in the Gangetic Plain*, DOI 10.1007/978-3-319-16124-2_9

Globally, contaminated drinking water is the chief source of chronic human intoxication (Smith et al. 2000; Kapaj et al. 2006; Rahman et al. 2009) that may result in skin ailments, such as hyper pigmentation and keratosis, and progress to cancer and ultimately death (WHO 1993). The current drinking water quality guideline for As is 10 µg/L (WHO 1993). Based on the data compilation from West Bengal and Bangladesh, (DPHE 1999) the average total As contents of fluvio-deltaic sediments is 15.9 mg/kg for onshore samples and 10.3 mg/kg for offshore samples (Ravenscroft et al. 2005). Sediment core analyses show that maximum As concentrations are noted for the fine-grained sediments at shallow depths (Nickson 1998; Paranis 1997; AAN 1999; DPHE 1999; Ahmed et al. 2004; Hasan et al. 2009). The As contaminated aquifers are pervasive within lowland organic rich, clayey deltaic sediments in the Bengal basin and locally within similar facies in narrow, entrenched river valleys within the Ganga Alluvial Plain (Acharyya and Shah 2004; Acharyya 2005).

Since the Middle Gangetic Plain is the most heavily populated area, so deterioration of health condition in this study is a matter of chief concern. Recent studies on the water quality in a major part of the Gangetic plains of India in the states of Uttar Pradesh and Bihar indicate the presence of elevated concentrations of As in drinking water wells (Acharyya and Shah 2004; Chakraborti et al. 2004; Acharyya 2005; Chauhan et al. 2009; Sankararamakrishnan et al. 2008; Srivastava et al. 2008). However, as seen from the above review of these cited literatures, water quality parameters and their relationship to the distribution of As has not been evaluated especially in this region. The present study has been attempted with the following objectives: (1) to quantify the As in the groundwater of the Buxar district, Bihar, India; (2) to trace the seasonal variations of As and associated major ions over two consecutive years of time, and (3) to understand the mechanism controlling the mobilization of As and its evolution.

9.2 Methodology

9.2.1 Study Area Description

The area taken for current study falls under two neighbouring districts of Bhojpur and Buxar of Bihar. The region lies just south to the River Ganga under middle Gangetic alluvial plain amid the flood prone belt of Sone-Ganga interfluves of Middle Ganga Plains (MGP). River Sone is a southern perennial tributary of Ganga and joins river Ganga at midway to Patna, capital of Bihar state (CGWB 2009). The study area is comprised of two elongated crescent shaped old river channels of Ganga ('Bhagar', local name), which is completely disconnected from main river course and are located opposite to each other. These channels are of different dimensions, the bigger one (situated west) is a 27 km long harbouring large number of villages while the smaller one is Gaura Bhagar, approximately 10 km long and harbouring comparatively lesser population. In southern part, Dharmawati river

(Dehra Nadi), meets Gaura Bhagar, which is an oxbow left over portion of meandered Ganga, that finally meets Ganga (CGWB 2009). A recent manmade embankment between abandoned river channel and River Ganga completely disconnects its link with main river. Between these two crescent shaped ponds Ganga first moves northward and then takes a southward loop, making this region active in respect to sedimentary processes (Saha et al. 2009).

The annual flood cycle of river Ganga spread new silt over entire area that is considered to be good for soil fertility, which in turn supports highly dense population. This area is known for surplus food production and intensive groundwater extraction to support the agriculture and drinking water need of sufficiently high population. Majority of geographical area about 94,000 km^2 of Bihar fall under Middle Ganga Plains (MGP) (Saha et al. 2009). Annual rain fall in study region is between 1,021 and 1,080 mm, and nearly 85 % of annual rainfall is due to Southwest monsoon (active between June and September). There is large fluctuation of temperature (5–45 °C). Important land uses for economic activities of local people in the study area are subsistence agriculture including main crops of paddy, wheat, maize, pulses and sugarcane. Arsenic contamination is posing a serious health crisis after reporting of several arsenic hotspots as there is no alternative source of drinking water supply other than groundwater extracted through local hand pumps.

9.2.1.1 Hydrogeological Settings

The study area typically lies on thick Holocene deposits and reported to be affected by elevated groundwater arsenic in south-eastern part of MGP (Central Ground Water Board 2010). The newer alluvium deposition is called Diara formation.

The channels are not deeply incised and exposed banks are those of the recent, aggrading floodplain system and the alluvial sequences in this area range from middle Pleistocene to Holocene. The sedimentation in the narrow entrenched Sone-Ganga interfluve was influenced by sea level fluctuation in Holocene causing formation of large fluvial lakes (Singh 2004). The older alluvium is present in the ground above flood level line in basin whereas newer alluvium is confined to the river channel and their flood plains. Both alluviums are distinct on the basis of colour and chemical composition. The newer alluvium is anoxic and composed of sand and silt-clay. The upper part is argillaceous (silty clay to sandy clay) and shallow part is made up of fine to medium sand with clay lenses (Saha et al. 2010). Lower alluvium is more weathered and oxidized, typically have yellowish colour which is indicative of oxic environment at the time of its deposition. Saha (2009) reported presence of quartz, muscovite, chlorite, kaolinite, feldspar, amphibole and goethite in newer alluvium of MGP.

The physiography of combined Bhojpur and Buxar districts are divided into two main parts, Northern Alluvium which has gentle slope toward north and flat plains of south. The selected abandoned river channels for present investigation are confined within newer alluvium entrenched channels and flood plains of Ganga river (Fig. 9.1). The detailed lithology of upto 300 m depth of this area is underlain by

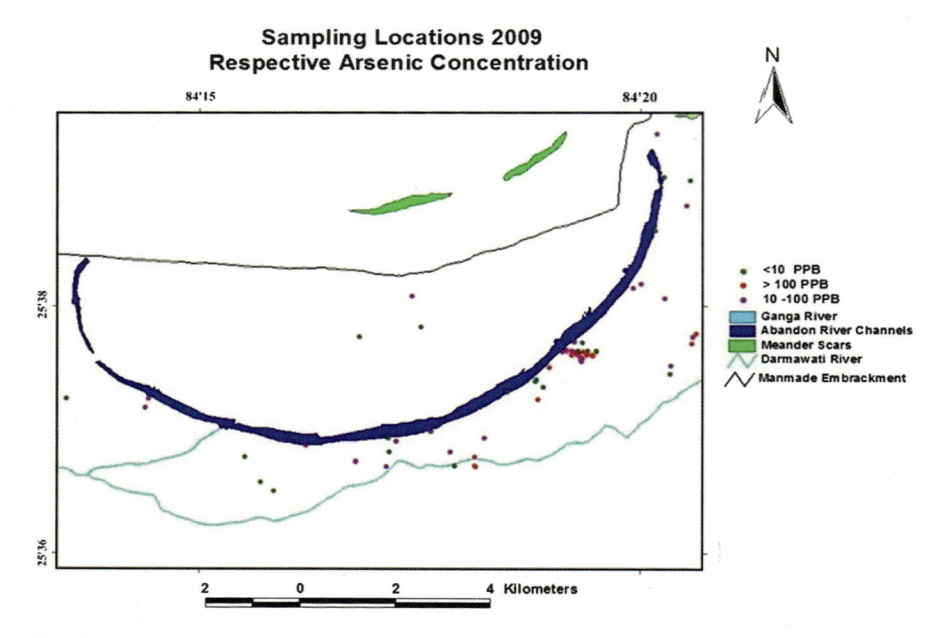

Fig. 9.1 Study area map showing sample location with their respective As concentration

multilayer sequence of aquifers alternating with aquitards like sandy-clay and clay represents a two-tier aquifer system. This typical two-tier system is in continuance across entire stretch of Ganga-Sone interfluve (Buxar to Patna). The upper 100–120 m depth is marked with shallow unconfined aquifers, where arsenic dissolution is reported. Deeper aquifers exist upto depth of 300 m which are separated by middle clay (light yellow sandy clay) of 20–30 m thickness. The upper aquifer system is referred as active channel deposits and marked as newer alluvium (Saha et al. 2009; CGWB 2006). Shallow aquifers in this region face groundwater fluctuation at moderate extent, typically 5.3 m in pre-monsoon and 3.78 m in post-monsoon. Groundwater flow is directed towards Ganga river with sluggish speed. Groundwater of both shallow and deeper aquifer system are of different age, approximated at 40 years and 3,000 years respectively, which suggest their unconfined and confined to semi-confined nature of aquifers and also establishes the role of middle clay as aquitards (Saha et al. 2009).

9.2.2 Materials and Methods

To understand the spatial and temporal variation in groundwater chemistry of the study area, groundwater samples were collected from Buxar district for two consecutive seasons representing pre-monsoon season in January 2009 ($n=52$) and post-monsoon seasons in June, 2009 ($n=69$). All the monitoring wells were drilled where groundwater extraction is being carried out by hand pumps or tube-wells. The information about the age of wells was collected from local people and the range varied from less than 1 year to 35 years, with an average of 6 years. Similarly, the depth information was collected from users and it varied from 40 to 185 m.

Alkalinity was measured in the field using a titration method. Three duplicate samples were collected and filtered using 0.45 μm Millipore membrane filters at each location. For analyses of major cations, 1 L of each sample was acidified by adding about 8 ml of ultra-pure nitric acid to bring pH less than 2. For NO_3^- analysis, H_3BO_3 acid was used as preservative (Kumar et al. 2009). The samples were stored at 4 °C until analysis. The analysis was done by standard techniques using Atomic Absorption Spectrophotometer (Shimadzu AAS) (for Na^+, K^+, Ca^{2+} and Mg^{2+}) and Spectrophotometer (Shimadzu, UV-1700) (for Cl^-, NO_3^- and SO_4^{2-}) (APHA et al. 2005). The precision of the chemical analyses was carefully defined by checking ion charge balance, which was generally found within ±10 %. To evaluate the potential relationship between the various physicochemical parameters the Pearson correlation coefficient was used. Factor analysis for groundwater samples was carried out by using "Statistical Package for Social Sciences (SPSS), version-16.00". "Principal component analysis" and "Varimax Rotation" were used for extracting and deriving factors, respectively.

9.3 Results and Discussion

9.3.1 General Expressions of Hydro-geochemical Data

The analytical results of groundwater samples of Buxar and their computed values have been summarized along with their standard deviation in Table 9.1. In general, groundwater was found alkaline in nature with an average value of more than 7.00. A slight increase observed in the post-monsoon with respect to the pre-monsoon indicates that dissolution has been enhanced due to high interaction between soil and rainwater as well as due to dilution from the influx of rainwater of lower alkalinity (Kumar et al. 2009). TDS and EC were found higher in the post-monsoon in terms of range but lower in their mean values. This pattern of data clearly shows that there are some areas affected with salt patch where additional leaching is taking place. On the other hand the higher average value of EC in the pre-monsoon suggests the enrichment of salt due to evaporation effect in the pre-monsoon followed by subsequent dilution through rainwater. Very high standard deviation in EC for the post-monsoon also suggests the spatial variability of leaching and dilution with recharging rainfall water, which can be further, linked with the local variation in point sources, soil type, multiple aquifer system and other agriculture related activities in the area.

In Buxar, dominant cations remain in the order of $Ca^{2+} > Mg^{2+} > Na^+ > K^+$ throughout the study period. Although, clear seasonal variations were observed both in minimum and maximum concentrations of these cations, but the difference in mean

Table 9.1 Comparative statistical summary of hydrogeochemical parameters of groundwater in the pre-monsoon and post-monsoon

Parameters	Pre-monsoon ($n=52$)		Post-monsoon ($n=69$)	
	Mean ± SD	Range	Mean ± SD	Range
Eh	−173 ± 59.3	−58.2–15.6	−165 ± 84.0	−38.7–129
pH	7.6 ± 0.1	7.9–8.2	7.1 ± 0.1	8.0–8.4
EC	417.9 ± 70.7	539.5–684.9	275 ± 197.5	514.1–1,201
TDS	201.5 ± 35.1	261.8–334.1	132 ± 97.7	249.2–592
Na^+	18.3 ± 8.3	27.8–46.0	14 ± 8.1	26.1–50.3
K^+	1.3 ± 1.0	3.4–5.1	1.9 ± 9.4	5.4–78.1
Mg^{2+}	20.2 ± 10.9	37.8–72.4	14.3 ± 8.4	31.2–60.2
Ca^{2+}	59.7 ± 25.0	104.4–156.7	41.6 ± 16.0	67.2–122.1
H_4SiO_4	15.2 ± 5.5	24.4–39.9	15.7 ± 3.6	23.9–33.4
HCO_3^-	167.4 ± 60.6	292–387.1	184.2 ± 81.8	390.8–524.6
NO_3^-	1.5 ± 9.0	10–44.0	0.3 ± 31.4	18.2–170.4
PO_4^{3-}	0.03 ± 0.4	0.8–1.4	0.016 ± 1.6	1.32–10.8
SO_4^{2-}	1.9 ± 11.6	13.1–44.8	1.7 ± 29.8	17.7–152.3
F^-	0.02 ± 0.1	0.2–0.6	0.006 ± 0.05	0.1–0.3
Cl^-	21.3 ± 13.8	49.5–79.7	4.2 ± 9.6	16.0–46.6
As (T)	1.7 ± 126.9	124.5–401	0.5 ± 39.1	73.5–945.3
As^{3+}	22.4 ± 100.5	199–325	10.9 ± 10.0	132.3–392.5
Fe^{2+}	0.19 ± 6.5	6.2–22.8	0.2 ± 7.5	7.3–24.7
Mn^{2+}	0.005 ± 0.3	0.4–1.1	0.003 ± 0.3	0.5–1.8
Zn^{2+}	0.01 ± 0.2	0.1–1.0	0.005 ± 0.2	0.1–1.0
Tz^+	3.8 ± 1.2	7.1–9.6	4.6 ± 1.3	7.5–11.3
Tz^-	3.6 ± 1.2	6.7–9.3	4.3 ± 1.4	7.6–11.4

All values are in mgL^{-1} except pH, EC (μScm^{-1}), ORP (mV), As(T) (μgL^{-1}) and As^{3+} (μgL^{-1})

value is not significant except for K^+ (Table 9.1). This may be due to the dynamic change among the different hydro-geochemical processes operating in the area. Despite the greater resistance of potassium silicate weathering, ions are released during weathering but they seem to be used up in the formation of secondary minerals. The abnormal concentration of potassium at few places may be due to urban pollution and fertilizer leaching.

Alkalinity of water is the measure of its capacity for neutralization. Bicarbonate represents the major sources of alkalinity. Bicarbonate is slightly higher in post-monsoon period indicating the contribution from carbonate weathering process. There was a slight variation observed in seasonal and spatial distribution of HCO_3^- but very significant at certain locations, may be due to contributions from carbonate lithology. The significant increase of Cl^- in the post-monsoon substantiates the high leaching of salt with percolating rain water. In general, in most of the natural waters, SO_4^{2-} is found in smaller concentrations than the Cl^-. This indicates addition of sulfate by the breakdown of organic substances of weathered soils, sulfate leaching, from fertilizers and other human influences.

Nutrients in the groundwater were found in the order of $NO_3^- > H_4SiO_4 > PO_4^{3-}$. The concentration of silica in groundwater samples showed some seasonal fluctuations. The possible explanation may be viewed in the perspective of the existence of alkaline environment, which enhances the solubility of silica and reveals secondary impact of silicate weathering. The average value of nitrate found higher in the post-monsoon, may be due to leaching of NO_3^- from fertilizers and biocides during irrigation of agriculture land. The PO_4^{3-} in the study area was very low, may be because of phosphate adsorption by soils as well as its limiting factor nature due to which whatever PO_4^{3-} is applied to the agricultural field is used by the plants (Holman et al. 2008). Wheat plantation in the area consumes more PO_4^{3-} which supports our observation (Kumar et al. 2008).

9.3.2 Graphical Representation of Hydro-geochemical Data

An overall characterization of hydro-geochemical data can be possible by knowing the hydro chemical facies of water, generally known as water type, using various plots like Durov diagram (1948), Piper (1944) tri-linear diagram, Schoeller diagram (1965), Stiff diagram and Radial plot. Figures 9.2, 9.3, 9.4 and 9.5 traces the seasonal variation in water type of groundwater samples collected in January '09 and

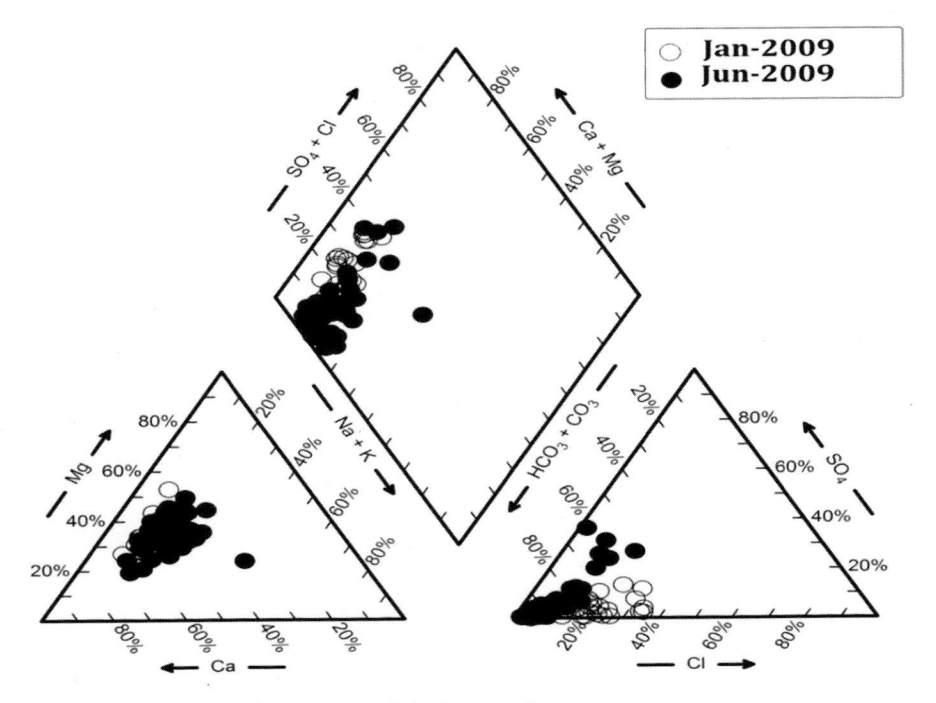

Fig. 9.2 Piper plot showing water type in both pre- and post-monsoon season

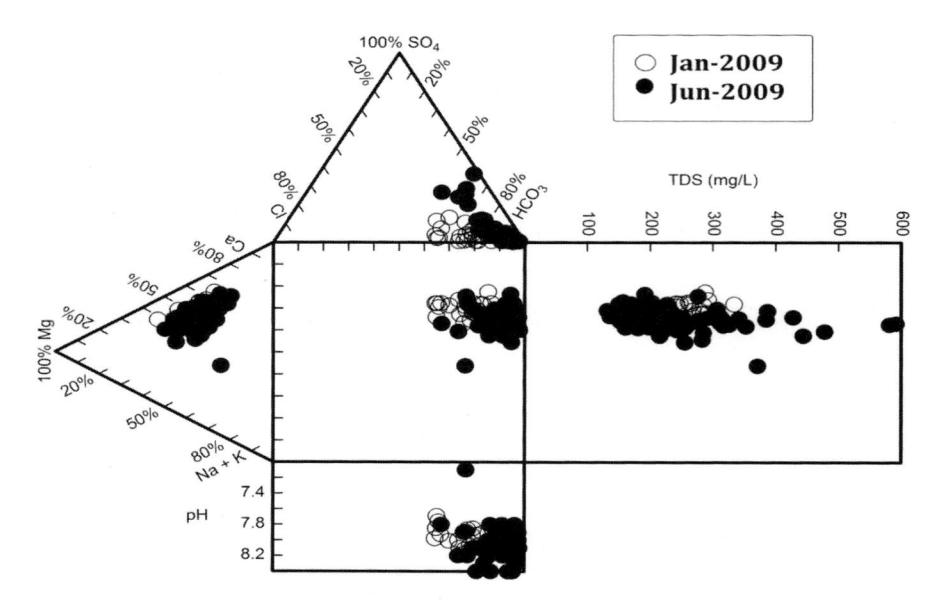

Fig. 9.3 Durov plot showing water type in Buxar district in both the seasons

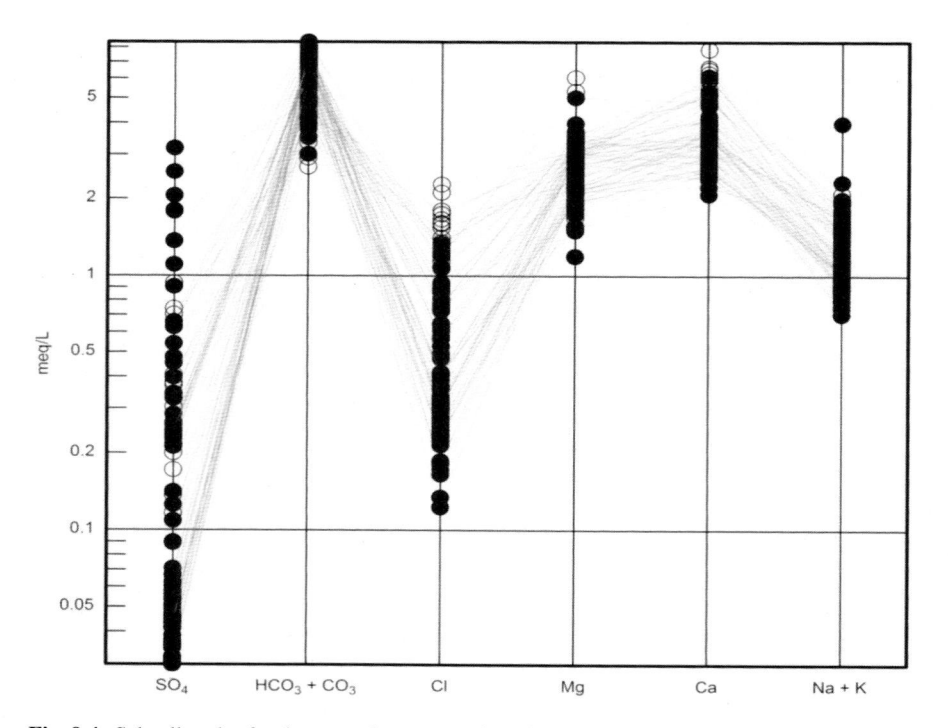

Fig. 9.4 Schoeller plot for the groundwater samples of Buxar district

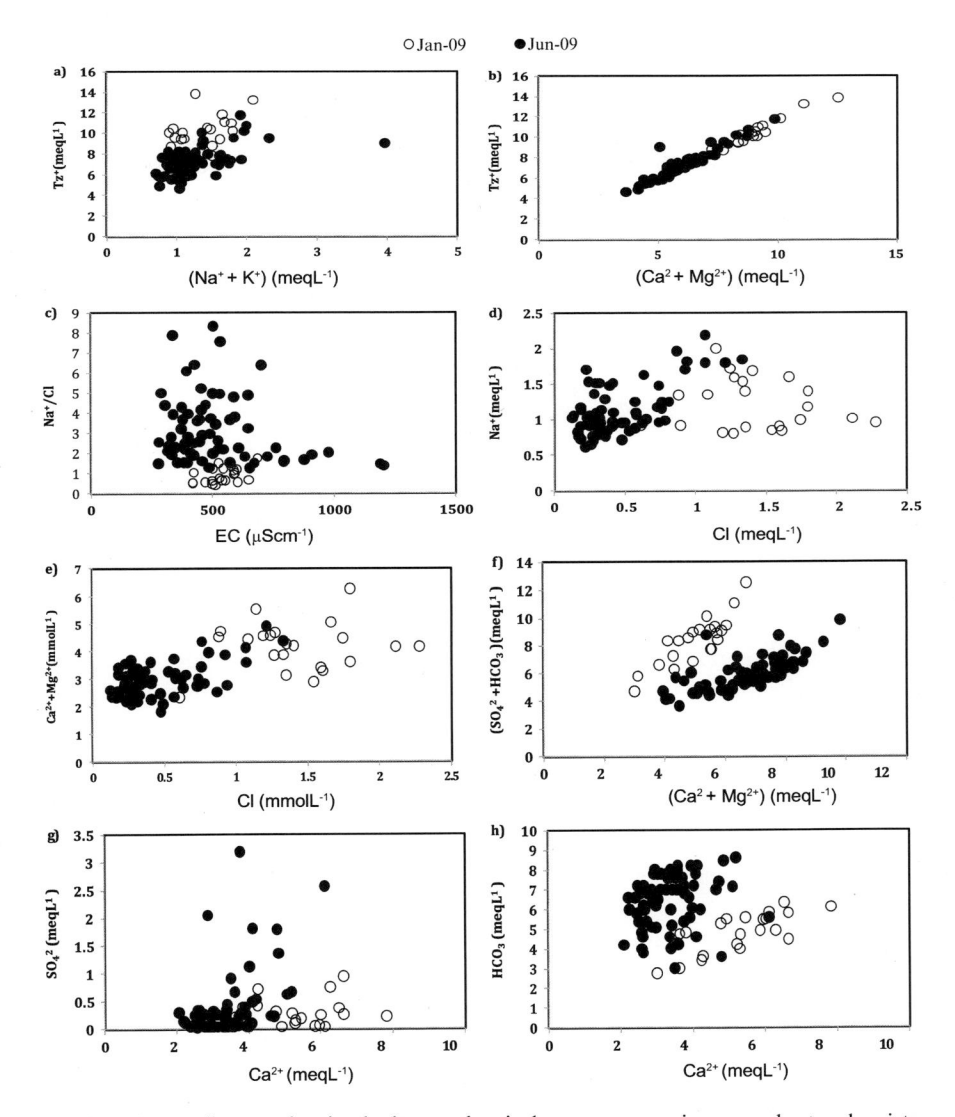

Fig. 9.5 Scatter diagram showing hydro-geochemical process governing groundwater chemistry in Buxar district

June '09. There is a little change in the hydro-geochemical facies of water, as a little shift towards bicarbonate and alkali metals domain can be observed in the post-monsoon with respect to the pre-monsoon samples. After plotting entire data set in the piper plot, the diamond field suggests that Ca^{2+}-Mg^{2+}-HCO_3^- type of water remain the major water type in the area throughout the year.

Durov representation in Fig. 9.3 also suggests that alkaline earths exceed alkali metals and weak acidic anions exceed strong acidic anions. Such water has temporary hardness. The positions of data points in this domain represent Ca^{2+}–Mg^{2+}–HCO_3^- water type. However, the presence of Ca^{2+}–Mg^{2+}–SO_4^{2-} type of waters as well as Na^+–HCO_3^- type of waters cannot be overruled. Such water has permanent hardness and does not deposit residual sodium carbonate in irrigation use. Thus, salinity is not an issue in the area and no problem for irrigation and drinking uses. As far as major ions are concerned, some waters are prone to deposit residual sodium carbonate in irrigation use and cause foaming problems.

Finally, the distribution of dataset on the Schoeller plot (Fig. 9.4) suggests that most of the sampling points exhibit temporary hardness i.e. Ca^{2+}–Mg^{2+}–HCO_3^- water type. This finding is in line with the previous work. In addition Schoeller plot suggests that it is Na^+ and Cl^- that varies a lot and differs most from sample to sample. Thus the impacts of irrigation return flow and landuse in the floodplain of Buxar is getting prominent. It also suggests that, while Ca^{2+}-Mg^{2+} and HCO_3^- seem to originate from geological processes, Na^+, Cl^- and sulfate are mainly anthropogenic induced that makes difference between the sampling locations.

9.3.3 Identification of Hydro-geochemical Process

The hydro-geochemical data are subjected to various conventional graphical plots in order to identify the hydro-geochemical processes and mechanisms in the aquifer region of study area.

Silicate weathering can be understood by estimating the ratio between Na^++K^+ and total cations (TZ^+). The relationship between Na^++K^+ and total cations (TZ^+) of the area indicates that the majority of the samples are plotted much above the $Na^++K^+=0.5TZ^+$ line (Fig. 9.5a) indicating the involvement of silicate weathering (Stallard and Edmond 1983). Furthermore, weathering of soda feldspar (albite) and potash feldspars (orthoclase and microcline) may contribute Na^+ and K^+ ions to groundwater. Feldspars are more susceptible for weathering and alteration than quartz in silicate rocks.

Further, $Ca^{2+}+Mg^{2+}$ vs total cations (TZ^+) plot of groundwater samples (Fig. 9.5b) has a linear spread mostly on or above 1:1 equiline ($Ca^{2+} + Mg^{2+}=TZ^+$) which indicates that Ca^{2+} and Mg^{2+} are major contributing cations to the groundwater of the area resulting from the weathering of carbonate minerals. In the groundwater of Buxar alkali metals are not as abundant as that of alkaline earth metals that may be due to its fixation in the formation of clay minerals. The source of $Ca^{2+}+Mg^{2+}$ could be weathering and dissolution of carbonate and sulphate minerals like calcite, dolomite and gypsum.

The molar ratio of Na^+/Cl^- for groundwater samples of the study area generally ranges from 0.03 to 3.92 (Fig. 9.5c). As most of the samples having Na^+/Cl^- molar ratio below 1 are found in the post-monsoon indicating halite dissolution in the post-monsoon, which is replaced by silicate weathering in the pre-monsoon at few locations where ratio is found above 1. The trend of EC vs Na^+/Cl^- scatter diagram of the groundwater samples shows that the trend is inclined, which indicates that evaporation may not be the major geochemical process controlling the chemistry of groundwater. The sodium versus chloride (Fig. 9.5d) plot indicates that most of the pre-monsoon samples lie slightly above the equiline. The excess of Na^+ is attributed from silicate weathering (Stallard and Edmond 1983) while the post-monsoon samples are lying below it, indicating that evaporation may have led to the addition of Cl^- in the post-monsoon due to water level rise which causes more salt dissolution from the soil. Na^+ concentration is also being reduced by ion-exchange. Hence Na^+ and Cl^- do not increase simultaneously. The samples in both the pre and post monsoon seasons do not show increasing trend of $Ca+Mg$ with salinity (Fig. 9.5e), indicating that reverse ion exchange is the dominant process.

The carbonates from various sources might have been dissolved and added to the groundwater system with recharging water during irrigation, rainfall or leaching and mixing processes. In $Ca^{2+}+Mg^{2+}$ vs $SO_4^{2-}+HCO_3^-$ scatter diagram (Fig. 9.5f), the points falling along the equiline ($Ca^{2+}+Mg^{2+}=SO_4^{2-}+HCO_3^-$) suggest that these ions have resulted from weathering of carbonates and silicates (Rajmohan and Elango 2004; Kumar et al. 2006, 2009). Most of the points, which are placed in the $Ca^{2+}+Mg^{2+}$ over $SO_4^{2-}+HCO_3^-$ side, indicate that carbonate weathering is the dominant hydro-geochemical process, while those placed below the 1:1 line are indicative of silicate weathering. The plot of $Ca^{2+}+Mg^{2+}$ versus $SO_4^{2-}+HCO_3^-$ suggests that most of the points in the post-monsoon seasons are placed in the $Ca^{2+}+Mg^{2+}$ over $SO_4^{2-}+HCO_3^-$ side, indicating that carbonate weathering is the dominant hydro-geochemical process.

Figure 9.5f also suggests that the dissolutions of calcite, dolomite and gypsum are the dominant reactions in a system. Ion exchange tends to shift the points to right due to an excess of $SO_4^{2-}+HCO_3^-$ (Cerling et al. 1989; Fisher and Mulican 1997). If reverse ion exchange is the process, it will shift the points to the left due to a large excess of $Ca^{2+}+Mg^{2+}$ over $SO_4^{2-}+HCO_3^-$. The plot of $Ca^{2+}+Mg^{2+}$ versus $SO_4^{2-}+HCO_3^-$ shows that most of the groundwater samples of the post-monsoon were below the 1:1 line except few samples which do indicate reverse-ion exchange that is the case with pre-monsoon samples. A possibility of calcite weathering by sulfuric acid is also there as indicated by Ca^{2+} and SO_4^{2-} ratio (Fig. 9.5g). Sulphuric acid may be produced by SO_x emission dissolved in rain from automobile and industrial sources.

The presence of carbonic acid and sulphuric acid enhances weathering reactions. If the weathering of carbonates is by carbonic acid, the equivalent ratio of dissolved Ca^{2+} and HCO_3^- in the groundwater resulting from calcite weathering is 1:2, whereas for dolomite weathering it is 1:4. If sulphuric acid is the weathering agent, then the $Ca^{2+}:SO_4^{2-}$ ratio is almost 1:1 (Das and Kaur 2001). In Ca^{2+} vs HCO_3^- scatter diagram (Fig. 9.5h), only few groundwater samples fall along the 1:2 line and most of them lie above the 1:1 equiline.

9.3.4 Distribution of Arsenic

In both the pre and post monsoon seasons, it was observed that As decreased with increase in HCO_3^- (Fig. 9.6a), the reason could be due to the fact that As and HCO_3^- compete with each other for adsorption at the mineral surfaces, e.g. Fe/Mn

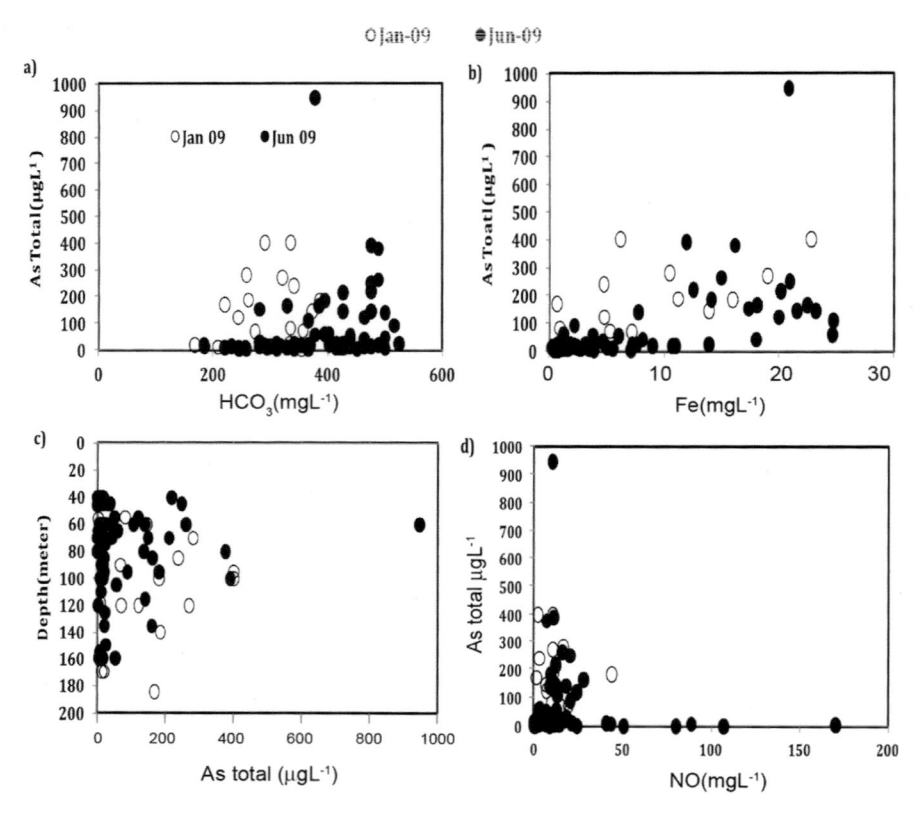

Fig. 9.6 Scatter plot showing relationship between (**a**) total As vs HCO_3^-, (**b**) total As vs Fe, (**c**) total As with depth and (**d**) As with NO_3^-

oxyhydroxides, clay minerals, and weathered mica. The relationship between As and Fe is shown in the plot (Fig. 9.6b). Only a few samples of post and pre monsoon fall close to the equiline, indicating a linear relationship between the samples. However majority of the samples do not show a linear relationship between As and Fe.

The plot of Depth versus As (Fig. 9.6c) shows that maximum As concentration is observed at shallow depths between 50 and 100 m. As decreases with an increase in depth of over 100–150 m. The reason could be the existence of Fe-oxyhydroxide layer and organic matter at shallower depths. The scatter plot between As vs NO_3^- (Fig. 9.6d) shows that As concentration in both the pre and post monsoon seasons are high, but the concentration of NO_3^- are low. This indicates that NO_3^- is a donor of oxygen for microbial degradation of organic matter, and facilitates As release.

Overall, these observations suggest that the high As concentrations in the lower catchment of the Ganges river are a result of multiple source areas and the likelihood of related mechanisms of mobilization across the entire flood plain. The As concentration <50 μg/L in the central portion and >50 μg/L in the Buxar district could be possibly due to an accumulation of coarser sediment along a Holocene course of the river or due to some local effects. This indicates that how depositional environment and geological age are important factors in controlling As mobilization.

The depth distribution of As shows similar observations reported by Bhattacharya et al. (2009) where a maximum As concentration was observed at a depth of 20–40 m where the active terminal electron accepting processes (TEAPs) drive the As mobilization (Bhattacharya et al. 2002a, 2006, 2009; McArthur et al. 2001; Mukherjee et al. 2006; Dhar et al. 2008; Zheng et al. 2004). High concentration of As is associated with low concentration of HCO_3^- in the well waters at depth between 10 and 20 m (Fig. 9.6a). The low concentration of HCO_3^- between 10 and 20 m depth corresponds to the most important anion species, which competes with As for adsorption sites at mineral surfaces (e.g., Fe/Mn oxyhydroxides, clay minerals, and weathered mica), consequently releasing As into the groundwater (Charlet et al. 2007). Moreover, the reduction of arseniferous Fe-oxyhydroxides may be coupled to precipitation of Fe^{2+} as $FeCO_3$ (Sracek et al. 2004; Bhattacharya et al. 2001) when anoxic conditions develop during microbial oxidation of sedimentary organic matter (Bhattacharya et al. 1997) and may be responsible for poor relationship between As and iron. This assessment is supported by poor correlation between As and NO_3^- (Fig. 9.6d), which reveals that apart from dissolved O_2, NO_3^- is thermodynamically favoured electron acceptor for microbial degradation of dissolved organic materials (Drever 1997) in the shallow aquifers (HP) of Buxar district. Biomediated reductive dissolution of hydrated iron oxide (HFO) by anaerobic heterotrophic Fe^{3+} reducing bacteria (IRB) play an important role in release of sorbed As to groundwater (Nickson et al. 1998).

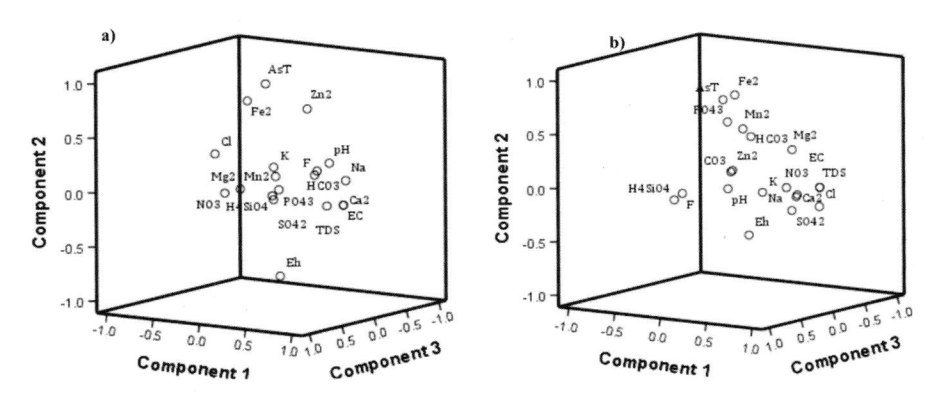

Fig. 9.7 3d plot PCA analysis of groundwater samples of (**a**) January and (**b**) June 2009

9.3.5 Statistical Analysis

In the month of January 2009 (Fig. 9.7a), i.e. the post-monsoon season, it is observed that NO_3^-, Mg^{2+} and Fe^{2+} have positive loading on the component one. The reason could be the existence of a reducing or anoxic condition after precipitation, high NO_3^- facilitates the decomposition of organic matter and a reducing condition. Component two has high loadings by EC and TDS, indicating that silicate and carbonate weathering are the major processes controlling the chemistry of water. PO_4^{3-} and Mn^{2+} show positive loadings on component three. In component four, total arsenic and arsenite show positive loadings, while H_4SiO_4 and F show negative loading, indicating the presence of anion exchange as a process for the release of As^{3+} in water. Component 5 has high loading by Cl^-, due to dilution by precipitation.

In the month of June 2009 (Fig. 9.7b), in the pre monsoon season, high positive loadings on component one is shown by EC, TDS, indicating a low water table, and concentration of the ions. Component two has high loadings by total arsenic and arsenite; F^- and HCO_3^- have high loading on component three. It indicates the presence of carbonate dissolution, which can be a source of F^- and HCO_3^-.

9.4 Conclusion

The groundwater quality depends mainly on composition of recharging water, the mineralogy and reactivity of the geological formations in aquifers, and the impact of human activities. Arsenic contamination in groundwater in the Ganga-Brahmaputra fluvial plains in India becomes major concern for India. As contamination depends upon Red-ox potential (Eh), adsorption/desorption, precipitation/dissolution, presence and concentration of competing ions, biological transformation, etc. Environmental conditions may also affect the geochemical mobility of certain constituents. As concentration decreased with increase in HCO_3^- may be due to the fact that As and HCO_3^- compete with each other for adsorption at the mineral surfaces. As and Fe did not have significant co-relation indicating their independent source. Maximum As concentration is observed at shallow depths between 50 and 100 m. As decreased with an increase in depth of over 100–150 m. The reason could be the existence of Fe-oxyhydroxide layer and organic matter at shallower depths. Low concentration of NO_3^- indicates that NO_3^- is a donor of oxygen for microbial degradation of organic matter, and facilitates As release. Depositional environment and geological age are important factors in controlling As mobilization.

Hydro-geochemical study showed that carbonate weathering is the main governing process controlling groundwater quality in the Buxar district. Groundwater in the district is mainly Ca^{2+}-Mg^{2+}-HCO_3^- type which remains the major water type in the area throughout the year. There is a little change in the hydro-geochemical facies of water, as a little shift towards bicarbonate and alkali metals domain can be observed in the post-monsoon with respect to the pre-monsoon samples. Salinity is not an issue in the area thus no problems for irrigation and drinking uses as far as major ions are concerned. Though hardness may be a concern which may create problem for irrigation purpose, it can also cause foaming problems.

References

AAN (1999) Arsenic contamination of groundwater in Bangladesh. Interim report of the research at Samta village. Report of the Asian Arsenic Network

Acharyya SK (2005) Arsenic levels in groundwater from quaternary alluvium in the Ganga plain and the Bengal basin, Indian subcontinent: insight into influence of stratigraphy. Gondwana Res 8:1–12

Acharyya SK, Shah BA (2004) Risk of arsenic contamination in groundwater affecting Ganga alluvial plain India. Environ Health Perspect 112:A19–A20

Ahmed KM, Bhattacharya P, Hasan MA, Akhter SH, Alam SMM, Bhuyian MAH, Imam MB, Khan AA, Sracek O (2004) Arsenic contamination in groundwater of alluvial aquifers in Bangladesh: an overview. Appl Geochem 19(2):181–200

APHA, AWWA, WEF (2005) Standard methods for examination of the water and wastewater, 21st edn. APHA, Washington, DC. ISBN 0-87553-047-8

Ben DS, Berner Z, Chandrasekharam D, Karmakar J (2003) Arsenic enrichment in groundwater of West Bengal, India: geochemical evidence for mobilization of As under reducing conditions. Appl Geochem 18:1417–1434

Bhattacharya P, Chatterjee D, Jacks G (1997) Occurrence of arsenic contaminated groundwater in alluvial aquifers from the Delta Plains, Eastern India: option for safe drinking water supply. Int J Water Res Dev 13:79–92

Bhattacharya P, Jacks G, Jana J, Sracek A, Gustafsson JP, Chatterjee D (2001) Geochemistry of the Holocene alluvial sediments of Bengal Delta Plain from West Bengal, India: implications on arsenic contamination in groundwater. In: Jacks G, Bhattacharya P, Khan AA (eds) Groundwater arsenic contamination in the Bengal Delta Plain of Bangladesh, Proceedings of the KTH-Dhaka University Seminar, University of Dhaka, Bangladesh KTH special publication, TRITA-AMI REPORT 3084, pp 21–40

Bhattacharya P, Frisbie SH, Smith E, Naidu R, Jacks G, Sarkar B (2002a) Arsenic in the environment: a global perspective. In: Sarkar B (ed) Handbook of heavy metals in the environment. Marcell Dekker, New York

Bhattacharya P, Jacks G, Ahmed KM, Khan AA, Routh J (2002b) Arsenic in groundwater of the Bengal Delta Plain aquifers in Bangladesh. Bull Environ Contam Toxicol 69:538–545

Bhattacharya P, Ahmed KM, Hasan MA, Broms S, Fogelström J, Jacks G, Sracek O, von Brömssen M, Routh J (2006) Mobility of arsenic in groundwater in a part of Brahmanbaria district, NE Bangladesh. In: Naidu R, Smith E, Owens G, Bhattacharya P, Nadebaum P (eds) Managing arsenic in the environment: from soil to human health. CSIRO Publishing, Melbourne

Bhattacharya P, Hasan MA, Sracek O, Smith E, Ahmed KM, von Brömssen M, Huq SMI, Naidu R (2009) Groundwater chemistry and arsenic mobilization in the Holocene flood plains in south-central Bangladesh. Environ Geochem Health 31:23–44

Central Ground Water Board (2010) Ground water in Bihar: issues and prospects, Ministry of Water Resources Govt. of India

Cerling TE, Pederson BL, Damm KLV (1989) Sodium–calcium ion exchange in the weathering of shales: implications for global weathering budgets. Geology 17:552–554

CGWB (2006) Ground water resources of National Capital Territory, Delhi. Ministry of Water Resources Govt. of India

CGWB (2009) West Bengal special issue. Quarterly journal of the Central Groundwater Board, Ministry of Water Resources. Bhu-Jal-Newz, 24(1):1–96. Available at http://cgwb.gov.in/documents/Bhujal_newz_24_1.pdf

Chakraborti D, Samanta MK, Rahman MM, Ahmed S, Chowdhury UK, Hossain MA, Mukherjee SC, Pati S, Saha KC, Dutta RN, Quamruzzaman Q (2004) Groundwater arsenic contamination and its health effects in the Ganga–Meghna–Brahmaputra plain. J Environ Monit 6:74N–83N

Charlet L, Chakraborty S, Appelo CAJ, Roman-Ross G, Nath B, Ansari AA, Lanson M, Chatterjee D, Mallik B (2007) Chemodynamics of an arsenic "hotspot" in a West Bengal aquifer: a field and reactive transport modeling study. Appl Geochem 22:1273–1292

Chauhan D, Nickson R, Iyengar L, Sankararamakrishnan N (2009) Groundwater geochemistry and mechanism of mobilization of arsenic into the ground water of Ballia district, Uttar Pradesh, India. Chemosphere 75(1):83–89

Das BK, Kaur P (2001) Major ion chemistry of Renuka lake and weathering processes, Sirmaur district, Himachal Pradesh, India. Environ Geol 40:908–917

Department of Public Health Engineering (DPHE) (1999) Groundwater studies for arsenic contamination in Bangladesh, Dhaka, Bangladesh. Main report on Phase I and volumes S1–S5

Dhar RK, Zheng Y, Stute M, van Geen A, Cheng Z, Shanewaz M, Shamsudduha M, Hoque MA, Rahman MW, Ahmed KM (2008) Temporal variability of groundwater chemistry in shallow and deep aquifers of Araihazar, Bangladesh. J Contam Hydrol 99:97–111

Drever JI (1997) The geochemistry of natural waters, 3rd edn. Prentice Hall, Englewood Cliffs

Durov SA (1948) Natural waters and graphical representation of their composition. Dokl Akad Nauk USSR 59:87–90

Fisher RS, Mulican IIIWF (1997) Hydrochemical evolution of sodium-sulfate and sodium-chloride groundwater beneath the Northern Chihuahuan desert, Trans-Pecos, Rexas, USA. Hydrogeol J 10(4):455–474

Harvey C, Swartz CH, Badruzzaman ABM, Keon-Blute NE, Yu W, Ashraf AM, Jay J, Niedam V, Beckie R, Brabander DJ, Oates PM, Ashfaque KN, Islam S, Hemond HF, Ahmed MF (2002) Arsenic mobility and groundwater extraction in Bangladesh. Science 298:1602–1606

Hasan MA, von Brömssen M, Bhattacharya P, Ahmed KM, Sikder AM, Jacks G, Sracek O (2009) Geochemistry and mineralogy of shallow alluvial aquifers in Daudkandi upazila in the Meghna flood plain, Bangladesh. Environ Geol 57:499–511. doi:10.1007/s00254-008-1319-8

Holman IP, Whelan MJ, Howden NJK, Bellamy PH, Willby NJ, Rivas-Casado M, McConvey P (2008) Phosphorus in groundwater: an overlooked contributor to eutrophication? Hydrol Process 22(26):5121–5127

Kapaj S, Peterson H, Liber K, Bhattacharya P (2006) Human health effects from chronic Arsenic poisoning – a review. J Env Sci Health Part A 41:1399–2428

Kumar M, Kumari K, Ramanathan AL, Rajinder S (2007) A comparative evaluation of groundwater suitability for irrigation and drinking purposes in two agriculture dominated districts of Punjab, India. Environ Geol 53:553–574

Kumar M, Kumari K, Ramanathan AL, Singh UK, Ramanathan AL (2008) Hydrogeochemical processes in the groundwater environment of Muktsar, Punjab: conventional graphical and multivariate statistical approach. Environ Geol 57:873–884

Kumar M, Ramanathan AL, Rao MS, Kumar B (2006) Identification and evaluation of hydrogeochemical processes in the groundwater environment of Delhi, India. Environ Geol 50:1025–1039

Kumar M, Ramanathan AL, Keshari AK (2009) Understanding the extent of interactions between groundwater and surface water through major ion chemistry and multivariate statistical techniques. Hydrol Process 23:297–310

McArthur JM, Ravenscroft P, Safiullah S, Thirlwall MF (2001) Arsenic in groundwater: testing pollution mechanisms for aquifers in Bangladesh. Water Resour Res 37:109–117

Mukherjee AB, Bhattacharya P, Jacks G, Banerjee DM, Ramanathan AL, Mahanta C, Chandrashekharam D, Chatterjee D, Naidu R (2006) Groundwater arsenic contamination in India: extent and severity. In: Naidu R, Smith E, Owens G, Bhattacharya P, Nadebaum P (eds) Managing arsenic in the environment: from soil to human health. CSIRO Publishing, Melbourne

Nickson RT, McArthur JM, Burgess WG, Ahmed KM, Ravenscroft P, Rahman M (1998) Arsenic poisoning of Bangladesh groundwater. Nature 395:338–349

Nickson RT, McArthur JM, Ravenscroft P, Burgess WG, Ahmed KM (2000) Mechanism of arsenic release to groundwater, Bangladesh and West Bengal. Appl Geochem 15:403–413

Nriagu JO, Bhattacharya P, Mukherjee AB, Bundschuh J, Zevenhoven R, Loeppert RH (2007) Arsenic in soil and groundwater: an introduction. In: Bhattacharya P, Mukherjee AB, Bundschuh J, Zevenhoven R, Loeppert RH (eds), Arsenic in soil and groundwater environment: biogeochemical interactions, health effects and remediation, trace metals and other contaminants in the environment, vol 9 (Ser ed Nriagu JO). Elsevier, Amsterdam

Paranis DS (1997) Principles of applied geophysics, 5th edn. Chapman & Hall, London. ISBN 0-412-64 080-5

Piper AM (1944) A graphical procedure in the geochemical interpretation of water samples. Trans Am Geophys Union 25:914–923

Rahman A, Vahter M, Smith AH, Nermell B, Yunus M, ElArifeen S, Persson LA, Ekström EC (2009) Arsenic exposure during pregnancy and size at birth: a prospective cohort study in Bangladesh. Am J Epidemiol 169:304–312

Rajmohan N, Elango L (2004) Identification and evolution of hydrogeochemical processes in the groundwater environment in an area of the Palar and Cheyyar River Basins, Southern India. Environ Geol 46:47–61

Ravenscroft P, Burgess WG, Ahmed KM, Burren M, Perrin J (2005) Arsenic in groundwater of the Bengal Basin, Bangladesh: distribution, field relations, and hydrogeological setting. Hydrogeol J 13:727–751

Saha D (2009) Arsenic groundwater contamination in parts of Middle Ganga Plain, Bihar. Curr Sci 97(6):753–755

Saha D, Sreehari SMS, Dwivedi SN, Bhartariya KG (2009) Evaluation of hydrogeochemical processes in arsenic contaminated alluvial aquifers in parts of Mid-Ganga Basin, Bihar, Eastern India. Environ Earth Sci 61:799–811

Saha D, Sahu S, Chandra PC (2010) Arsenic-safe alternate aquifers and their hydraulic characteristics in contaminated areas of Middle Ganga Plain, Eastern India. Environ Monit Assess 175:331–348

Sankararamakrishnan N, Chauhan D, Nickson RT, Iyengar L (2008) Evaluation of two commercial field test kits used for screening of groundwater for arsenic in northern India. Sci Total Environ 40:162–167

Schoeller (1965) Hydrodynamique lans lekarst (ecoulemented emmagusinement). Actes Colloques Doubronik, I, AIHS et UNESCO

Smith AH, Lingas EO, Rahman M (2000) Contamination of drinking water by arsenic in Bangladesh: a public health emergency. Bull World Health Organ 78(9):1093–1103

Sracek O, Bhattacharya P, Jacks G, Gustafsson JP, von Brömssen M (2004) Behavior of arsenic and geochemical modeling of arsenic enrichment in aqueous environments. Appl Geochem 19(2):169–180

Srivastava AK, Govil PC, Tripathi RM, Shukla RS, Srivastava RS, Vaish DP, Nickson RT (2008) Groundwater for sustainable development: problems, perspectives and challenges. In: Bhattacharya P, Ramanathan AL, Mukherjee AB, Bundschuh J, Chandrasekharam D, Keshari AK (eds) Initial data on arsenic in groundwater and development of a state action plan, Uttar Pradesh, India. Taylor & Francis/A.A. Balkema, Rotterdam, pp 271–281

Stallard RF, Edmond JM (1983) Geochemistry of Amazon, the influence of geology and weathering environment on the dissolved load. J Geophys Res 88:9671–9688

Stumm W, Morgan JJ (1981) Aquatic chemistry: an introduction emphasizing chemical equilibria in natural waters. Wiley, New York

WHO (World Health Organization) (1993) Guidelines for drinking water quality, vol 1, 2nd edn, Recommendations. WHO, Geneva

Zheng Y, Stute M, van Geen A, Gavrieli I, Dhar R, Simpson HJ, Ahmed KM (2004) Redox control of arsenic mobilization in Bangladesh groundwater. Appl Geochem 19:201–214

Chapter 10
Chemical Characteristics of Arsenic Contaminated Groundwater in Parts of Middle-Gangetic Plain (MGP) in Bihar, India

Sanjay Kumar Sharma, AL. Ramanathan, and V. Subramanian

10.1 Introduction

Groundwater is the primary source of water for domestic, agricultural and industrial uses in many countries. Water resources of good quality are becoming an important issue being vital to health, safety and socio-economic development of man. The rising demands for hygienic water often cannot be met by surface water supplies. This has led to increased dependence on groundwater resources in many parts of the world. Hence suitable quantity and quality of groundwater become a more crucial alternative resource to meet the drastic increase in social and agricultural development and to avoid the expected deterioration of groundwater quality due to heavy abstraction for miscellaneous uses. Water quality and its management have received added attention in developing countries. Further intensive use for irrigation makes groundwater a critical resource for human activities. Thus water remains the principal driver for development in South Asia.

The shallow aquifers (~50 m depth) in the Gangetic alluvial deposits encounter intensive groundwater draft. The middle Ganga plain (MGP) covering about 89 % geographical area of Bihar (~94,000 km^2) holds potential alluvial aquifers. The tract is known for surplus food production and intensive groundwater extraction for drinking, irrigation, and industrial uses. Chemical quality plays a significant role in groundwater resource management in Bihar as the entire drinking and a

---

S.K. Sharma (✉) • AL. Ramanathan
School of Environmental Sciences, Jawaharlal Nehru University, New Delhi 110067, India
e-mail: sanju.jnu@gmail.com

V. Subramanian
School of Environmental Sciences, Jawaharlal Nehru University, New Delhi 110067, India

Amity Institute of Environmental Sciences, Amity University,
Sector 125, Noida 201313, UP, India

major part of irrigation consumption is extracted from Quaternary aquifers (Saha 1999). Chemical composition of groundwater is controlled by many factors that include composition of precipitation, geological structure and mineralogy of the watersheds and aquifers, and geological processes within the aquifer (Andre et al. 2005). However, in a groundwater system of alluvial nature, anthropogenic influences may mask the normal chemical alteration trends. The impact of agricultural practices and domestic waste disposal pose a major threat to quality of shallow groundwater.

Groundwater arsenic contamination in fluvial plains of India and Bangladesh and its consequences to the human health have been reported as one of the world's biggest natural groundwater calamities to mankind. Arsenic (As) occurrence in natural waters is the result of natural or anthropogenic sources, natural sources being the major cause of environmental As problems (Smedley and Kinniburgh 2002). Natural As in groundwater at concentrations above the drinking water standard of 10 µg/L is not uncommon. Extensive spread of As-contaminated areas in the Bengal basin and parts of the Ganga plain, which are virtually free of industrial, mining, or thermal water activities, and reports of similar problems in other deltas and floodplains suggest that As-contamination in groundwater in flood-delta region is geogenic. Recent studies have linked the occurrence of arsenic in aquifers with regional geology and hydrogeology, the most serious contamination being in alluvial/fluvial or deltaic plains (Mandal et al. 1996; Bhattacharya et al. 1997, 2004, 2009; Nickson et al. 2000; Acharyya 2002; Ahmed et al. 2004; Nath et al. 2008; Chauhan et al. 2009; Kumar et al. 2010; Bhowmick et al. 2013). The most common process that explains the As mobilization is the reductive dissolution of iron and oxides/hydroxide minerals and subsequent release of As into the groundwater, which occurs under anaerobic conditions (Bhattacharya et al. 1997; Nickson et al. 2000; Zheng et al. 2004; Mukherjee et al. 2008). Holocene alluvial sediment with slow flushing rate is one of the significant contributors to As enrichment in groundwater (Nordstrom 2002). The As-contaminated areas, in the Ganga plain, are exclusively confined to the Newer Alluvium (Holocene).

The investigated area is located in the middle Ganga plain, amid the flood-prone belt of Sone-Ganga interfluve region. The Quaternary deposits within 300 m below ground in As affected areas in Sone–Ganga interfluve is two-tier aquifer system as revealed by hydrostratigraphic analysis. The upper part of the shallow aquifer (within ~50 m bgl) is affected by groundwater As contamination (Saha 2009). Tracts of As-polluted areas are reported from adjoining parts of eastern Uttar Pradesh (U.P.), and northern parts of Jharkhand states, which are located in the middle section of the Ganga alluvial plain and of the deltaic Bengal basin located in lower Gangetic plain.

The paper intends to evaluate the natural and human induced chemical processes in the shallow alluvial aquifers and characterize the groundwater in parts of Bhojpur district in Bihar. A detailed periodic investigation was carried out to appraise the groundwater chemistry of As-contaminated aquifers. This study also investigates the mechanisms of As mobilization from Quaternary sedimentary aquifers in parts of MGP and its enrichment in the local groundwater.

10.2 Materials and Methods

10.2.1 Site Description

The study was carried out in two rural residential areas namely Semaria Ojha Patti and Kastulia in northern Bhojpur district of Bihar, India (Fig. 10.1). Bhojpur district lies between 25°10′ to 25°40′N and from 83°45′ to 84°45′E inhabiting population of 22,33,415 with population density 903 km^{-2}. The study area is located in the middle-Gangetic alluvial plain amid the floodprone belt of Sone-Ganga interfluve region at about 8 km south of the Ganga river. Sone river, a right-hand perennial tributary of the Ganga, flows along the eastern boundary of the study area whereas the northern, southern and western boundaries are marked by the Ganga course. This region has some of the most fertile soil found in the country as new silt is continually laid down by the rivers every year. Annual normal rainfall in this part of MGP is 1,061 mm. The period of general rains (southwest monsoon) usually starts in third week of June with a sudden rise of relative humidity to over 70 % and a fall in temperature between about 5 °C and 8 °C from about 35 °C. The rainfall

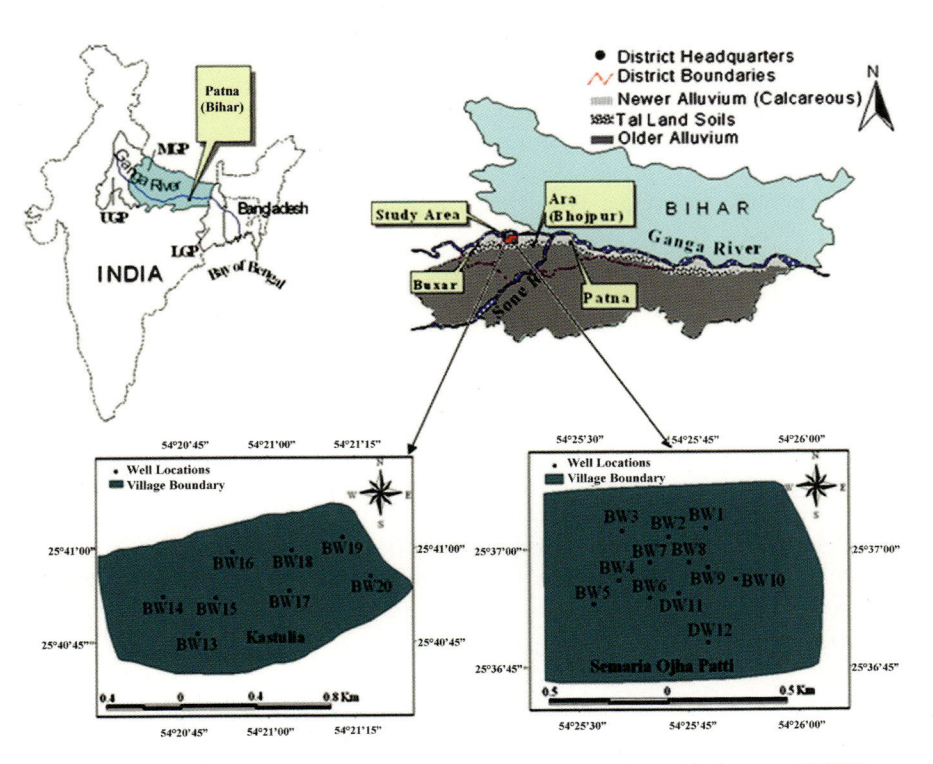

Fig. 10.1 Study area maps showing sampling location and its physiographic features. *UGP* Upper Gangetic Plain, *MGP* Middle Gangetic Plain, *LGP* Lower Gangetic Plain

continues with intermissions till the end of September or the early part of October providing over 85 % of the annual rainfall.

Semaria Ojha Patti (~4 km^2 in area) and Kastulia (~3 km^2 in area), with population of 3,500 inhabitants and 1,500 inhabitants, respectively, are remote agricultural regions located at nearly 30 km from Ara, the district's headquarter lying between two major important cities Patna and Buxar in the MGP in Bihar. Important land uses for economic activities to the local people in the study area are subsistence agriculture and pastoral production. Numerous mechanical bore-wells were found in the area drilled by farmers for irrigation. The depth of these wells ranges from 30 to 60 m yielding few litres to 30 L/s of water. The main cultivation for livelihood sustenance is focused on paddy, wheat, maize, pulses, and some cash crops (e.g. sugarcane) which are water intensive. The irrigation demand is largely met by the extraction from shallow aquifers in the area. There are no potable water supply networks in these rural localities; hence people rely mostly on groundwater extracted through hand-pumps and dug-wells.

10.2.2 Hydrogeological Settings

The MGP is predominantly alluvial deposits of Quaternary age. Most parts of the MGP are characterized by a rapidly filling and aggradational regime. The channels are not deeply incised in this area, and exposed bank sediments are those of the recent, aggrading floodplain system and the alluvial sequences in this area range in age from Middle Pleistocene to Holocene (Sinha et al. 2005). The sedimentation in the narrow and entrenched Sone-Ganga interfluve region of the active middle-Gangetic flood plain was influenced by sea level fluctuation during the Holocene, causing increased aggradations and formation of large fluvial lakes and swamps (Singh 2001). A higher sediment supply in the MGP may be a function of, apart from its greater proximity to the Himalayan front in the north, a higher crustal shortening rate and a higher average Holocene uplift rate (Lave and Avouac 2000). The newer alluvium, as distinct from the older unit, is characteristically unoxidised and consists of sand and silt-clay, which were mainly deposited in a fluvial and fluvio-lacustrine setting (Acharya 2005). The upper succession of the newer alluvium is predominantly argillaceous (silty clay or sandy clay), followed by fine to medium sand with occasional clay lenses forming the shallow aquifer (Saha et al. 2010). The As is generally associated with many types of mineral deposits and in particular those containing sulphide minerals such as arsenopyrite. However, elevated concentrations are sometimes found in fine grained argillaceous sediments and phosphorites (Thornton 1996).

X-ray Diffractrometry results of As-safe older alluvial and As-contaminated newer alluvial soil samples from the MGP reveals presence of minerals such as quartz, muscovite, chlorite, kaolinite, feldspar, amphibole and goethite (Shah 2008). The study area is confined within newer alluvium entrenched channel and floodplain of the Ganga river (Fig. 10.1). The area is underlain by a multi-layer sequence of sand (aquifer) alternating with aquitards like sandy-clay and clay, down to depth of 300 m. In the Sone-Ganga interfluves region, the Quaternary deposits within 300 m below ground exhibit two-tier aquifer system. The shallow aquifer system is confined within 120–130 m depth, followed by a laterally continuous 20–30 m thick clay/sandy-clay zone forming the aquitard. The deeper aquifer system exists below this aquitard, which continues down to 240–260 m below ground (Saha 2009). The shallow aquifer system is referred as active channel deposit and marked as newer alluvium. The groundwater in deep aquifer remains under confined to semi-confined condition in the affected Bhojpur district as revealed by the pumping test carried out by Saha (2009). The deep aquifer can, therefore, be developed for safe community based drinking water supply in the area.

10.2.3 Sampling, Field Measurements and Laboratory Analyses

A total of 20 groundwater samples were collected from 20 shallow wells (≤55 m depth) located in residential houses within the study areas based on general survey of water quality from the occupants of the respective houses during August 2006–May 2009. Sampling locations were chosen on the basis of availability of wells and in a manner such that the whole study area is uniformly covered. The wells' locations are shown in Fig. 10.1. Some wells were selected from the houses where residents reported symptoms of arsenic poisoning. This was with a view of evaluating the quality and safety of the water being consumed, from elemental composition and possible health implication perspectives. The samples were collected from hand-pumps and few active dug-wells. First, the water was left to run from the sampling source (hand-pump) for 4–6 min to pump out the volume of water standing in the casing before taking the final sample and then water samples were collected in pre-cleaned, acid sterilized high density 250 ml polypropylene bottles. The physical parameters namely pH, Eh, EC and TDS were measured in-situ by portable corresponding digital instruments. Further, water samples were filtered through 0.45 μm Millipore membrane filter papers and immediately preserved in portable ice-box to minimize chemical alteration until they were brought at the earliest possible to the laboratory where the samples were kept at 4 °C in a thermostatic cold chamber until further processing and analyses. About 100 ml of sample was acidified with ultrapure HNO_3 (2 % v/v) to pH ~2 to stabilize the dissolved metals and rest of the portion was kept unacidified. Additional samples were collected in chromic-acid-washed brown glass bottles and filtered with glass-fibre (GF/C) filter papers for the analysis of dissolved organic carbon (DOC).

Cations were analyzed in the acidified fraction, whereas the unacidified fraction was used for the analysis of anions. Bicarbonate content was determined following the potentiometric titration method (Clescerl et al. 1998). Anions and cation (NH_4^+) were analyzed by Dionex ICS-90 Ion Chromatograph using the specific analytical columns "IonPac AS12A (4×200 mm)" for anions and "IonPac CS12A

(4×250 mm)" for cations. Dissolved silica (H_4SiO_4) and DOC were determined from the unacidified water samples by the molybdosilicate method (Clescerl et al. 1998) and the vial digestion method using COD reactor, respectively; finally, absorbance measurements were carried out using a Cecil Spectrophotometer (model no. 594) for both the parameters. Metals such as Na, K, Ca, Mg, Fe and Mn were measured by Shimadzu AA-6800 Atomic Absorption Spectrophotometer (AAS) in Flame mode of analysis, whereas total dissolved As were quantified using AAS coupled with on-line hydride generator (HG). Samples were added with 20 % (w/v) KI (2 mL in 20 mL sample) and kept for 15 min reaction period for the reduction of As(V) to As(III) prior to the analysis. Further, As(III) was reduced to arsine (AsH_3) by on-line HG using 1 % (w/v) $NaBH_4$ in 0.1 % (w/v) NaOH and 5 M HCl as buffer solution. The inclusion of on-line HG generally increases the sensitivity of detection and reduces the possible interferences from the sample matrix. Merck-certified standard solutions and AnalR grade chemicals were used for quality control during analyses. Milli-Q water was used during processing and analyses of the samples. Standards for major and trace elements were analyzed as unknown intermittently to check precision as well as accuracy of the analysis and the precision was ≤5 %. The AAS was programmed to perform analyses in triplicates and RSD was found within ±3 % in most of the cases reflecting the precision of the analysis. The computed ionic balance errors for the analytical database were observed within a standard limit of ±5 % attesting good quality of the data.

10.3 Results and Discussion

10.3.1 Groundwater Chemistry

Chemical characteristics of groundwater as observed during investigation period in the study area are statistically presented in Table 10.1. Groundwaters are characterized as nearly neutral to moderately alkaline with pH ranging 7.5–8.3 (mean 7.9). The lower redox potential (Eh) recorded in groundwater wells (−41 mV to −15 mV; mean ~ −23 mV) suggest that groundwater in the studied area of MGP was under moderately reducing condition perhaps induced by frequent flooding in the region. The moderately reducing aquifer condition in the area is further confirmed by considerable presence of other redox-sensitive elements such as Fe and Mn discussed in next section. Significant variations in chemical composition indicate high salinity in many of the groundwaters, with total dissolved solids (TDS) ranging 375–1,087 mg/L (mean 602 mg/L) and conductivity (EC) in the range 581–1,643 µS/cm (mean 943 µS/cm). Higher EC implies intensive mineralization in groundwater. Irrigation increases the salinity of irrigation-return-flow from three to ten times that of the applied water (Todd 1980). Since irrigation is the primary user of water, irrigation-return-flow may be the major cause of groundwater pollution in the study area. Further recycling of saline groundwater for irrigation may result in a progressive increase in soil and groundwater salinity.

Table 10.1 Statistical chemical characteristics of groundwater in parts of middle-Gangetic plain in the state of Bihar, India

Parameters	Unit	Minimum	Maximum	Average	Std. Dev.
Depth	m	15.24	54.86	27.36	13.4
pH	pH unit	7.5	8.3	7.9	0.3
Eh	mV	−41	−15	−23	6.2
E.C.	µS/cm	581	1,643	943	269
TDS	mg/L	375	1,087	602	179
Na^+	mg/L	26.95	94	37.1	16.7
K^+	mg/L	4.51	119.91	20.99	32.4
Mg^{2+}	mg/L	20.31	41.93	31.67	7.4
Ca^{2+}	mg/L	49.19	119.62	78.26	19.9
HCO_3^-	mg/L	253.49	519.6	386.63	72
F^-	mg/L	0.22	1.37	0.47	0.3
Cl^-	mg/L	8.48	155.23	44.64	41.1
NO_3^-	mg/L	3.03	133.8	34.55	32.8
PO_4^{3-}	mg/L	0.27	1.36	0.65	0.3
SO_4^{2-}	mg/L	3.99	94.56	27.56	23.4
H_4SiO_4	mg/L	27.35	41	34.49	3.8
DOC	mg/L	0.98	5.25	3.22	1.2
NH_4^+	mg/L	0.2	4.08	1.49	1.1
Fe	mg/L	0.227	4.728	1.037	1
Mn	mg/L	0.042	0.231	0.099	0.1
As_T	µg/L	21.96	190.21	83.57	46.8
pCO_2	atm	−2.67	−1.7	−2.21	0.3

Major ion composition showing a wide range of variations in the parts of middle-Gangetic alluvial aquifers is dominated by HCO_3^- (253.49–519.60 mg/L, mean 386.63 mg/L) and in lower proportion by Cl^- (8.48–155.23 mg/L, mean 44.64 mg/L). Concentrations of NO_3^- (3.03–133.80 mg/L, mean ~34.5 mg/L) and SO_4^{2-} (3.99–94.56 mg/L, mean 27.56 mg/L) were relatively low in the samples, except for some locally high occurrences, contributing 0.75–13.61 % (mean 5.64 %) and 1.27–23.33 % (mean 6.33 %), respectively to the total anions. The dominance of HCO_3^- accounting for 50.23–93.33 % (mean ~75 %) of the total anions, classifies the area as a recharge zone (Subba Rao 2007). The elevated HCO_3^- reflects an apparent lithological control of major ion composition in groundwaters. Hence, this infers a dominance of mineral dissolution in the aquifers (Stumm and Morgan 1996). High HCO_3^- concentrations are typical features of many of these groundwaters that can be attributed to the dissolution of carbonate via biodegradation of organic matter under local reducing condition and reaction of silicates with carbonic acid. The origin of Cl^- derives mainly from the non-lithological sources and can also be contributed, especially, from the surface sources through the domestic wastewaters, septic tanks, irrigation-return flows and chemical fertilizers (Hem

1991). The Cl⁻ contribution of 3.88–28.14 % (mean ~13 %) to the total anions in groundwater of the study area may be attributed to the influences of irrigation return-flows and chemical fertilizers. The agricultural activity is intensive and long-term, and no other sources are evident in the area. The NO_3^- can be a direct indicator of anthropogenic contaminants from agricultural land uses and Cl⁻ also indicates the same source because salinization due to seawater intrusion or brines from deep aquifers would not be expected in the newer alluvial regime.

Natural NO_3^- sources are minor which include low concentration soil NO_3^- from decay of sparse, natural vegetation and nitrogen-bearing rock units. Indeed, 25 % of the groundwater samples exceeds drinking water limit of 45 mg/L NO_3^- (BIS 2003). In natural conditions, mainly from the non-lithological sources, the concentration of NO_3^- does not exceed 10 mg/L in the water (Ritzi et al. 1993). The observed higher concentrations exceeding 10 mg/L, in 80 % of the monitored locations, reflect the man-made pollution (Hem 1991), apparently due to application of fertilizers targeted for higher crop yields in the study area. The shallow aquifers in Ganga Alluvial Plain at places are known for high NO_3^- load (Handa 1983), often exceeding the limit for human consumption (45 mg/L) prescribed by BIS (2003). The SO_4^{2-} is likely to be contributed by multiple sources, such as atmospheric deposition, agricultural activities and pyrite oxidation (N'egrel and Pauwels 2004). In general, high SO_4^{2-} concentrations may be derived either by sulphide or SO_4^{2-} weathering (Stallard and Edmond 1987). Because the shallow aquifers in the area were recharged from the irrigated agricultural field and were not affected by seawater, it might also possibly be inferred that SO_4^{2-} dissolved in groundwaters were derived from agricultural activities and decomposition of organic matter. Higher SO_4^{2-} concentrations in some cases may partly be attributed to pyrite oxidation. High levels of Cl⁻ and SO_4^{2-} generally observed together with high NO_3^- levels in shallow groundwaters of agricultural areas are reported to be derived from fertilizers (Chae et al. 2004). Thus the observed higher concentrations of Cl⁻, SO_4^{2-} and NO_3^- are an indication of man-made source, as also reported by Hem (1991).

The PO_4^{3-} concentrations, on the other hand, are fairly low (0.27–1.36 mg/L, mean 0.65 mg/L) accounting for 0.09–0.63 % (mean 0.25 %) to the total anions. Under natural conditions, the dissolved phosphate (PO_4^{3-}) concentration should not exceed 0.5 mg/L, as the solubility of phosphate mineral is limiting (Subramanian 1984). High concentrations in some wells are apparently due to anthropogenic influences including use of fertilizers in copious quantities to increase agricultural output. Contamination sources from household activities, such as sewage and large-scale livestock feedlots also appear to contribute to the contamination, because the area is devoid of proper sewage system. The concentration of F⁻ in groundwaters varied significantly from 0.22 to 1.37 mg/L (mean 0.47 mg/L), contributing 0.12–0.87 % (mean 0.32 %) to the total anions. The F⁻ in drinking water has a narrow optimum concentration range in relation to human health. While a moderate concentration range between 0.7 and 1.2 mg/L prevents dental caries, higher concentrations are responsible for dental and skeletal fluorosis (Marimon et al. 2007). Mere 5 % of the samples exceeded drinking water quality limit of 1.0 mg/L recommended by BIS (2003). Elevated concentration in some of the groundwater samples could be attributed to the weathering of F⁻-bearing minerals such as micas and amphiboles in the rocks.

The concentrations of major cations such as Ca^{2+}, Mg^{2+}, Na^+ and K^+ ions in the groundwater samples varied between 49.19–119.62 mg/L, 20.31–41.93 mg/L, 26.95–94.00 mg/L and 4.51–119.91 mg/L, with mean values of 78.26, 31.67, 37.10 and 20.99 mg/L, respectively. Their concentrations (on an equivalent basis) represent 28.34–53.07 %, 20.93–39.11 %, 12.38–27.60 % and 1.03–22.80 % with mean values ~46 %, ~31 %, ~18 % and ~5 % of the total cations, respectively. The observed excess of Na^+ over K^+ is because of the greater resistance of K^+ to chemical weathering and its adsorption on clay minerals restricting its mobility. Although K^+ is the least dominant cation in the groundwater of the study area, its concentration in most of the samples were detected often modest to high for the study period. Dissolution of K-bearing minerals and ion-exchange reactions may be involved. But high K^+ concentrations in the basin cannot exclusively be linked to groundwater-rock interaction as its natural occurrence in higher concentrations is rare and may be the result of increased K-rich fertilizer used for agricultural land uses. The occurrence of K^+ in groundwater, especially in higher concentrations reflects manure spreading and fertilization on the surface as well as a short flow path and infiltration length of the shallow aquifers in the area. The results suggest that K^+ is the potential indicator of anthropogenic impacts on groundwater, namely agricultural practices in the recharge area of the aquifers. High K^+ concentration in several groundwater samples also seems to be derived from domestic effluents. The elevated concentrations of Ca and Mg may be attributed to the dissolution of carbonate minerals. Dissolved silica (H_4SiO_4) concentration in groundwater is most conveniently related to interaction with solid phases, as it enters the system during the weathering of silicate minerals. The groundwater has high concentrations of H_4SiO_4 ranging between 27.35 and 41.00 mg/L (mean 34.49 mg/L). The range reflects both high groundwater temperatures and reaction of fine-grained (e.g. clays) and easily weathered (e.g. feldspar) silicate minerals in the aquifers. Further, H_4SiO_4 shows dominance over other nutrients such as NO_3^- and PO_4^{3-}.

10.3.1.1 Weathering Processes and Evolution of Groundwater Chemistry

Reactions between CO_2 and carbonate and silicate minerals, containing the common alkali and alkaline earth metals (Na, K, Mg and Ca), produce the bicarbonate and carbonate anions that neutralize the positive cation charges and generate alkalinity. The computed effective partial pressure of carbon dioxide (pCO_2) in groundwater samples range from $10^{-1.70}$ to $10^{-2.67}$ atm (mean $10^{-2.21}$ atm), which are higher than that in the atmosphere ($10^{-3.50}$ atm) (Table 10.1). The trend indicates that groundwater is in disequilibrium with the atmosphere. The groundwater pCO_2 in the study area are higher than those reported in case of surface waters (Sharma and Subramanian 2008). This increase in pCO_2 would typically be counterbalanced by mineral weathering reactions in the aquifer. Thus groundwater pCO_2, significantly higher (upto $~10^{-1.7}$ atm) than atmospheric as well as surface waters, could possibly be caused by biologic activity leading to the decay of organic matter in the sediments through sequence of redox reactions including reductive dissolution of Fe(III)–oxyhydroxides in the solid phase and root respiration in the vadose zone during recharge. The observed high pCO_2 values suggest that, in general, redox reactions in the aquifers are responsible for the generation of CO_2 which, in turn, results in the elevated level of HCO_3^- in the groundwater samples, as observed in the study area.

In order to understand the foremost functional sources of major dissolved ions in the groundwater, the chemical data of groundwater samples of the study area are discussed below taking into account stoichiometric relations among dissolved ions during weathering process. The predominance of Ca^{2+} and Mg^{2+} cations and high $(Ca^{2+} + Mg^{2+})/(Na^+ + K^+)$ ratio is the characteristic feature of prevailing carbonate weathering in the basin (Sharma and Subramanian 2008). The average mill equivalent ratio of $(Ca^{2+} + Mg^{2+})/(Na^+ + K^+)$ worked out to be 3.66 (range 1.06–5.65). The observed higher ratio values indicate the dominance of carbonate weathering and suggest that chemical composition of groundwater in the parts of middle-Gangetic alluvial aquifers is controlled by carbonate lithology of the Ganga basin. Lower $(Ca^{2+} + Mg^{2+})/(Na^+ + K^+)$ ratios obtained in few cases suggest additional source of Ca^{2+} and Mg^{2+} ions, which might have been balanced by Cl^- and SO_4^{2-} and/or partly supplied by silicate weathering in addition to the carbonate weathering. The Ca^{2+}/Mg^{2+} molar ratio was calculated for further justification. In case of dolomite dissolution, the Ca^{2+}/Mg^{2+} molar ratio equals to 1 whereas, the higher ratio indicates the source from calcite (Maya and Loucks 1995). The observed higher Ca^{2+}/Mg^{2+} ratios (>1) varying between 1.10 and 1.82 (mean 1.49) suggest dissolution of calcite favoured over dolomite in the aquifers.

Most of the samples (75 %) have Na^+/Cl^- molar ratio >1, which indicate that Na^+ is released from silicate weathering (Meybeck 1987). If it is so, the groundwater would have HCO_3^- as the most dominant anion (Rogers 1989), as observed in the study area. This also suggests that higher concentration of alkalies is from sources other than precipitation. At few locations, the ratio was found less than 1 which implies for halite dissolution as the source of Na^+, indicating the influence of secondary evaporate deposits in the aquifers. However, the Cl^--enriched samples reveal that surplus chlorides are likely to result in aqueous species such as $CaCl_2$ and $MgCl_2$. The $(Na^+ + K^+)/TZ^+$ ratio is an index to assess the contribution of the cations via silicate weathering (Stallard and Edmund 1987). The $(Na^+ + K^+)/TZ^+$ ratios ranging from 0.15 to 0.48 (mean 0.24) suggest inputs from the weathering of aluminosilicates such as Na- and K-feldspars and/or from soils. Thus the results, taking into account hydrogeochemical processes discussed above, reveal that groundwater quality in the study area is primarily affected by natural mineralization with secondary inputs from anthropogenic sources in terms of major dissolved constituents. The Piper plot of major ions (in %milli equivalents) shown in Fig. 10.2 allows identification of groundwater chemical types and presence of chemical evolutionary trends that may exist during processes of water-rock interaction in the study area.

Groundwater in the study area is of predominantly Ca-HCO_3 type, reflecting the initial impacts of water-rock interaction in the soil zone and shallow bedrock environments. The occurrence of Cl^- in the groundwater cannot be explained

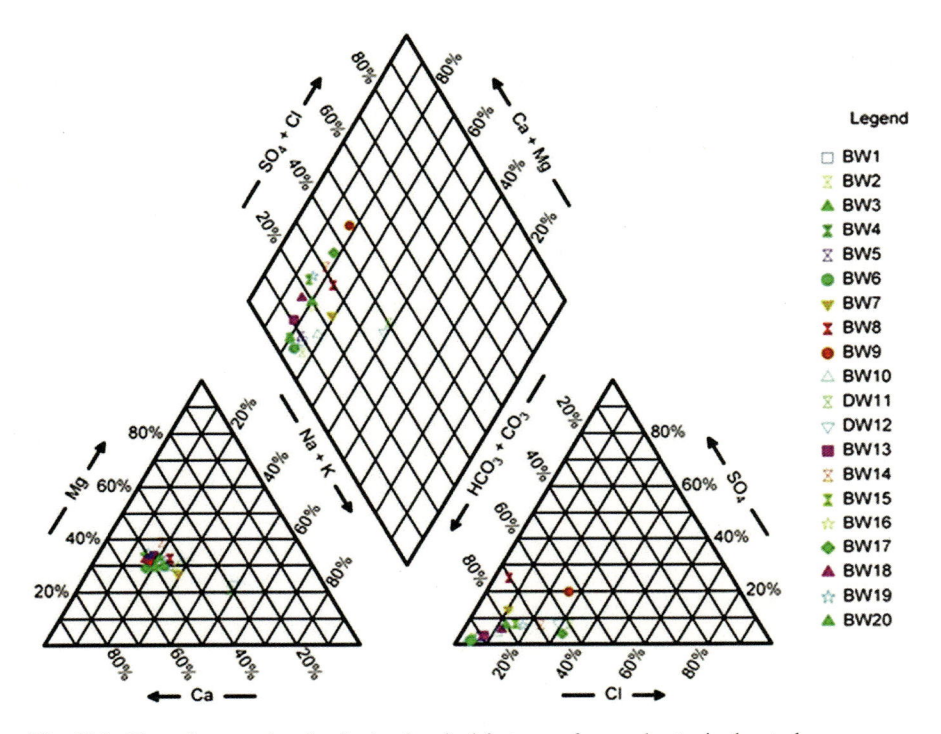

Fig. 10.2 Piper diagram showing hydrochemical features of groundwater in the study area

through water-rock interaction, since it derives mainly from a non-lithological source and it also tends to prefer to form compounds with alkalis rather than Ca^{2+} and Mg^{2+}. Therefore, it may be inferred that Cl^- has been consumed in the formation of alkali chlorides. Further, the observed alkali concentrations in most of the samples are much higher than those required for the formation of chlorides. It is, therefore, logical to expect the existence of aqueous species such as alkali bicarbonates in the groundwater. The corollary to this situation is presented by the Cl-enriched samples in which surplus chlorides are likely to result in aqueous species such as $CaCl_2$ and $MgCl_2$ as discussed above. The diamond shaped field (Fig. 10.2) further indicates that with very few exceptions, no cation-anion pair exceeds 50 % of the total ions which, in turn, reflects that the chemical properties are dominated by alkaline earths ($Ca^{2+} + Mg^{2+}$) and weak acids ($HCO_3^- + CO_3^{2-}$). Thus it may be concluded that the groundwater in the parts of middle-Gangetic alluvial plain is primarily of bicarbonate alkaline earth type. The rock-water interaction processes tend to release cations (especially Ca^{2+} and Na^+) and anions (especially HCO_3^-) with raising pH condition. Subsequent ion-exchange reactions may modify the groundwater chemistry, while redox condition may lead to the dissolution of minor and trace elements in groundwater. Solutes which are subject to these processes may have been derived from water-rock interaction and/or from leachates coming with the rainfall recharge.

10.3.2 Source and Distribution of Redox-Sensitive Parameters

The redox-sensitive characteristics of groundwater are presented in Table 10.1. The concentration of DOC in groundwater varied considerably from 0.98 mg/L to as high as 5.25 mg/L, with a mean value of 3.22 mg/L. Abandoned channels, swamps and active channels in MGP are perennially or seasonally water filled where the aquifers are presumed to be enriched in biomass and thereby organic carbon (Shah 2008). During recharging of aquifers, organic carbon of surface water is also transported into the aquifers upto a certain depth depending on hydrological conditions. The reported high DOC concentration in these shallow alluvial aquifers is attributed to the sedimentary biomass contribution via microbial degradation in the aquifers and surficial contribution during groundwater recharge. Saline ammonia (NH_4^+) is a useful indicator for sewage pollution. Another possible source of NH_4^+ is through natural anaerobic decomposition of sedimentary organic matter. The concentrations of NH_4^+ are fairly low to moderate with few exceptions, ranging between 0.20 and 4.08 mg/L, with an average of 1.48 mg/L. Mobility of As could be affected by the redox reactions involving either aqueous or adsorbed As (Manning and Goldberg 1997). The total dissolved As (As_T) concentrations in the shallow aquifers in the study area significantly varied from 21.96 to 190.21 µg/L (mean 83.57 µg/L), which far exceed drinking water guideline of 10 µg/L (BIS 2003; WHO 2011).

Thick sequences of young sediments are quite often the sites of high groundwater As concentration. The variability may be attributed both to the variations in the As content of the soils, rocks and the aquifer minerals that serve as the proximate source of As in the groundwater and to the variable extent of processes that release or sequester As from or to solid phases. The exceptionally elevated As concentrations at few locations could be attributed to the local dissolution of Fe(III)–oxyhydroxides in perhaps more reducing condition. Reducing conditions of the aquifers are also reflected by sizeable presence of Fe and Mn in most of the groundwater samples ranging between 0.23 and 4.73 (mean 1.04) and 0.04–0.23 (mean 0.10), respectively. The As_T and other important redox parameters such as Fe, PO_4^{3-}, DOC and NH_4^+ related to its release into groundwater were plotted against depth of the wells monitored (Fig. 10.3). Variations in concentration of these redox characteristics with the depth indicated some trend, generally high values being found in the shallowest wells upto the depth of ~20 m with maximum concentrations at ~15 m below the ground level and then declined towards the deeper part of the shallow aquifers. These variations may be attributed to varying redox states at shallow depths within the aquifer. The trend suggests that microbial degradation of sedimentary organic matter in the shallow aquifers under anoxic condition is perhaps responsible for the release and mobilization of As in groundwater. Correlation between As and Fe or Mn is often observed in groundwater because, under certain conditions, the presence of Fe/Mn oxyhydroxides could lead to desorption of As, Fe and Mn. In order to infer their role in mobilization of As into groundwater, the As concentrations were plotted against selected physico-chemical parameters showing meaningful correlations. The highest As concentrations, in general, were detected in groundwaters with pH values exceeding 8. This may be attributed to As desorption from the

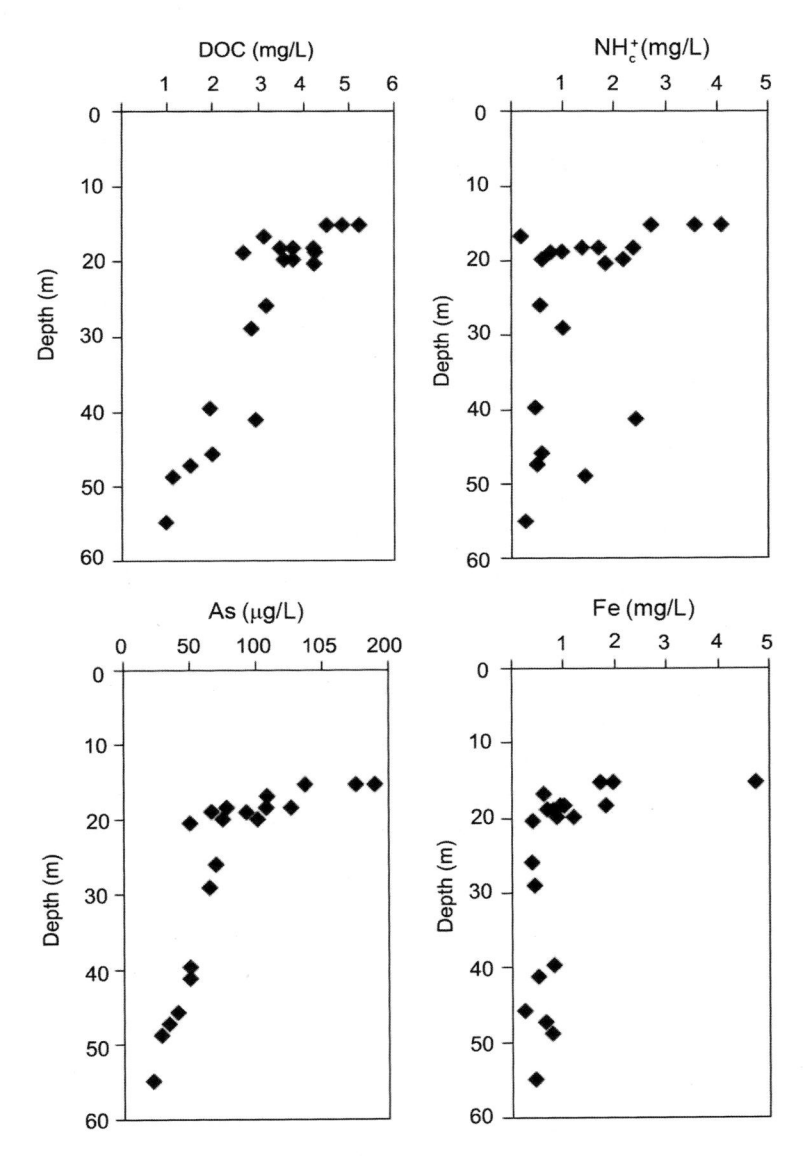

Fig. 10.3 Scatter plots of redox sensitive characteristics of groundwater with respect to well-depths depicting their vertical enrichment profile

oxide surfaces at pH above 8.0 (Smedley and Kinniburgh 2002). The positive correlation between pH and As justifies the desorption processes at higher pH values ($r=0.62$; $p<0.01$; Fig. 10.4a). A significant positive correlation between As and Fe concentrations in the groundwater ($r=0.80$; $p<0.01$; Fig. 10.4b) suggests that As is released in groundwater due to reductive dissolution of Fe(III)–oxyhydroxides. The As and Mn were negatively correlated ($r=-0.46$; $p<0.01$), suggesting their

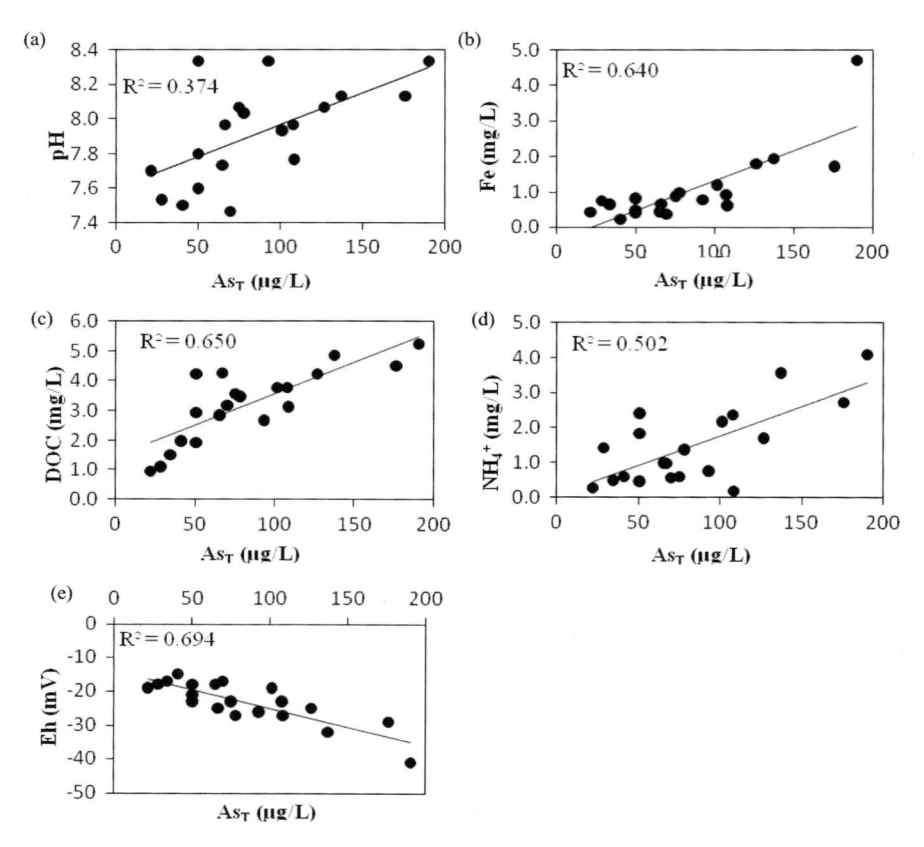

Fig. 10.4 Relationship of Arsenic (As) with selected hydrogeochemical characteristics of groundwater in the study area [(**a**) As_T vs pH; (**b**) As_T vs Fe; (**c**) As_T vs DOC; (**d**) As_T vs NH_4^+; (**e**) As_T vs Eh]

different sources in the wells, while generally low concentrations of PO_4^{3-} detected in the groundwater do not lead to any conclusion regarding its role in the mobilization and enrichment of As in the aquifers.

It has also been confirmed by the experimental desorption of As sorbed to mineral surfaces by PO_4^{3-} (Manning and Goldberg 1997), that phosphorus in groundwater cannot contribute to As pollution. No clear correlation between As and SO_4^{2-} was observed in present study, which suggests that pyrite oxidation is not the basic mechanism for As mobilization in these aquifers. The relationship between As and HCO_3^- is also well documented because this ion can play an important role in the mobilization of As through the competition for adsorption sites and through the formation of arseno-carbonate complexes. However absence of any correlation between these two parameters nullifies the hypothesis for the present case.

Strong positive correlations of As with DOC ($r=0.81$; $p<0.01$) and NH_4^+ ($r=0.71$; $p<0.01$) signify their role in mobilizing As through microbial degradation of organic matter in the aquifer sediments (Fig. 10.4c, d). Significant negative correlation between Eh and As ($r=-0.73$; $p<0.01$) supports the phenomenon of As mobilization in the moderately reducing aquifers impacted by frequent flooding in the area (Fig. 10.4e). Tremendous amount of groundwater pumped out especially during dry seasons leading to flow in of monsoonal rain and local surface water loaded with surficial organic matter, which upon reduction, dissolves Fe-oxides bearing As and thus may trigger the release of As from sediments in the study area. The positive correlation of As with Fe and elevated HCO_3^- concentrations in groundwater are similar to those recorded over broader geographical areas (McArthur et al. 2001). Thus the results from present study support the hypothesis that desorption of As accompanied with reductive dissolution of Fe(III)–oxyhydroxides in the aquifer sediments is the principal mechanism by which As is released to the groundwater. Presence of elevated As concentration despite Fe concentration being low at few locations may be attributed to the precipitation of Fe-carbonates (Nickson et al. 2000). Slow groundwater flow also contributes to an elevated As concentration of water (Smedley et al. 2003). The aquifers in newer alluvial section of MGP have low hydraulic gradient with less fluctuation suggesting a sluggish flow regime in the area (Saha 2009). Hence groundwater As enrichment in the study area may also be partly attributed to the influence of longer groundwater residence time due to its limited flow in the alluvial aquifers of MGP. Based on the above discussion, it can be explicitly figured out that mobilization and enrichment of As in this part of middle-Gangetic plain is due to the influence of the sediments' geochemical and hydrogeological characteristics of the alluvial aquifers with partial inputs from anthropogenic sources.

10.4 Conclusions

Weathering, mineral dissolution, leaching and other processes related to anthropogenic activities generally control groundwater chemistry in this part of MGP. Groundwater is of predominantly Ca-HCO$_3$ type produced mainly by the dissolution of carbonate minerals, which reflects the influence of their host rock in the study area. The predominance of Ca^{2+} and Mg^{2+} cations and high $(Ca^{2+} + Mg^{2+})/(Na^+ + K^+)$ ratio in the groundwater indicate the prevailing carbonate weathering in the aquifers. The elevated concentrations of NO_3^- and Cl^- ions are caused by intensive agricultural practices in the basin. Thus it can be concluded that the chemical solutes of groundwater samples are influenced by anthropogenic inputs in addition to the natural chemical weathering. The hydrogeochemical characteristics of the alluvial aquifers govern the mobility of As in groundwater. Positive correlations of As with Fe, NH_4^+ and DOC, as well as the negative correlation with Eh, typically characterizes anoxic aquifers where reductive dissolution of Fe-phases and release of surface bound As is the principal source of dissolved As in groundwater of the middle-Gangetic alluvial plain. The As contamination in groundwater presents significant threats to health of the local communities in the area. Nevertheless, groundwater is being widely used for drinking and other domestic works due to the lack of potable water supply networks or other clean water resources in these rural localities. Noticeable variations with well-depth were observed with higher As concentrations being found in the shallowest wells. Therefore, from mitigation and policy point of view, installation of deep wells in the affected villages can provide a reliable source of drinking water.

Acknowledgements The University Grant Commission (UGC) is gratefully acknowledged for awarding research fellowship (to the first author) which has been helpful for carrying out the present study.

References

Acharyya SK (2002) Arsenic contamination in groundwater affecting major parts of southern West Bengal and parts of western Chhattisgarh: source and mobilization process. Curr Sci 82:740–744

Acharya SK (2005) Arsenic trends in groundwater from quaternary alluvium in the Ganga plain and the Bengal basin, Indian subcontinent: insights into influences of stratigraphy. Gondwana Res 8(1):55–66

Ahmed KM, Bhattacharya P, Hasan MA, Akhter SH, Alam MA, Bhuyian H et al (2004) Arsenic enrichment in groundwater of the alluvial aquifers in Bangladesh: an overview. Appl Geochem 19:181–200

Andre L, Franceschi M, Pouchan P, Atteia O (2005) Using geochemical data and modelling to enhance the understanding of groundwater flow in a regional deep aquifer, Aquitaine Basin, south-west of France. J Hydrol 305:40–62

Bhattacharya P, Chatterjee D, Jacks G (1997) Occurrence of arsenic contamination of groundwater in alluvial aquifers from Delta Plain, Eastern India: option for safe drinking supply. Int J Water Resour Dev 13:79–92

Bhattacharya P, Welch AH, Ahmed KM, Jacks G, Naidu R (2004) Arsenic in groundwater of sedimentary aquifers. Appl Geochem 19(2):163–167

Bhattacharya P, Aziz Hasan M, Sracek O, Smith E, Ahmed KM, von Bromssen M, Huq SMI, Naidu R (2009) Groundwater chemistry and arsenic mobilization in the Holocene flood plains in south-central Bangladesh. Environ Geochem Health 31:23–43

Bhowmick S, Nath B, Halder D, Biswas A, Majumder S, Mondal P, Chakraborty S, Nriagu J, Bhattacharya P, Iglesias M, Roman-Ross G, Guha Mazumder D, Bundschuh J, Chatterjee D (2013) Arsenic mobilization in the aquifers of three physiographic settings of West Bengal, India: understanding geogenic and anthropogenic influences. J Hazard Mater 262:915–923

BIS (2003) Indian standard: drinking water–specifications (first revisions). Bureau of Indian Standard, New Delhi

Chae GT, Kim K, Yun ST, Kim KH, Kim SO, Choi BY, Kim HS, Rhee CW (2004) Hydrogeochemistry of alluvial groundwaters in an agricultural area: an implication for groundwater contamination susceptibility. Chemosphere 55:369–378

Chauhan VS, Nickson RT, Chauhan D, Iyengar L, Sankararamakrishnan N (2009) Groundwater geochemistry of Ballia district, Uttar Pradesh, India and mechanism of arsenic release. Chemosphere 75:83–91

Clescerl LS, Greenberg AE, Eaton AD (eds) (1998) Standard methods for the examination of water and wastewater, 20th edn. American Public Health Association, American Water Works Association, Water Environment Federation, Washington, DC

Handa BK (1983) The effect of fertilizers use on groundwater quality in the phreatic zone with special reference to Uttar Pradesh. In: Proceedings of the seminar on assessment development and management of groundwater research. Central Groundwater Board, New Delhi

Hem JD (1991) Study and interpretation of the chemical characteristics of natural water. Scientific Publishers, Jodhpur

Kumar P, Kumar M, Ramanathan AL, Tsujimura M (2010) Tracing the factors responsible for arsenic enrichment in groundwater of the middle Gangetic Plain, India: a source identification perspective. Environ Geochem Health 32:129–146

Lave J, Avouac JP (2000) Active fold of fluvial terraces across the Siwaliks Hills, Himalayas of central Nepal. J Geophys Res 105(B3):5735–5770

Mandal BK, Roy Choudhury T, Samanta G, Basu GK, Chowdhury PP, Chandra CR, Lodh D, Karan NK, Dhar RK, Tamili DK, Das D, Saha KC, Chakroborti D (1996) Arsenic in groundwater in seven districts of West Bengal, India—the biggest arsenic calamity in the world. Curr Sci 70:976–986

Manning BA, Goldberg S (1997) Adsorption and stability of arsenic(III) at the clay-mineral water interface. Environ Sci Technol 31:200–201

Marimon MPC, Knöller K, Roisenberg A (2007) Anomalous fluoride concentration in groundwater – is it natural or pollution? A stable isotope approach. Isotopes Environ Health Stud 43(2):165–175

Maya AL, Loucks MD (1995) Solute and isotopic geochemistry and groundwater flow in the Central Wasatch Range, Utah. J Hydrol 172:31–59

McArthur JM, Ravencroft P, Safiullah S, Thirlwall MF (2001) Arsenic in groundwater: testing pollution mechanism for sedimentary aquifers in Bangladesh. Water Resour Res 37:109–117

Meybeck M (1987) Global chemical weathering of surficial soils estimated from river dissolved loads. Am J Sci 287:401–428

Mukherjee A, von Bromssen M, Scanlon BR, Bhattacharya P, Fryar AE, Hasan Md A, Ahmed KM, Chatterjee D, Jacks G, Sracek O (2008) Hydrogeochemical comparison and effects of overlapping redox zones on groundwater arsenic near the Western (Bhagirathi sub-basin, India) and Eastern (Meghna sub-basin, Bangladesh) margins of the Bengal Basin. J Contam Hydrol 99:31–48

N'egrel P, Pauwels H (2004) Interaction between different groundwaters in Brittany catchments (France): characterizing multiple sources through strontium and sulphur isotope tracing. Water Air Soil Pollut 151:261–285

Nath B, Stüben D, Basu Mallik S, Chatterjee D, Charlet L (2008) Mobility of arsenic in West Bengal aquifers conducting low and high groundwater arsenic. Part I: Comparative hydrochemical and hydrogeological characteristics. Appl Geochem 23:977–995

Nickson RT, McArthur J, Ravenscroft P, Burgess W, Ahmed KM (2000) Mechanism of arsenic release to groundwater, Bangladesh and West Bengal. Appl Geochem 15:403–413

Nordstrom DK (2002) Worldwide occurrences of arsenic in ground water. Science 296:2143–2145

Ritzi RW, Wright SL, Mann B, Chen M (1993) Analysis of temporal variability in hydrogeochemical data used for multivariate analyses. Ground Water 31:221–229

Rogers RJ (1989) Geochemical comparison of groundwater in areas of New England, New York and Pennsylvania. Ground Water 27:690–712

Saha D (1999) Hydrogeological framework and groundwater resources of Bihar. Central Ground Water Board, Patna

Saha D (2009) Arsenic groundwater contamination in parts of middle Ganga plain, Bihar. Curr Sci 97(6):753–755

Saha D, Sarangam SS, Dwivedi SN, Bhartariya KG (2010) Evaluation of hydrogeochemical processes in arsenic-contaminated alluvial aquifers in parts of Mid-Ganga Basin, Bihar, Eastern India. Environ Earth Sci 61(4):799–811

Shah BA (2008) Role of quaternary stratigraphy on arsenic contaminated groundwater from parts of middle Ganga Plain, UP–Bihar, India. Environ Geol 35:1553–1561

Sharma SK, Subramanian V (2008) Hydrochemistry of the Narmada and Tapti Rivers, India. Hydrol Process 22:3444–3455

Singh IB (2001) Proxy records of neotectonics, climate changes and anthropogenic activity in the late Quaternary of Ganga plain. In: Proceedings of the national symposium on role of earth sciences in integrated development and societal issues, 2–4 November 2001, Lucknow, India. Special Publication 65, vol 1. Geological Survey India, Calcutta

Sinha R, Tandon SK, Gibling MR, Bhattacharjee PS, Dasgupta AS (2005) Late Quaternary geology and alluvial stratigraphy of the Ganga basin. Himal Geol 26(1):223–240

Smedley PL, Kinniburgh DG (2002) A review of the source, behaviour and distribution of arsenic in natural waters. Appl Geochem 17:517–568

Smedley PL, Zhang M, Zhang G, Luo Z (2003) Mobilisation of arsenic and other trace elements in fluviolacustrine aquifers of the Huhhot Basin, Inner Mongolia. Appl Geochem 18:1453–1479

Stallard RF, Edmond JM (1987) Geochemistry of the amazon: weathering chemistry and limits to dissolved inputs. J Geophys Res 92:8293–8302

Stumm W, Morgan JJ (1996) Aquatic chemistry. Wiley, New York

Subba Rao N (2007) Groundwater quality as a factor for identification of recharge zones. Environ Geosci 14:79–90

Subramanian V (1984) River transport of phosphorous and genesis of ancient phosphorites. Geol Survey India (Special Publication) 17:11–15

Thornton I (1996) Sources and pathways of arsenic in the geochemical environment: health implications. Geol Soc Lond (Special Publications) 113:153–160

Todd DK (1980) Groundwater hydrology. Wiley, New York

WHO (2004) Guidelines for drinking-water quality, vol 1, 3rd edn. World Health Organization, Geneva

WHO (2011) Guidelines for drinking water quality, 4th edn. World Health Organization, Geneva

Zheng Y, Stute M, van Geen A, Gavriela I, Dhar R, Simpson H, Schlosser P, Ahmed KM (2004) Redox control of arsenic mobilization in Bangladesh groundwater. Appl Geochem 19:201–214

Chapter 11
An Insight into the Spatio-vertical Heterogeneity of Dissolved Arsenic in Part of the Bengal Delta Plain Aquifer in West Bengal (India)

Santanu Majumder, Ashis Biswas, Harald Neidhardt, Simita Sarkar, Zsolt Berner, Subhamoy Bhowmick, Aishwarya Mukherjee, Debankur Chatterjee, Sudipta Chakraborty, Bibhash Nath, and Debashis Chatterjee

11.1 Introduction

Naturally occurring, carcinogenic, arsenic (As) is omnipresent in hydrological systems, and is considered as the most serious abiotic contaminant of groundwater in several parts of the world (Smedley and Kinniburgh 2002; Chatterjee et al. 2005; Charlet et al. 2007; Mukherjee et al. 2008a; Neumann et al. 2010 and references therein). Holocene aquifers of south-east Asia (mostly shallow, <50 m) often contain high As groundwater. The groundwater is predominantly used for irrigation and domestic purposes, e.g., cooking, drinking and bathing (Bhattacharya et al. 1997; Bhattacharyya et al. 2003a; Charlet et al. 2007). In south-east Asia, As-rich

---

S. Majumder
ICEE, Department of Environmental Management, University of Kalyani,
Kalyani, Nadia 741235, West Bengal, India

Department of Chemistry, University of Girona, Campus Montilivi, s/n. 17071, Girona, Spain

Department of Chemistry, University of Kalyani, Kalyani, Nadia
741235, West Bengal, India

A. Biswas • S. Sarkar • D. Chatterjee (✉)
Department of Chemistry, University of Kalyani, Kalyani, Nadia
741235, West Bengal, India
e-mail: debashis.chatterjee.ku@googlemail.com

H. Neidhardt • Z. Berner
Institute of Mineralogy and Geochemistry, Karlsruhe Institute of Technology, KIT,
Karlsruhe D-76128, Germany

S. Bhowmick
Department of Chemistry, University of Girona, Campus Montilivi, s/n. 17071, Girona, Spain

Department of Chemistry, University of Kalyani, Kalyani, Nadia 741235, West Bengal, India

groundwaters are often found in alluvial plains of regional rivers (Fendorf et al. 2010). Prolonged consumption of groundwater with elevated levels of As may cause a formidable threat to human health and millions of people are now at risk (Bhattacharyya et al. 2003a; Chatterjee et al. 2010; Nath et al. 2008a). Arsenic contamination in groundwater and related health issues is considered as the greatest mass poisoning in human history (Smith et al. 2000).

In the Bengal Delta Plain (BDP), the spatial distribution of groundwater As is often heterogeneous and patchy (contaminated areas interspersed with non-contaminated areas) (Charlet et al. 2007; Nath et al. 2008b; Biswas et al. 2011). Arsenic content in aquifer sediments (1–22 mg/kg, average 6.9 mg/kg) is generally low (Nath et al. 2008c). The level of As in sediments depend on their texture, grain size and mineralogy (Pal et al. 2002; Nath et al. 2005). In surface soils, As concentration is also low (0.4–2 mg/kg) (Kabata-Pendius and Pendius 2001). Arsenic content in the deltaic river bed sediments (1.3–5.6 mg/kg) and their waters (0.33–3.83 µg/L) is also low (Datta and Subramanian 1998; Stüben et al. 2003), whereas extensively exploited alluvial and/or deltaic aquifers (<50 m) are contaminated with high As (>10 µg/L) groundwater. The occurrence of As-rich groundwater is, therefore, difficult to explain by As content in the BDP sedimentary environment. Over the years, it has become evident that instead of a local high As source in the Holocene sediment, several hot-spots of As-rich (>150 µg/L) groundwater have been identified in various parts of the BDP (Bhattacharyya et al. 2003b; Nath et al. 2005; Mukherjee et al. 2008b; Bhowmick et al. 2013).

In the present study, we investigated spatial scale heterogeneity of As in a deltaic aquifer of the BDP (Chakdaha block, Nadia district, West Bengal, India) where a large number of high As wells were previously identified (RGNDWM 2001; Jana 2004; IFCPAR 2004). The study initiated with groundwater collection (mostly from hand-operated drinking water wells in rural areas) and testing to demonstrate the degree of As heterogeneity. A statistical evaluation of (hydro)chemical features was carried out to explain the factors regulating the groundwater geochemical processes. Finally, the inter-relationship between prevailing land use pattern and groundwater chemistry of surveyed wells was studied to decipher local (hydro)chemical environment and to demonstrate their relationship to justify typical spatial heterogeneity of As distribution in the aquifer.

A. Mukherjee
Department of Basic Sciences and Humanities, Hooghly Engineering & Technology College, Hooghly 712103, West Bengal, India

D. Chatterjee
Heritage Institute of Technology, East Kolkata Township, Chowbaga Road, Anandapur, Kolkata 700107, West Bengal, India

S. Chakraborty
Department of Chemistry, Kanchrapara College, Kanchrapara 743145, West Bengal, India

B. Nath
School of Geosciences, University of Sydney, Sydney, NSW 2006, Australia

11.2 Materials and Methods

11.2.1 Study Area

The study area (latitude 23° 00′ 20″–23° 05′ 20″ N and longitude 88° 31′ 40″–88° 49′ 00″ E) is located in Nadia district, approximately 65 km north of Kolkata, West Bengal, India (Fig. 11.1), and encompasses an integral part of the BDP. The area is the main geomorphic component of the world's largest fluvio-deltaic system consisting of Ganges, Brahmaputra and Meghna rivers.

A large part of the area (>70 %) is characterized by shallow (10–50 m) groundwater wells (mostly hand operated). Apart from these, a few motorized large-diameter deep tube wells (depth >100 m) for community water supply are also available. A large number of shallow motorized irrigation wells (10–50 m) aid the practice of sustainable agriculture all-year round. These shallow wells significantly support Boro rice cultivation (summer paddy during February–June). The gross crop area is ~1.0 million ha with a cropping intensity of 140 % (PHED 1993). The climate is tropical, hot and humid (temperature 16–42 °C; average relative humidity >65 %) with annual rainfall ranging between 1,295 and 3,945 mm (Bhattacharyya and Chatterjee 2001).

11.2.2 Sampling and Analytical Techniques

Groundwater samples ($n = 181$) were collected from tube wells in acid washed pre-cleaned polyethylene (PE) bottles (capacity, 100–200 mL). Prior to sampling, tube wells were flushed for several minutes to discharge standing volume of water (about

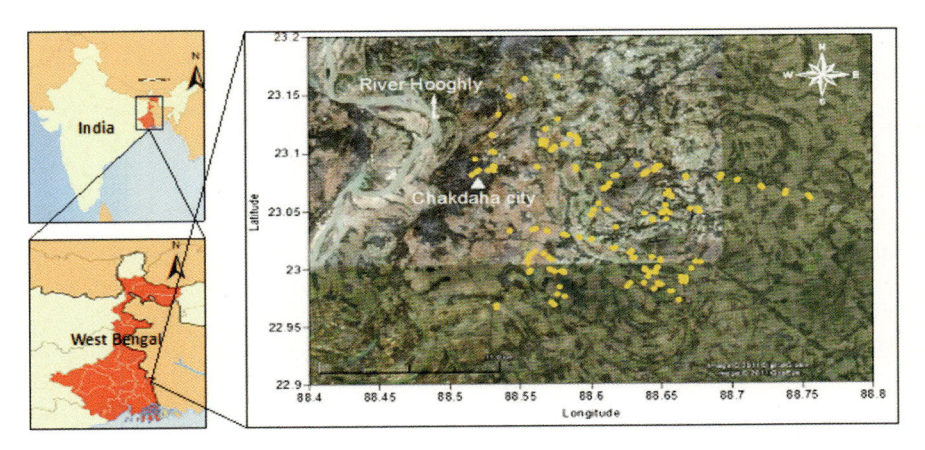

Fig. 11.1 Map of the study area in Chakdaha block, Nadia district (West Bengal, India) showing locations where groundwater samples were collected from tube wells

40–60 L). This was done to get the fresh water from the aquifer. Water samples were filtered (0.45 μm cellulose nitrate membrane filter) on site and divided into two groups. First group was acidified with HNO_3 (0.2 % v/v, Suprapur, MERCK, Germany) for the analysis of cations and As using double focussing High Resolution Inductively Coupled Plasma Mass Spectrometer (HR-ICP-MS VG AXIOM, VG Elemental) in a clean room. The second group was left un-acidified for anion analysis with the help of Ion Chromatograph (IC, DIONEX ICS 1000) with a separating column (DIONEX AS 4 SC). A few field parameters pH, Eh, electrical conductivity (EC), temperature (using WTW multimeters) and HCO_3^- (titrating with 0.2 M H_2SO_4) were measured on site.

11.2.3 Statistical and Thermodynamic Calculations

The software package Statistica 6.0 (StatSoft Inc., USA) was used to evaluate the geochemical variation in groundwater. The variation of the groundwater chemical composition was assessed by factor analysis (Liu et al. 2003; Nath et al. 2008b). Principal Component Analysis (PCA) was used to derive factor loadings limiting maximum eigen value to 1.0. To facilitate the interpretation of factor loadings, the factor axis was rotated by the normalized varimax method, i.e., extracted factors were rotated until the variance of factor loadings become maximum. A loading close to ±1.0 indicates a strong relationship between the factors and the variables, while a loading close to zero explains a weak relationship between the factors and the variables (Davis 1986).

Saturation indices ($SI = \log [IAP\ K_T^{-1}]$) were calculated with PHREEQC version 2.15 (Parkhurst and Appelo 1999) with IAP as the ion activity product and K_T as the equilibrium solubility constant of a mineral phase at ambient temperature.

11.3 Results

11.3.1 Groundwater Chemistry

The chemical composition of the groundwater samples that were collected from tube wells over a ~300 km² area is summarized in Table 11.1. The results indicate a large variation in the chemical composition of the groundwater and also high variations in respective As concentrations (mean 52 μg/L, range 0.31–333 μg/L). It is worth to note that As concentration exceeded, in a large number of samples, the Indian threshold value for As in drinking water of 50 μg/L ($n=63$, 35 % of total 181 samples) and the WHO guideline value of 10 μg/L in a total of 132 samples (i.e., 73 % of the surveyed tube wells). The values of electrical conductivity are moderate (mean 707 μS/cm, range 200–1,514 μS/cm) and pH is circum-neutral (mean 7.12,

Table 11.1 Statistical summary of water quality parameters ($n=181$) in groundwater from tube wells

Parameters	Min	Max	Mean	Median	SD (±)
Depth (m)	7.0	271	39	24	43
pH	5.95	7.99	7.12	7.12	0.37
Eh (mV)	−238	56	−109	−112	37
EC (µS/cm)	200	1,514	707	672	187
Temp (°C)	25	31	27	26	0.74
As (µg/L)	0.31	333	52	34	62
Fe (µg/L)	8.0	19,408	3,415	2,467	3,647
DOC (mg/L)	BDL	21	2.1	1.7	1.8
Na (mg/L)	4.4	112	22	18	15
Mg (mg/L)	5.4	43	23	22	6.4
K (mg/L)	0.77	45	3.8	3.0	3.8
Ca (mg/L)	20	178	100	99	26
Li (µg/L)	0.42	15	4.1	3.0	2.9
B (µg/L)	9.6	86	29	26	12
Al (µg/L)	2.4	913	22	11	75
P (µg/L)	2.2	3,196	628	406	651
V (µg/L)	0.03	15	0.74	0.11	1.9
Cr (µg/L)	0.04	39	0.76	0.13	3.5
Mn (µg/L)	5.3	2,524	321	263	284
Co (µg/L)	0.02	9.4	0.27	0.10	0.87
Ni (µg/L)	0.26	26	1.1	0.71	2.2
Cu (µg/L)	0.39	99	2.6	1.1	8.4
Zn (µg/L)	5.2	6,964	272	33	805
Rb (µg/L)	0.06	8.2	2.2	1.9	1.5
Sr (µg/L)	103	656	295	267	109
Mo (µg/L)	0.21	8.4	1.1	0.88	0.84
Cd (µg/L)	0.01	0.52	0.07	0.05	0.05
Sb (µg/L)	0.30	0.50	0.40	0.40	0.06
Ba (µg/L)	28	592	176	163	86
Tl (µg/L)	0.02	0.71	0.09	0.06	0.09
Pb (µg/L)	0.08	78	1.9	0.82	6.18
Th (µg/L)	0.10	0.30	0.20	0.10	0.10
U (µg/L)	0.01	12	0.60	0.03	1.7
HCO_3^- (mg/L)	200	588	425	422	50
F^- (mg/L)	0.01	1.6	0.12	na	0.24
Cl^- (mg/L)	0.67	147	23	14	25
NO_3^- (mg/L)	BDL	13	1.6	1.2	2.1
PO_4^{3-} (mg/L)	0.11	4.0	0.71	0.48	0.68
SO_4^{2-} (mg/L)	0.14	50	7.3	4.6	8.3

Note: *BDL* below detection limit, *na* not available

range 5.95–7.99). Eh values are low (mean −109 mV, range −238 to 56 mV), while values of dissolved oxygen (DO) are low to very low (<0.1 mg/L). Redox sensitive elements (such as Fe and Mn) are slightly elevated while NO_3^- is low to very low (43 % of samples are below detection limit). SO_4^{2-} concentrations have a mean value of 7.3 mg/L (range 0.14–50 mg/L), whereas Cl^- concentrations have a mean value of 23 mg/L (range 0.67–147 mg/L). The distributions of Cl^- and SO_4^{2-} are depth dependent and generally decrease with depth (Fig. 11.2a, b). Highest concentrations of Cl^- and SO_4^{2-} are only restricted to shallow aquifers (15–50 m) while

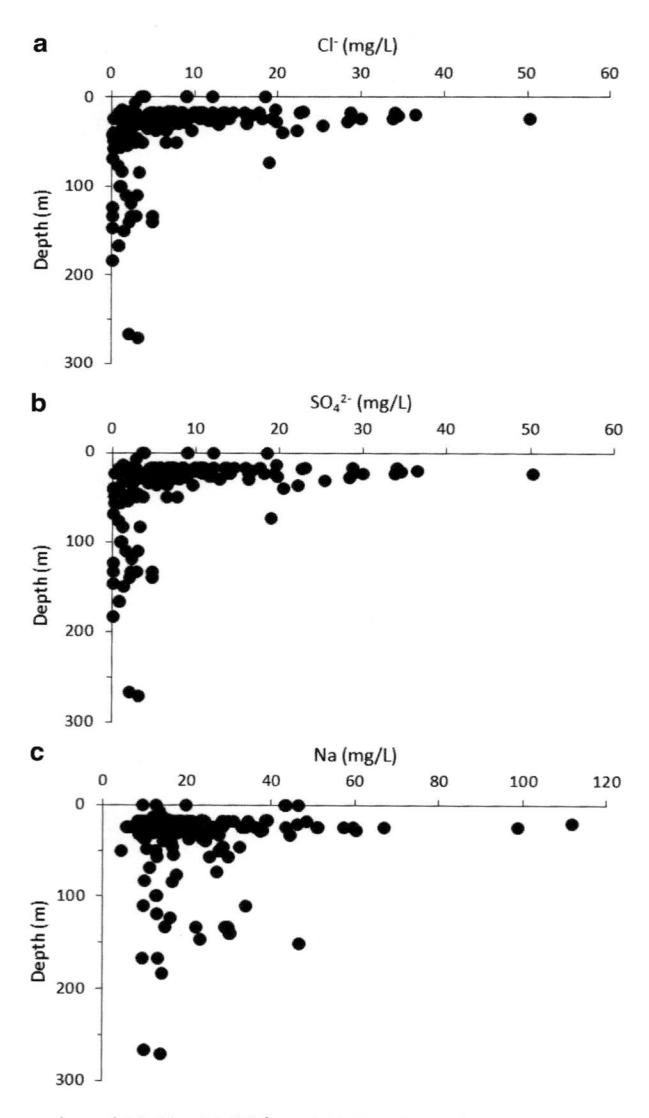

Fig. 11.2 Scatter plots of (**a**) Cl^-, (**b**) SO_4^{2-} and (**c**) Na with respect to depth

mostly remain close to the detection limit in the deeper aquifers. P concentrations (mean 628 µg/L, range 2.2–3,196 µg/L) in the surveyed wells are considerably higher than the general trends in the BDP (BGS and DPHE 2001). The variation of P concentration in the groundwater follows a trend similar to that of As concentrations. Highest P (>1 mg/L) and As (up to 332 µg/L) concentrations are found in the shallow wells (<50 m), while deeper wells show relatively lower P concentrations.

The dominant anion is HCO_3^- (mean 425 mg/L, range 200–588 mg/L) while cation is Ca (mean 100 mg/L, range 20–178 mg/L) (Table 11.1). Molar ratios of Mg and Ca are variable and ranging from 0.16 to 0.58 (mean 0.39). Among alkali metal concentrations, Na (mean 22 mg/L, 4.4–112 mg/L) is much higher than K (mean 3.8 mg/L, range 0.77–45 mg/L) and their molar ratios are also variable (mean 0.12, range 0.02–0.96). Na concentration is generally high in the shallow aquifers together with high Cl^- concentrations (Fig. 11.2a, c). The Piper plot shows that the groundwater is predominantly of $Ca-HCO_3^-$ type (Fig. 11.3). Zinc, Sr and Ba show relatively higher values, while rest of the elements show low values.

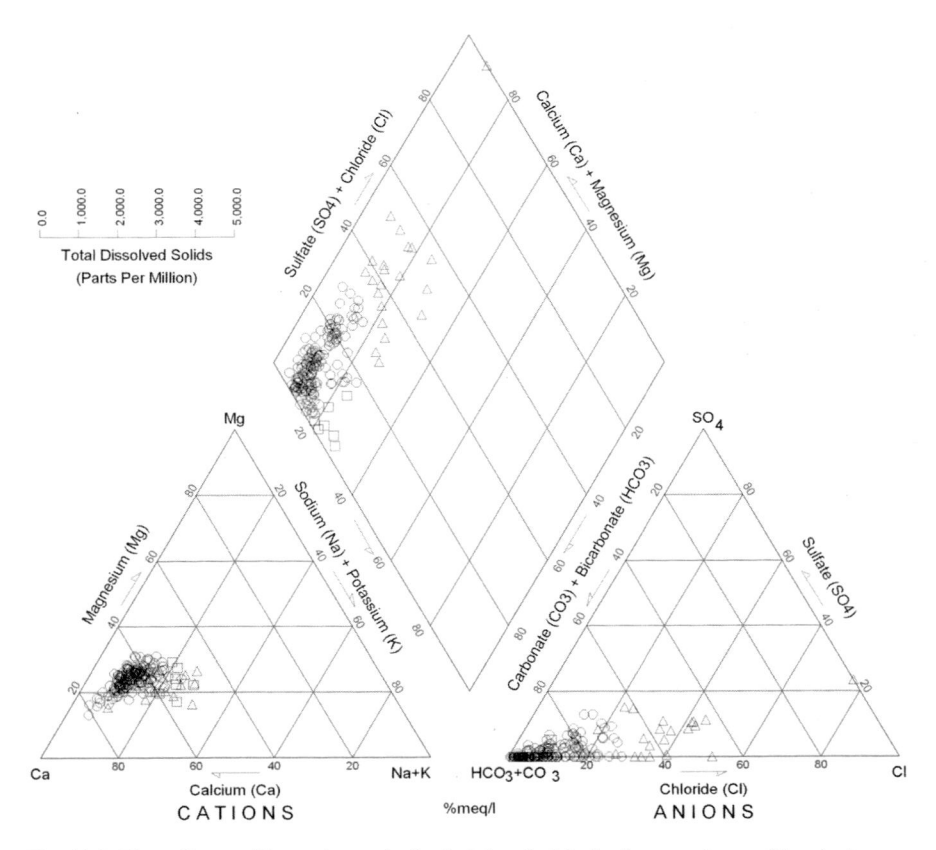

Fig. 11.3 Piper diagram illustrating main (hydro)chemical facies in groundwater. The *circles* represent the predominant $Ca-HCO_3$ type waters, *triangles* represent high Na – high Cl^- type waters and the *squares* represent high Na – low Cl^- type waters.

11.3.2 Depth and Spatial Distribution of Arsenic

Depth distribution of As shows a large variation (Fig. 11.4a) with a high concentration prevailing in the near-surface aquifer. Arsenic in the shallow aquifer is most contaminated where 41 % of the tube wells (40 out of 98 wells) exceeded the Indian national standard and 83 % of the tube wells (80 out of 98 wells) exceeded the WHO guideline value. Only a few wells ($n=18$ out of 98 wells) have As concentrations below the WHO guideline value. This suggests that even in the highly contaminated zone As concentrations are varying. Groundwater in the deep (81–150 m) and very deep (>150 m) aquifers and the intermediate aquifers contain low to very low As concentrations (Table 11.2). It is astounding to observe that dissolved As concentration in some of the deep wells are exceeding both the WHO guidelines and the Indian national drinking water standard.

The present investigation emphatically highlights the bewildering nature of the spatial distribution of As concentrations in groundwater of the study area. The spatial distribution pattern indicates that the contaminated wells are surrounded by safe

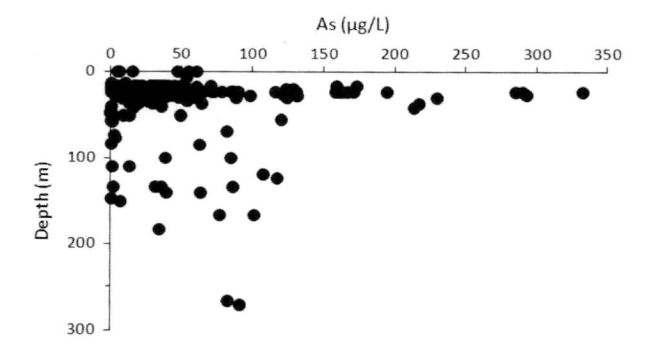

Fig. 11.4 Scatter plot of arsenic concentrations with respect to depth

Table 11.2 Depth profile of groundwater arsenic at different concentration ranges

As concentration ranges	Near-surface aquifer (<20 m)	Shallow aquifer (21–50 m)	Intermediate aquifer (51–80 m)	Deep aquifer (81–150 m)	Very deep aquifer (>150 m)
<10 µg/L	38.8 %	36.7 %	14.3 %	8.2 %	2 %
($n=49$)	($n=19$)	($n=18$)	($n=7$)	($n=4$)	($n=1$)
10–50 µg/L	30.4 %	57.9 %	2.9 %	7.2 %	1.4 %
($n=69$)	($n=21$)	($n=40$)	($n=2$)	($n=5$)	($n=1$)
50–150 µg/L	12.8 %	55.3 %	10.6 %	12.8 %	8.5 %
($n=47$)	($n=6$)	($n=26$)	($n=5$)	($n=6$)	($n=4$)
>150 µg/L	12.5 %	87.5 %	None	None	None
($n=16$)	($n=2$)	($n=14$)			

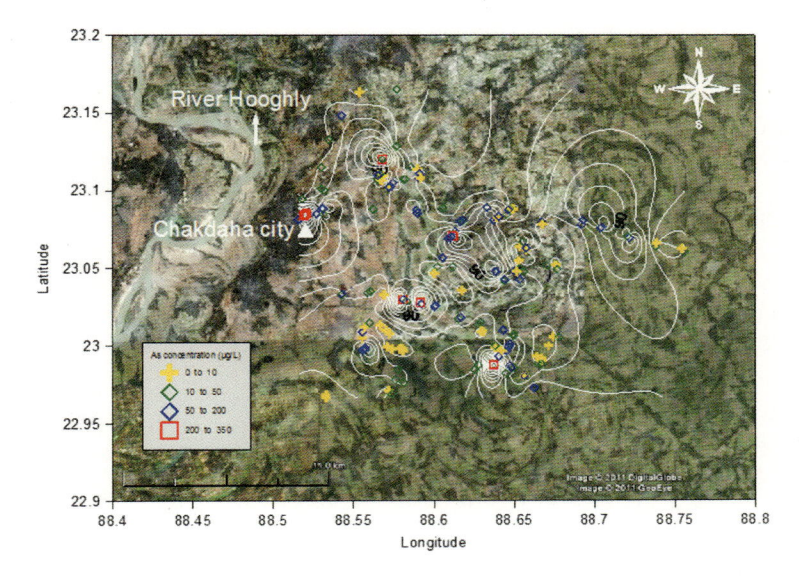

Fig. 11.5 Spatial distributions of As concentrations in groundwater

Table 11.3 Arsenic distribution in relation to local land use characteristics

Land use characteristics	Number of wells	% of wells	As (µg/L)	Mean As (µg/L)	As ≥50 µg/L (number of wells)	% of affected wells
Sanitation	74	41	0–217	44	26	35
Sanitation and surface water bodies	50	28	0–290	49	17	34
Surface water bodies	22	12	0–285	60	9	41
Agricultural fields	35	19	0–333	67	12	34

wells (Fig. 11.5). This is more pronounced in the low lying areas in the south and south-east of the study area.

11.3.3 Heterogeneity of Arsenic in Relation to Land Use Characteristics

Land use survey was conducted during sampling campaign. Arsenic distribution in four major land use characteristics is shown in Table 11.3. The highest number of wells (41 %) is located adjacent to sanitations (or sanitary installations) followed by sanitation and surface water bodies (28 %), agricultural fields (19 %) and surface water bodies (12 %). Mean As concentration varies among the selected land use characteristics (Table 11.3 and Fig. 11.6a). Tube wells located adjacent to agricultural fields have the highest mean As concentrations (67 µg/L). It is interesting to note that the number of contaminated wells (As ≥50 µg/L) are associated with sanitations ($n=26$) followed by sanitation and surface water bodies ($n=17$), agricultural fields ($n=12$) and surface water bodies ($n=9$). However, tube wells located near the surface water bodies are mostly affected when percentage of affected wells (As ≥50 µg/L) is considered (Table 11.3 and Fig. 11.6a).

Average depth and age distribution of tube wells with respect to land use characteristics are plotted in Fig. 11.6b. The figure illustrates that the contaminated wells are generally young (4–7 years) and located adjacent to surface water bodies and agricultural fields. Relatively older wells are located adjacent to sanitary installations

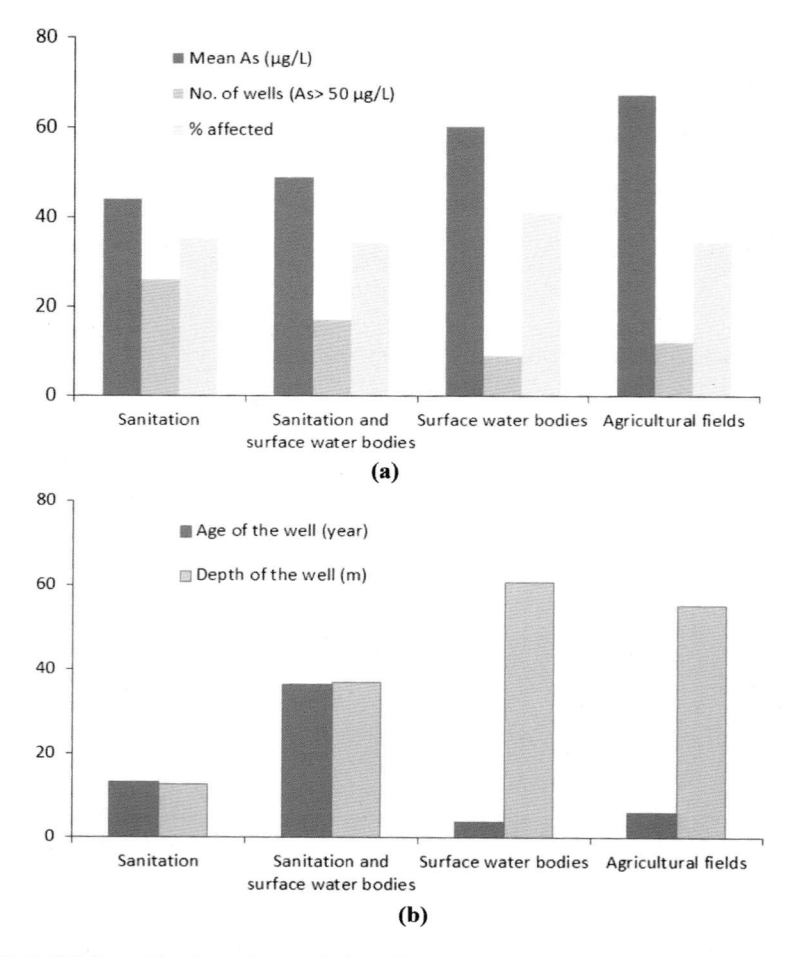

Fig. 11.6 Relation of land use characteristics with (**a**) As concentrations and (**b**) age and depth of tube wells

and sanitation and surface water bodies. The figure also indicates that deeper wells are located adjacent to surface water bodies and agricultural fields, whereas, shallower wells are located adjacent to sanitary installations and sanitation and surface water bodies.

11.3.4 Factor Analysis

Factor analysis aggregated the initial 21 variables into four factors (Table 11.4). Factor 1 has loadings >0.5 for Fe, V, Cr, Co and Ni. All these trace metals are typically associated with Fe-oxyhydroxide as well as mafic minerals (silicates). Factor 2 exhibited higher loadings for EC, Na, Mg, Ca, Cl$^-$ and SO$_4^{2-}$. The combination of these variables into a common factor corresponds to the major dissolved ionic load of the groundwater. This can be explained by the interaction of groundwater with aquifer materials and rocks of the catchment area. Factor 3 shows higher loading for As and PO$_4^{3-}$. Their similar algebraic sign suggests that the mobilization of both elements is coupled by the same processes. Factor 4 contains high loadings for Al, Co, Zn, Cd and Pb.

Table 11.4 Factor loadings (varimax normalized) of water quality parameters

Variables	Factor 1	Factor 2	Factor 3	Factor 4
pH	0.014	−0.158	−0.424	0.014
Eh	−0.118	−0.201	−0.291	−0.046
EC	−0.091	**0.884**	0.242	0.056
HCO$_3^-$	−0.234	0.477	0.491	0.232
As	0.261	−0.113	**0.708**	0.041
Fe	**0.753**	0.014	0.487	0.223
Na	0.029	**0.763**	0.078	0.175
Mg	−0.039	**0.810**	0.289	0.067
K	0.125	0.471	−0.279	−0.057
Ca	−0.010	**0.749**	0.218	0.054
Al	0.409	0.024	0.116	**0.789**
V	**0.836**	0.176	−0.091	0.183
Cr	**0.855**	0.004	0.163	0.198
Co	**0.709**	0.018	0.132	**0.639**
Ni	**0.831**	−0.024	0.083	0.139
Zn	−0.062	−0.014	0.052	**0.838**
Cd	0.462	0.105	−0.037	**0.679**
Pb	0.296	0.079	0.041	**0.852**
Cl$^-$	0.109	**0.789**	−0.092	−0.141
SO$_4^{2-}$	0.141	**0.691**	−0.483	−0.008
PO$_4^{3-}$	0.468	−0.065	**0.601**	0.060

Note: Marked (bold) loadings are >0.5

Table 11.5 Saturation indices of major mineral phases in groundwater

Minerals	Saturation indices (SI)			
	Agricultural fields	Sanitation	Sanitation and surface water bodies	Surface water bodies
Aragonite ($CaCO_3$)	0.15	0.09	0.09	0.27
Calcite ($CaCO_3$)	0.29	0.24	0.24	0.42
Dolomite [$CaMg(CO_3)_2$]	0.30	0.12	0.12	0.53
Hydroxyapatite [$Ca_5(PO_4)_3OH$]	1.49	1.29	1.65	1.88
Goethite (FeOOH)	2.83	2.62	2.68	3.05
Hematite (Fe_2O_3)	7.66	7.23	7.34	8.09
Vivianite [$Fe_3(PO_4)_2, 8H_2O$]	0.98	0.90	1.23	0.75
Siderite ($FeCO_3$)	0.93	0.86	0.89	0.89
Rhodochrosite ($MnCO_3$)	0.16	0.03 (Equilibrium)	0.13	0.10
Gibbsite [$Al(OH)_3$]	1.43	1.33	1.37	1.18

11.3.5 Mineral Saturation Index

Groundwaters of the surveyed wells are saturated with respect to carbonate (aragonite, calcite, dolomite, siderite and rhodochrosite), oxide (hematite), hydroxide (gibbsite), oxy-hydroxide (goethite) and phosphate (hydroxyapatite, vivianite) minerals (Table 11.5). Saturation of these minerals varies with land use characteristics (i.e., sanitations, sanitation and surface water bodies, agricultural fields and surface water bodies). Calculated SI values of 0.03 indicate that rhodochrosite is the only mineral that is close to equilibrium with the groundwater collected from tube wells located adjacent to sanitary installations. Saturation index computation further suggests that carbonate, oxide-hydroxide and phosphate minerals are significant contributors for the geochemical evolution of groundwater in tube wells located adjacent to agricultural fields and surface water bodies. Hence, local recharge from agricultural fields and surface water bodies is likely to play an important role in controlling As release in groundwater.

11.4 Discussion

The (hydro)chemistry of the surveyed wells indicate that the groundwater is partially enriched with dissolved As and other redox sensitive elements (such as Fe and Mn). The distribution of As is highly heterogeneous with some areas more contaminated than others. These results confirm regional As distribution trends within the BDP (van Geen et al. 2003; Zheng et al. 2004; Polizzotto et al. 2006; Bhattacharya et al. 2007; Mukherjee and Fryar 2008; Fendorf et al. 2010; Mukherjee et al. 2011).

Most significant observation is the occurrence of high As concentrations in shallow wells where Na and Cl$^-$ concentrations are high. Cl$^-$ in groundwater is probably derived from sanitary systems (Jacks et al. 1999; McArthur et al. 2012). Such local pollution sources were earlier reported in the BDP by Chatterjee et al. (2010). In the deeper part of the aquifer, Na concentrations are varying and Cl$^-$ concentrations remain stable. The combinations of high Na and high Cl$^-$, and high Na and low Cl$^-$ groundwater suggest the occurrence of local anthropogenic pollution sources. High Na (and low Cl$^-$) groundwater may also be caused due to ion exchange processes (Mukherjee and Fryar 2008), facilitating groundwater with low alkalinity, low Cl$^-$ and high alkaline earth metals. Low SO_4^{2-} concentration in some wells tends to be associated with strongly reducing groundwater. The lack of correlation between dissolved As and SO_4^{2-} suggests removal of SO_4^{2-} from groundwater as part of the bacterial reduction of SO_4^{2-} within the aquifer (Radu et al. 2005; Gault et al. 2005; Lowers et al. 2007; Mukherjee et al. 2008b).

The (hydro)chemical features and land use associations explain the importance of possible local anthropogenic influence on groundwater quality as well as heterogeneity in As distribution pattern. Higher concentrations of NO_3^-, SO_4^{2-} and Cl$^-$ in groundwater are associated with local sanitary installations as well as surface water bodies. The site specificity of the tube wells in relation to sanitation is further conformed when high Na groundwater is found in tube wells located adjacent to sanitary installations. Higher concentrations of PO_4^{3-} and HCO_3^- are primarily linked with agricultural fields and surface water bodies. Relatively higher mean HCO_3^- concentrations advocate surface input of dissolved organic material from agricultural fields and surface water bodies. Phosphate input into the groundwater from agricultural fields and ponds is common. During field investigations, the location of the pond is often observed in between agricultural fields and habitations where surface run-off from agricultural fields takes place (Biswas et al. 2011; Sahu et al. 2011). Mean As concentration is relatively higher in tube wells near agricultural fields and surface water bodies where PO_4^{3-} (run-off from agricultural fields) and HCO_3^- (bio-indicator and an end product of degraded natural organics) are high and dissolved organic carbon (DOC), NO_3^- and SO_4^{2-} are relatively low. This supports the influence of the combination of Fe-oxyhydroxide reduction and PO_4^{3-} competitive exchange processes for the occurrence of high As in groundwater (Biswas et al. 2011). In contrast, mean As concentration is relatively low in the tube wells located adjacent to sanitary installations and sanitation and surface water bodies, where SO_4^{2-}, NO_3^-, Cl$^-$ and Fe are relatively high and PO_4^{3-} and HCO_3^- are relatively low and these groundwater compositions suggest Fe-oxyhydroxide reduction as the principal process for As release.

Spatio-vertical heterogeneity of As concentrations is an important issue in groundwater of the BDP. The present study suggests that this heterogeneity may be a matter of aquifer perturbation. In the absence of primary sources (e.g., pyrite and arsenopyrite), secondary metal oxides/hydroxides of Fe and Mn along with micas are considered to be important. These are often dominant sources as they are abundant in the BDP sediments (Swartz et al. 2004; Polizzotto et al. 2006; Chakraborty et al. 2007; Nath et al. 2008b). In addition, the fluvial geomorphology and the land use pattern are also important factors that influence the local release and distribution of As in the groundwater.

Saturation of carbonate minerals (such as aragonite, calcite, dolomite, siderite and rhodochrosite) and high values of Ca, Mg, Fe and HCO_3^- in groundwater with circum-neutral pH is an indication of carbonate mineral dissolution. This often indicates that carbonate reactions are important in influencing groundwater (hydro) chemistry in the surveyed wells. Saturation index values for siderite are virtually unaffected amongst the various land use characteristics around the wells, reflecting that the influence of these local features on the thermodynamic equilibrium of siderite is not sufficient. Groundwater is commonly super-saturated with respect to hematite and saturated with respect to siderite. This suggests that the precipitation of Fe(III) phases from the aqueous phase is thermodynamically favourable and therefore likely to regulate As concentration in the groundwater. Saturation index values indicate that vivianite is saturated in most cases where As concentrations vary largely (mean 49–67 µg/L). Saturation indices of rhodochrosite and gibbsite are also important because the formation of rhodochrosite and gibbsite can regulate As concentration in groundwater. These minerals can act as a sink for As by co-precipitation with Mn(II) and Al(III) ions from the anoxic groundwater (Roman-Ross et al. 2006; Bhowmick et al. 2013). Contrastingly, these processes can further accelerate the rate of reduction of metal(s) oxide/hydroxides by shifting the equilibrium and increasing dissolved As concentration in the groundwater (Biswas et al. 2011). However, when the minerals are in equilibrium (rhodochrosite in case of sanitary installations), the process gets slowed and As release is possibly hindered. This further reinforces that the combined release mechanisms are important and operated by various local features around the well site.

Heterogeneous distribution (spatial and vertical variability) of As can be linked with multiple geochemical processes. Nevertheless, ion exchange and ion concurrence with PO_4^{3-} from anthropogenic sources (fertilizers) together with the complex deltaic environment may also have some impact on the As spatio-vertical heterogeneity.

11.5 Conclusion

The present study demonstrates that the groundwater chemistry of As affected aquifers in Chakdaha block, Nadia district, West Bengal, is predominantly under anoxic condition. Critical evaluation of the groundwater chemical composition reveals that Fe-oxyhydroxide reduction and As mobilization may not be the only process to explain high As concentrations in the studied groundwater. The major finding of our study is that the land use patterns (surface water bodies, sanitations and agriculture-aquifer-sediment interactions) are likely to have a strong influence on the release and distribution of As in local shallow aquifers. The heterogeneous distribution of these factors can suitably explain the corresponding heterogeneity of As concentrations in the groundwater. The competitive exchange of ions can also influence As distribution when tube wells are located adjacent to agricultural fields. However, the extent and inter-connectivity of all the processes vary considerably from place to place and from depth to depth. The calculated SI values also suggest that there may be a relation between tube well site characteristics (i.e., land use pattern) and mineralogical compositions influencing the thermodynamic equilibrium of the sediment-water interaction and release of As. The release mechanism of As is complex and no unique or individual mechanism can be universally applied to explain such heterogeneous distribution of As in the studied groundwater aquifers.

Acknowledgements The authors acknowledge the financial support under the DFG and BMZ Programme for Research Co-operation with Developing Countries (Programms zur Zusammenarbeit mit Wissenschaftlern in Entwicklungsländern, project Stu 169/37-1) to carry out the research work. The support of DST (New Delhi) under FIST and PURSE programme to the Department of Chemistry, University of Kalyani is also duly acknowledged. SM would also like to thank the Erasmus Mundus External Cooperation Window (EMECW-Action II) programme, EURINDIA for providing the doctoral research fellowship.

References

BGS & DPHE (2001) Kinniburgh DG, Smedley PL (eds) Arsenic contamination of groundwater in Bangladesh. British Geological Survey WC/00/19, Keyworth

Bhattacharya P, Chatterjee D, Jacks G (1997) Occurrence of arsenic-contaminated groundwater in alluvial aquifers from delta plains, Eastern India: options for safe drinking water supply. J Water Resour Dev 13:79–92

Bhattacharya P, Welch AH, Stollenwerk KG, McLaughlin MJ, Bundschuh J, Panaullah G (2007) Arsenic in the environment: biology and chemistry. Sci Total Environ 379:109–120

Bhattacharyya R, Chatterjee D (2001) Technique for arsenic removal from groundwater utilizing geological options – an innovative low cost remediation. Int Assoc Hydrol Soc 275:391–396

Bhattacharyya R, Chatterjee D, Nath B, Jana J, Jacks G, Vahter M (2003a) High arsenic groundwater: mobilization, metabolism and mitigation – an overview in the Bengal Delta Plain. Mol Cell Biochem 253:347–355

Bhattacharyya R, Jana J, Nath B, Sahu SJ, Chatterjee D, Jacks G (2003b) Groundwater As mobilization in the Bengal Delta Plain, the use of ferralite as a possible remedial measure – a case study. Appl Geochem 18:1435–1451

Bhowmick S, Nath B, Halder D, Biswas A, Majumder S, Mondal P, Chakraborty S, Nriagu J, Bhattacharya P, Iglesias M, Roman-Ross G, Guha Mazumder D, Bundschuh J, Chatterjee D (2013) Arsenic mobilization in the aquifers of three physiographic settings of West Bengal, India: understanding geogenic and anthropogenic influences. J Hazard Mater 262:915–923

Biswas A, Majumder S, Neidhardt H, Halder D, Bhowmick S, Mukherjee-Goswami A, Kundu A, Saha D, Berner Z, Chatterjee D (2011) Groundwater chemistry and redo processes: depth dependent arsenic release mechanism. Appl Geochem 26:516–525

Chakraborty S, Wolthers M, Chatterjee D, Charlet L (2007) Adsorption of arsenite and arsenate onto muscovite and biotite mica. J Colloid Interface Sci 309:392–401

Charlet L, Chakraborty S, Appelo CAJ, Roman-Ross G, Nath B, Ansari AA, Lanson M, Chatterjee D, Basu Mallik S (2007) Chemodynamics of an As "hotspot" in a West Bengal aquifer: a field and reactive transport modeling study. Appl Geochem 22:1273–1292

Chatterjee D, Roy RK, Basu BB (2005) Riddle of arsenic in groundwater of Bengal Delta Plain – role of non-inland source and redox traps. Environ Geol 49:188–206

Chatterjee D, Halder D, Majumder S, Biswas A, Bhattacharya P, Bhowmick S, Mukherjee-Goswami A, Saha D, Maity PB, Chatterjee D, Nath B, Mukherjee A, Bundschuh J (2010) Assessment of arsenic exposure from groundwater and rice in Bengal Delta Region, West Bengal, India. Water Res 44:5803–5812

Datta DK, Subramanian V (1998) Distribution and fractionation of heavy metals in the surface sediments of the Ganges-Brahmaputra-Meghna river system in the Bengal Basin. Environ Geol 36:93–101

Davis JC (1986) Statistics and data analysis in geology. Wiley, New York

Fendorf S, Michael HA, van Geen A (2010) Spatial and temporal variations of groundwater arsenic in south and south East Asia. Science 328:1123–1127

Gault AG, Islam FS, Polya DA, Charnock JM, Boothman C, Chatterjee D, Lloyd JR (2005) Microcosm depth profiles of arsenic release in a shallow aquifer, West Bengal. Mineral Mag 69:855–863

IFCPAR (Indo-French Centre for the Promotion of Advance Research) (2004) Report on "Geochemistry of Ganga sediment and groundwater: arsenic mobilization". University of Kalyani, Nadia, West Bengal, India and University of Grenoble, France

Jacks G, Sefe F, Carling M, Hammar M, Letsamao P (1999) Tentative nitrogen budget for pit latrines – Eastern Botswana. Environ Geol 8:199–203

Jana J (2004) Genesis of arseniferrous groundwater in the Bengal Delta Plain in West Bengal, Eastern India. Unpublished Ph.D. thesis, University of Kalyani, India

Kabata-Pendius A, Pendius H (2001) Trace elements in soils and plants, 3rd edn. CRC Press, Boca Raton

Liu CW, Lin KH, Kuo YM (2003) Application of factor analysis in the assessment of groundwater quality in a Blackfoot disease area in Taiwan. Sci Total Environ 313:77–89

Lowers HA, Breit GN, Foster AL, Whitney J, Yount J, Uddin MN, Muneem AA (2007) Arsenic incorporation into authigenic pyrite, Bengal Basin sediment, Bangladesh. Geochim Cosmichim Acta 71:2699–2717

McArthur JM, Sikdar PK, Hoque MA, Ghosal U (2012) Waste-water impacts on groundwater: Cl/Br ratios and implications for arsenic pollution of groundwater in the Bengal Basin and Red River Basin, Vietnam. Sci Total Environ 437:390–402

Mukherjee A, Fryar AE (2008) Deeper groundwater chemistry and geochemical modeling of the arsenic affected western Bengal basin, West Bengal, India. Appl Geochem 23:863–892

Mukherjee A, Bhattacharya P, Savage K, Foster A, Bundschuh J (2008a) Distribution of geogenic arsenic in hydrologic systems: controls and challenges. J Contam Hydrol 99:1–7

Mukherjee A, von Bromssen M, Scanlon BR, Bhattacharya P, Fryar AE, Hasan MA, Ahmed KM, Chatterjee D, Jacks G, Sracek O (2008b) Hydrogeochemical comparison and effects of over-lapping redox zones on groundwater arsenic near the western (Bhagirathi sub-basin, India) and Eastern (Meghna sub-basin, Bangladesh) margins of the Bengal Basin. J Contam Hydrol 99:31–48

Mukherjee A, Fryar AE, Scanlon BR, Bhattacharya P, Bhattacharya A (2011) Elevated arsenic in deeper groundwater of western Bengal basin, India: extents and controls from regional to local-scale. Appl Geochem 26:600–613

Nath B, Berner Z, Basu Mallik S, Chatterjee D, Charlet L, Stüeben D (2005) Characterization of aquifers conducting groundwaters with low and high arsenic concentrations: a comparative case study from West Bengal, India. Mineral Mag 69:841–853

Nath B, Sahu SJ, Jana J, Mukherjee-Goswami A, Roy S, Sarkar MJ, Chatterjee D (2008a) Hydrochemistry of arsenic-enriched aquifer from rural West Bengal, India: a study of the arsenic exposure and mitigation option. Water Air Soil Pollut 190:95–113

Nath B, Stuben D, Basu Mallik S, Chatterjee D, Charlet L (2008b) Mobility of arsenic in West Bengal aquifers conducting low and high groundwater arsenic. Part I: comparative hydrochemical and hydrogeological characteristics. Appl Geochem 23:977–995

Nath B, Berner Z, Chatterjee D, Basu Mallik S, Stüben D (2008c) Mobility of arsenic in West Bengal aquifers conducting low and high groundwater arsenic. Part II: comparative geochemical profile and leaching study. Appl Geochem 23:996–1011

Neumann RB, Ashfaque KN, Badruzzaman ABM, Ali MA, Shoemaker JK, Harvey CH (2010) Anthropogenic influences on groundwater arsenic concentrations in Bangladesh. Nat Geosci 3:46–52

Pal T, Mukherjee PK, Sengupta S (2002) Nature of arsenic pollutants in groundwater of Bengal Basin – a case study from Baruipur area, West Bengal, India. Curr Sci 82:554–561

Parkhurst DL, Appelo CAJ (1999) Users guide to PHREEQC (version 2): a computer program for speciation, batch–reaction, one–dimensional transport, and inverse geochemical calculation. United States Geological Survey Water Resources Investigations report, 99–4259

PHED (Public Health and Engineering Directorate) (1993) National Drinking Water Mission project on arsenic pollution on groundwater in West Bengal. Final report, Steering Committee on As investigation. Government of West Bengal, India

Polizzotto ML, Harvey CF, Li G, Badruzzman B, Ali A, Newville M, Sutton S, Fendorf S (2006) Solid-phases and desorption processes of arsenic within Bangladesh sediments. Chem Geol 228:97–111

Radu T, Subacz JL, Philippi JM, Barnett MO (2005) Effects of dissolved carbonate on arsenic adsorption and mobility. Environ Sci Technol 39:7875–7882

RGNDWM (Ministry of Rural Water and Development, Government of India) (2001) Phase-II report. University of Kalyani, Nadia

Roman-Ross G, Cuello GJ, Turrillas X, Fernández-Martínez A, Charlet L (2006) Arsenite sorption and co-precipitation with calcite. Chem Geol 223:328–336

Sahu SJ, Nath B, Roy S, Mandal B, Chatterjee D (2011) A laboratory batch study on arsenic sorption and desorption on guava orchard soils of Baruipur, West Bengal, India. J Geochem Explor 108:157–162

Smedley PL, Kinniburgh DG (2002) A review of the source, behaviour and distribution of arsenic in natural waters. Appl Geochem 17:517–568

Smith AH, Lingas EO, Rahaman M (2000) Contamination of drinking water by arsenic in Bangladesh: a public health emergency. Bull World Health Organ 78:1093–1103

Stüben D, Berner Z, Chandrasekharam D, Karmakar J (2003) Arsenic enrichment in groundwater of West Bengal, India: geochemical evidence for mobilization of As under reducing conditions. Appl Geochem 18:1417–1434

Swartz CH, Blute NK, Badruzzman B, Ali A, Brabander D, Jay J, Besancon J, Islam S, Hemond HF, Harvey CF (2004) Mobility of arsenic in a Bangladesh aquifer: inferences from geochemical profiles, leaching data, and mineralogical characterization. Geochem Cosmichim Acta 68:4539–4557

van Geen A, Zheng Y, Versteeg R, Stute M, Horneman A, Dhar R, Steckler M, Gelman A, Small C, Ahsan H, Graziano JH, Hussain I, Ahmed KM (2003) Spatial variability of arsenic in 6000 tube wells in a 25 km^2 area of Bangladesh. Water Resour Res 39:1140–1151

Zheng Y, Stute M, van Geen A, Gavrieli I, Dhar R, Simpson HJ, Schlosser P, Ahmed KM (2004) Redox control of arsenic mobilization in Bangladesh groundwater. Appl Geochem 19:201–214

Chapter 12
Surface Generated Organic Matter: An Important Driver for Arsenic Mobilization in Bengal Delta Plain

S.H. Farooq and D. Chandrasekharam

12.1 Introduction

An access to safe drinking water is primary human need, but at many places around the globe this primary requirement is not fulfilled and millions of people are forced to drink contaminated water. On a larger scale, various elements such as As, F, V, U, Pb, Hg etc. are contaminating the available drinking water resources (WHO 2006). Among all the above mentioned elements, As is posing a major threat and several countries around the globe are facing problem of As contamination in the groundwater. The problem of As contamination is most severe in the South-East Asian countries (Bhattacharya et al. 1997; Chandrasekharam et al. 2001; McArthur et al. 2004; Nickson et al. 1998; Smedley and Kinniburgh 2002; van Geen et al. 2008). Elevated As concentrations have been reported at places in Nepal, India, Bangladesh, Myanmar, Cambodia, Vietnam and Thailand etc. It is estimated that about 200 million people living in Asia are exposed to various health risks due to consumption of As contaminated water (Sun 2004). In Bengal delta plain alone more than 50 million people are routinely exposed, thus it is described as the worst case of mass-poisoning in the history of mankind (Ahsan et al. 2009; Chakraborti et al. 2003). The WHO (2011) permissible limit of As in drinking water is 10 µg/L, while the Indian permissible limit is five times higher (50 µg/L). In Bengal delta plain As concentrations more than 1,000 µg/L has been reported at several places. Still it appears that the extent of the contamination is not fully known and many new affected areas are getting discovered on regular basis. Apart from West Bengal,

S.H. Farooq (✉)
School of Earth, Ocean and Climate Sciences,
IIT Bhubaneswar, Bhubaneswar, Odisha 751007, India
e-mail: hilalfarooq@iitbbs.ac.in

D. Chandrasekharam
Department of Earth Sciences, IIT Bombay, Powai, Mumbai 400076, Maharashtra, India

higher As concentrations have been reported from other Indian states including Bihar, Uttar Pradesh, Assam, Jharkhand, Chattisgarh and Madhya Pradesh (Acharyya et al. 2005; Ahamed et al. 2006; Bhattacharjee et al. 2005; Chakraborti et al. 1999, 2003; Das et al. 1996; Paul and Kar 2004).

In spite of a wealth of information available on occurrence of As in groundwater, no definite source of As in ground water is well established. However, most of the researchers agree that in Bengal delta plain, As is released from the sediments that have coating of As on it (Smedley and Kinniburgh 2002; Bhattacharya et al. 2002 and references therein). Three models namely, oxidation, phosphate ion-exchange, and reduction have been proposed to explain "how As is releasing from the sediments and contaminating the groundwater". Oxidation model suggests that due to huge withdrawal of groundwater, water table drops and some part of the aquifer which was earlier under saturated condition gets exposed to the atmospheric gases. Under such conditions As-rich pyrite breaks down and releases arsenic (Das et al. 1996; Mallick and Rajagopal 1996). The phosphate ion-exchange model suggests that the arsenic anions sorbed on aquifer minerals are displaced into solution by competitive exchange by phosphate anions, sourced from over-application of fertilizer to agricultural fields (Acharyya et al. 1999).

The most accepted reduction model proposes that the naturally occurring As sorbed on iron oxyhydroxide (FeOOH) remobilizes when reducing condition develops (Bagla and Kaiser 1996; Bhattacharya et al. 1997, 2002; Islam et al. 2004; McArthur et al. 2004; Nickson et al. 2000; Stüben et al. 2003). Currently, this is the most accepted model for Bengal delta and the South East Asia. Reduction model considers an internal organic matter source (peat layer) to be responsible for driving a sequence of reactions which mobilize arsenic (McArthur et al. 2001; Ravenscroft et al. 2001). However, over the time, several other organic matter sources (i.e. organic matter rich clay layer, seepage from petroleum-derived hydrocarbon etc.) have been suggested (Rowland et al. 2006). Despite its conjectural importance, significant quantities of organic matter that remains available on the surface of Bengal delta (explained later), has not been discussed until recently. Different surface-derived organic compounds (more often represented in the form of a composite parameter as dissolved organic carbon (DOC)) that form due to decomposition of this organic matter can have a significant role in As mobilization from the sediments. Thus, a study has been taken to evaluate the potential of surface-derived organic carbon (in terms of DOC) on mobilization of As from the sediments.

12.2 Concept of Investigation

A careful observation of various As affected areas in South-East Asia clearly indicates that all affected areas are paddy cultivating areas, and different varieties of rice are grown throughout the year. Additionally, the problem is more severe in areas where jute production is also high (i.e. Bengal delta plain). This leads to the idea that there must be some relationship between paddy cultivation, jute production and As contamination of groundwater. The cultivation of rice (paddy cultivation) and processing of jute (to obtain jute fibre from the plant) produces significantly large quantities of DOC on the surface of Bengal delta.

In Bengal delta, the organic carbon production sites can broadly be categorized into two classes, (i) paddy fields and (ii) jute decomposing ponds.

(i) **Production of organic carbon in paddy fields:** In paddy fields huge quantity of organic matter remains available which when decayed forms significant quantity of DOC. The DOC produces mainly because farmer follows traditional way to cultivate paddy. In this method, the harvested crop is cut from the middle of stem, the remaining half of the stem and roots are ploughed back for the next cultivation. During the monsoon season, when the fields are filled with rain water, the remains of previous crop (roots and half of the stem) start to decompose and produce large quantities of particulate organic carbon (POC) and DOC (Chandrasekharam 2008). This way significant quantity of POC and DOC forms in the paddy cultivating fields of Bengal delta.

(ii) **Production of organic carbon in jute decomposing ponds:** Jute decomposing ponds act as point source for supplying organic carbon to Bengal Delta. Like paddy cultivation, here also, traditional processing technique is followed to obtain jute fibre from the jute plants. This involves retting of the jute plant in open ponds for several weeks, which generates significantly large quantities of DOC. This again leaves significant quantity of POC and DOC on the surface of Bengal delta.

The POC and DOC thus produced in paddy fields and jute decomposing ponds, ultimately moves downward with the percolating water. POC molecules being larger in size get clogged in the top sediments while DOC moves down. On its way to water table DOC molecules react with mineral surfaces and modify their sorption behaviour. Since West Bengal and Bangladesh rank among the top jute and rice (paddy) producing areas in the world, the quantities of decomposable organic matter available on the surface of Bengal delta can be imagined.

12.3 Material and Methods

12.3.1 Field Site

The study was conducted in the Nabipur block (24°14′,-24°17′ N; 88°31′-88°43′ E) of Murshidabad district of West Bengal, India (Fig. 12.1). Murshidabad is one of the most severely As affected district. Apart from paddy cultivation, it is also one of the main jute producing districts of West Bengal and the jute fibres from the jute plants are obtained by the traditional processing method (as described earlier).

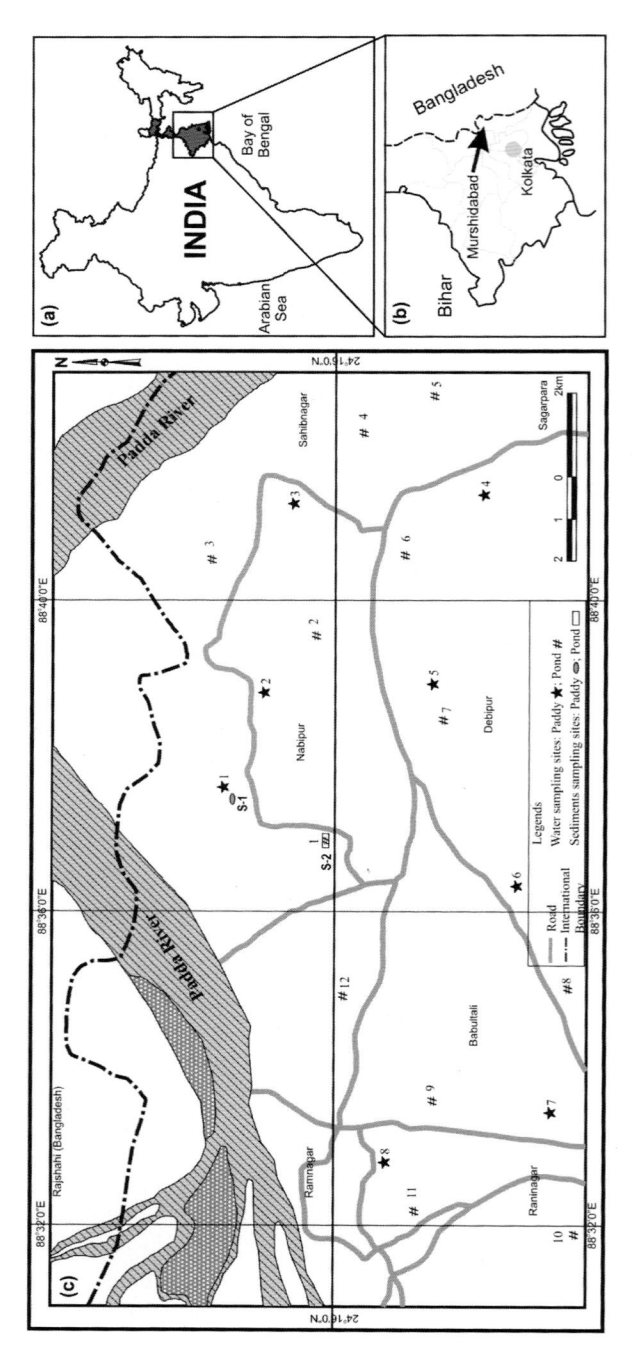

Fig. 12.1 Map showing the locations of water and sediment sample collection sites

12.3.2 Collection and Analysis of Water Samples

To determine the DOC content in the water standing in paddy fields and jute decomposing ponds, water samples from eight paddy fields and 12 jute decomposing ponds were collected (Fig. 12.1). Samples were filtered with 0.45 µm polycarbonate filters in the field and stored at a low temperature in amber coloured glass bottles to minimize the photo-oxidation of DOC. DOC concentrations in the collected water samples were determined by Shimadzu total organic carbon analyzer (TOC-5000). The procedure involves a high temperature oxidation prior to IR detection of CO_2. The detection limit was 0.1 mg/L and the precision ranged between 2 and 3 %.

12.3.3 Collection and Analysis of Sediment Profile

To determine the effect of percolating DOC on arsenic mobilization, sediment cores (up to aquifer depth, 9 m) were recovered from a paddy field and from the bottom of a jute decomposing pond (in Fig. 12.1, S1: paddy; S2: pond). To recover the undisturbed bulk sample Cable Percussion Technique was applied. The core was sliced into smaller segments at every 7.5 cm interval ($\approx$3 in.). This way, 84 sediment samples were collected from paddy field and 72 from the jute decomposing pond. The samples were immediately packed into resaleable plastic bags and purged with nitrogen gas to minimize the oxidation due to exposure to the atmosphere. In the laboratory, samples were freeze dried and stored at low temperature until further analysis.

Elemental composition of all collected sediments was determined by ED-XRF (Spectra 5000, Atomica). Arsenic content was measured using a Pd primary filter, which led to a detection limit of $\approx$1 mg/kg. To control the quality of data, a number of selected Certified Reference Materials (CRMs) such as GXR-2, GXR-5, Soil-5, SL-1, SCO-1, SDO-1 (Park City, Utah, USA) etc. have been repeatedly analyzed. In general, precision (better than 5 %) was calculated from repeated measurements of the standard material and accuracy (better than 10 %) was checked by including CRMs (GXR-2, GXR-5 and Soil-5). Total C content of the sediments was measured by Carbon-Sulphur-Analyser (CSA 5003, Leybold Heraeus, Germany), while the inorganic C content was determined by a Carbon-Water-Analyser (CWA 5003, Leybold Heraeus, Germany). The total organic C (TOC) content of the sediments was calculated by subtracting inorganic C from the total C content. Constituent minerals (i.e. quartz, plagioclase, k-feldspar, kaolinite/chlorite etc.) in the sediments were identified by X-ray diffraction (XRD) analysis (Krisalloflex D500, Siemens, Germany) at 40 kV and 25 mA. CuKα-1 radiation was used at angles between 3° and 63°. The semi-quantitative evaluation of the spectra was done based on the calibration curves obtained from different samples with known mineral composition produced at the IMG, KIT, Karlsruhe. Grain size of the bulk sediment samples was determined by using a Laser-granulometer (Hydro 2000G).

Approximately 0.5 g of dry bulk sample material was initially disaggregated with an ultrasonic probe using water as dispersant medium. Reported grain sizes were grouped into clay (<2 μm), silt (2–63 μm) and sand (>63 μm).

12.3.4 Column Leaching Experiment

To determine the effect of percolating DOC on the sediments, conditions similar to those existing in Bengal delta (i.e. natural conditions) were created in the laboratory through the column experiment. Separate column experiments were run for paddy field and pond profiles.

For paddy column experiment, based on the chemical composition of the samples (XRF) and grain size analysis, the 9-m long sediment core has been divided into two zones, i.e. Zone 1 and 2 (explained in Sect. 12.4.2). However, on the similar basis the sediment core collected from the jute decomposing pond has been divided into three zones: i.e. Zone 1, 2 and 3 (explained in Sect. 12.4.2). For paddy column experiment, representative material was prepared from the samples of each zone and used as the infill of two interconnected columns (30 cm each), arranged in the same order as the zones occur in the core (material from Zone 1 in Column 1AP, from Zone 2 in Column 2AP; P denotes paddy). Another identical set of columns (set B; labelled 1BP, and 2BP; B denotes blank) was prepared and run in parallel. Similarly, for pond column experiment the representative material was filled in three interconnected columns (1AJ, 2AJ and 3AJ; J denotes jute), arranged as described above and an identical parallel set of columns containing columns 1BJ, 2BJ and 3BJ was prepared and run in parallel. The columns were connected in such a way, that the leaching solution successively passed through both columns for paddy and all the three columns for pond column leaching experiment (Fig. 12.2a, b).

The leaching solution (i.e. DOC solution) for paddy column experiment was made by decomposing paddy plants and for pond column experiment by decomposing the jute plants separately, in the laboratory under controlled conditions. Separate decomposition and leaching with the respective DOC solution was necessary as the quality (e.g. molecular size and functional groups) and thus reactivity of DOC depends mainly upon the composition of source/parent material, medium of decomposition, temperature and rate of decomposition, and the presence of microbial communities.

By decomposing the respective material under comparable environment, it is reasonable to assume that the DOC produced in laboratory will be nearly identical to that of the DOC produced in a paddy field and in the jute decomposing ponds of West Bengal, and will behave similarly to that produced under natural conditions in Bengal delta.

Based on average DOC concentration measured in the paddy fields and jute decomposing ponds, both the paddy column set (AP) and pond column set (AJ) were leached with a solution containing 100 mg/L of respective DOC, while the columns of set B (BP and BJ) were leached with plain water. Set B of columns was

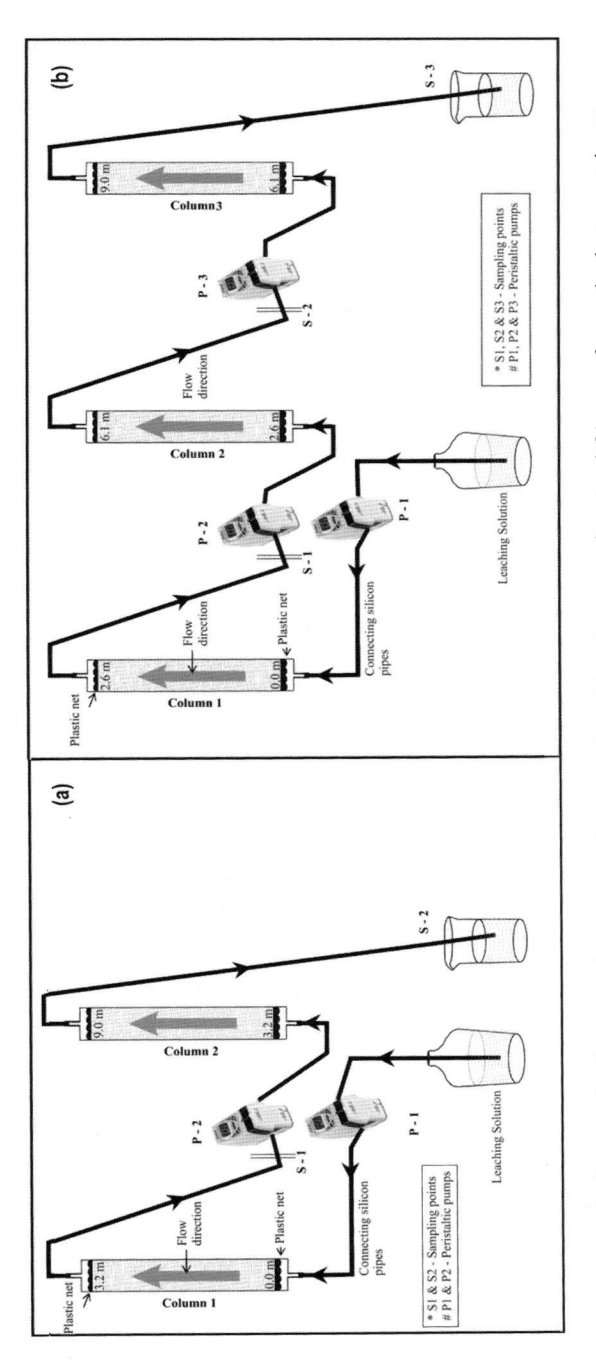

Fig. 12.2 Schematic diagram of column experiment set-up: (**a**) set-up for paddy column experiment and (**b**) set-up for pond column experiment

used as reference, allowing to clearly determine the effect of DOC on As mobilization. Leachates were sampled at the exits of each column on daily basis and were analyzed for As content. The column experiment for paddy and pond profile was run continuously for a period of 41 days and 42 days respectively.

12.4 Results and Discussions

12.4.1 Water Samples

The DOC content in paddy field ranges between 56 and 61 mg/L, except in one sample where DOC content as high as 128 mg/L is registered (Table 12.1). Similarly, in jute ponds also, DOC concentration ranges around the narrow limit of 60–70 mg/L and in few samples higher concentrations were registered (maximum conc. 153 mg/L). The average DOC concentration registered in paddy field and jute decomposing pond is 67.2 mg/L and 74 mg/L, respectively.

The DOC concentrations in paddy fields and in jute decomposing ponds are controlled by several factors such as time of sampling, availability/quantity of decomposable organic matter, rate of decomposition and the amount of rainfall. All these factors vary widely both in space and time; thus, no single concentration (neither mean nor highest) can be considered as representative for the entire Bengal delta. Specifically for paddy field, the other factors that may also influence DOC concentrations include (i) the variety of rice cultivated in previous cropping, as different plant varieties have different decomposition rate and (ii) different agricultural fields usually do not have boundaries of similar height; one with higher boundary can accumulate more rain water that can act as diluting agent, when compared to others with lower boundary (Farooq et al. 2010). Similarly, in case of jute ponds, the main controlling factor for DOC content in pond water is the quantity of plant matter (jute plant) dumped in the pond for decomposition. The quantity placed in each pond depends primarily on the number of ponds present in the area and the harvest yield of the jute crop. There is no fixed ratio of pond size to the amount of plant matter placed. Thus the concentration of DOC depends completely on the wish of the pond owner (Farooq et al. 2012). Keeping all these factors in mind, in this study, we took 100 mg/L of DOC in the leaching solution for both the paddy and pond column leaching experiments. Further, this will help in comparing the As mobilization data from both column leaching experiments quickly and easily.

Table 12.1 DOC concentrations (mg/L) in different paddy fields and jute decomposing ponds

Sample no.	1	2	3	4	5	6	7	8	9	10	11	12
DOC cons. in paddy fields	58.9	56.0	60.9	61.1	57.3	128	56.5	58.1	–	–	–	–
DOC conc. in jute ponds	62.3	77.1	69.4	61.0	62.6	61.8	92.7	58.6	67.7	153	62.1	60.2

12.4.2 Elemental Composition and Grain Size Distribution Along the Sediment Profiles

The depth-wise variation of As and Fe content in paddy and pond sediments profile is shown in Fig. 12.3 (a: paddy profile, b: pond profile). Paddy field profile clearly shows two zones (Fig. 12.3a). Zone 1 is enrichment zone where the As concentrations are well above the background concentrations (10 mg/kg) for the Bengal delta sediments (Norra et al. 2005). This is mainly because of the continuous influx of As

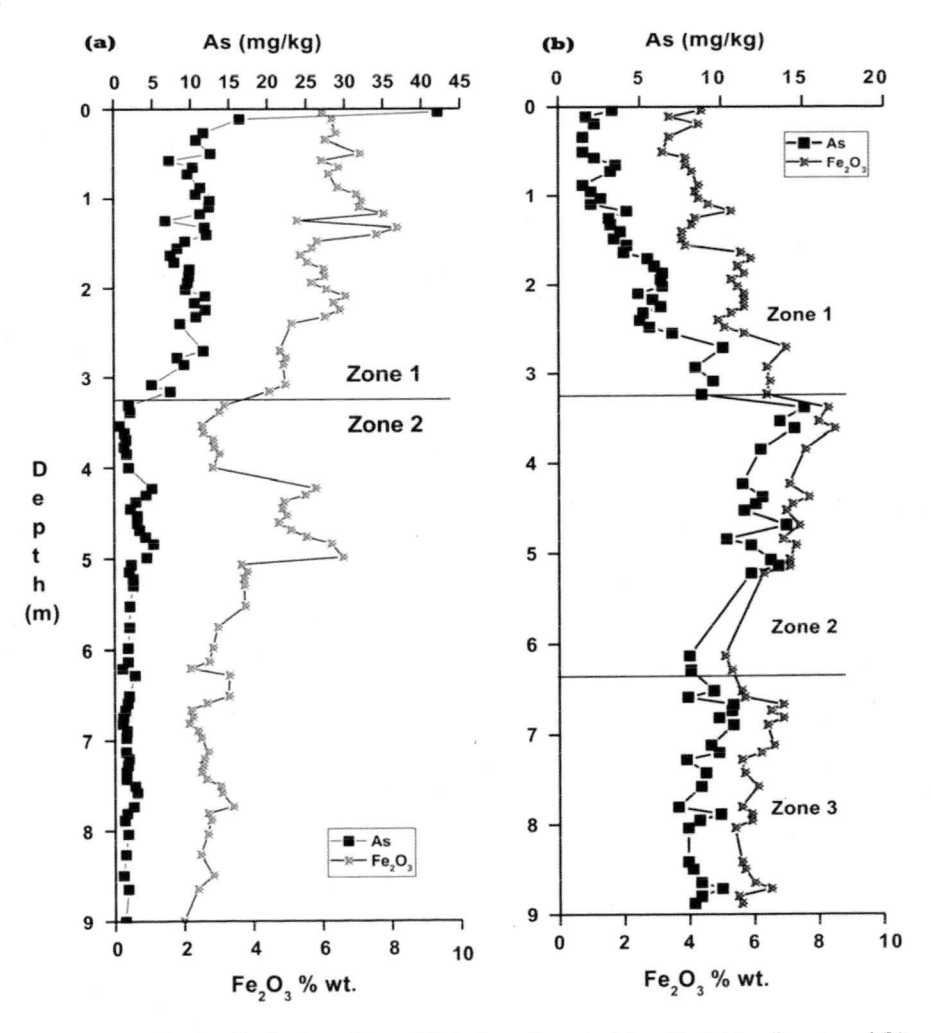

Fig. 12.3 Depth-wise distribution of As and Fe in the sediments: (**a**) paddy field sediments and (**b**) pond sediments

through contaminated irrigation water. The As supplied through irrigation water accumulates itself on to the sediments and also gets adsorbed on the iron plaque formed at the roots of paddy plants (Ahsan et al. 2009; Dittmar et al. 2007; Norra et al. 2005; van Geen et al. 2006).

Around 50 % of roots of paddy plant extend in first 10–15 cm below the ground surface (Mathan and Natesan 1988) and thus, the major fraction of unusually higher As concentration at this depth seems to be contributed by the accumulation of As in root zones (Otte et al. 1995). The presence of dominating finer particles in Zone 1 sediments additionally facilitates As adsorption by providing abundant sites where As loosely binds itself (Ona-Nguema et al. 2005). Various studies have shown that the concentration of As in top sediments would have been much higher if a mechanism for continuous removal of As does not exists. The removal mechanism may involve reduction induced mobilization or diffusion of accumulated As into floodwater or transportation of As to the lower horizons (Dittmar et al. 2007). Zone 2 sediments do not show higher As concentrations and As concentration near background concentration (at deeper level) has been registered. This may be because of the fact that all the As that mobilizes from top sediments of Zone 1 has been attenuated within the lower portion of Zone 1 itself. The same is true, not just for As but also for other redox sensitive elements including Fe, Zn, Cu etc.

The elemental composition of sediment profile below the jute decomposing pond clearly show three zones, (i) Zone 1 (0–2.6 m), which has As and Fe concentrations less than their background concentrations in Bengal delta sediments; (ii) Zone 2 (2.6–6.1 m), in which the concentrations of the same elements are higher than their respective background values, and (iii) Zone 3 with a composition similar to the background (Fig. 12.3b). Such a trend indicates the mobilization/transfer of As from Zone 1 to Zone 2 and in case of jute decomposing ponds this is mainly due to the effect of percolating DOC (explained later).

Mineralogy of paddy profile shows that Zone 1 is dominated by clay minerals which make up to 66 % (average 48.7 %) of the total minerals present. Additionally, kaolinite (av. 9.8 %) and other clay minerals are also significantly higher in this zone as compared to Zone 2, where quartz (av. 28.4 %), plagioclase (av. 11.5 %) and K-feldspar (av. 14.5 %) are the dominant minerals (Table 12.2). Concentrations of clay minerals in jute sediment profile is 39.5 %, 56.6 % and 47.4 %, in Zone 1, 2 and 3, respectively. Grain size analysis of paddy profile shows that Zone 1 is dominated by silt (average 44.9 %), while Zone 2 is dominated by sand (average 57.2 %) sized grains (Table 12.3). In case of jute decomposing pond profile, the clay sized particles show a general increase with depth (4–57 %), sand size particles a general decrease with depth (47–1 %) and the silt content vary within a narrow range (38–54 %). However, the presence of sand laminations (in form of higher sand and lower clay content) is also noticed at different depths throughout the profile. Clay patches are common feature of any deltaic morphology and in Bengal delta and jute decomposing ponds are located on these clay patches so that they can hold rain water for longer time. This makes these ponds more useable for jute decomposition.

Table 12.2 Abundance of minerals (wt %) in different zones along the sediment profile

Site	Depth zone	Minerals								Phyllo silicates/clay minerals (%)
		Quartz (%)	Plagioclase (%)	K-feldspar (%)	Dolomite (%)	Calcite (%)	Organic matter (%)	Kaolinite (%)		
Paddy	Zone 1 (0–3.2 m)	6–30 (av. 19.4)	2–16 (av. 4.6)	0–19 (av. 6.0)	0–2 (av. 1.2)	3–12 (av. 9.9)	0.2–1.7 (av. 0.4)	5–21 (av. 9.8)		26–66 (av. 48.7)
	Zone 2 (3.2–9.0 m)	6–47 (av. 28.4)	0–27 (av. 11.5)	1–22 (av. 14.5)	1–6 (av. 2.0)	2–12 (av. 6.2)	0.1–0.7 (av. 0.2)	4–18 (av. 7.5)		10–57 (av. 29.8)
Pond	Zone 1 (0–2.6 m)	14–45 (av. 25.1)	1–12 (av. 6.1)	0–20 (av. 8.7)	1–4 (av. 1.3)	3–12 (av. 8.6)	0.1–1.2 (av. 0.4)	0–30 (av. 10.2)		13–50 (av. 39.5)
	Zone 2 (2.6–6.1 m)	11–22 (av. 16.8)	0–6 (av. 2.5)	0–7 (av. 2.1)	0–1 (av. 0.8)	2–12 (av. 9.9)	0–2.1 (av. 0.6)	7–16 (av. 10.5)		42–70 (av. 56.6)
	Zone 3 (6.1–9.0 m)	8–27 (av. 18.9)	2–8 (av. 4.6)	0–12 (av. 5.3)	0–3 (av. 1.3)	10–12 (av. 11.2)	0.3–1 (av. 0.5)	7–22 (av. 10.8)		29–59 (av. 47.4)

Table 12.3 Grain size distribution along the sediment profile

Site	Depth zone	Clay (0.063–2.0 μm)	Silt (2–63.0 μm)	Sand (63–2,000 μm)
Paddy	Zone 1	8. 5–50.0	35.3–55.7	2.2–56.2
	(0–3.2 m)	(av. 33.0)	(av. 44.9)	(av. 22.1)
	Zone 2	0.4–48.2	2.6–55.0	3.3–97.0
	(3.2–9.0 m)	(av. 11.3)	(av. 31.6)	(av. 57.2)
Pond	Zone 1	3.9–41.7	37.8–53.8	7.4–58.3
	(0–2.6 m)	(av. 24.4)	(av. 46.8)	(av. 28.7)
	Zone 2	40.2–49.7	41.7–50.8	1.2–17.4
	(2.6–6.1 m)	(av. 43.5)	(av. 47.2)	(av. 9.2)
	Zone 3	38.4–57.4	38.1–48.6	4.1–16.0
	(6.1–9.0 m)	(av. 45.7)	(av. 44.9)	(av. 9.4)

12.4.3 Column Experiments

The results of column leaching experiment conducted with paddy field sediments are discussed under the sub-heading paddy profile and jute decomposing ponds under the sub-heading pond profile.

(i) *Paddy Profile*

In column leaching experiment conducted with the Zone 1 sediments, for first 4 days higher fluctuation in As concentration has been registered and from 5th to 9th day a continuous increase in As concentration in the leachates has been noticed. However, As shows quite a constant mobilization from 10th day till the end of the experiment (Fig. 12.4a).

In Zone 1 sediments, around 80–90 % of the As is associated with mobile, loosely bound, organically bound and Fe-Mn bound fractions (Farooq et al. 2010). In first few days, mobile and loosely bound As fractions were mobilized while with the development of stronger reducing condition As bound to Fe-Mn phases also mobilized. Once the reducing condition developed appropriately the As mobilization became stable and continued with almost the same rate throughout the experiment. Thus, from 10th day onwards till the end of the experiment not much variation in As concentration was found. The reducing conditions in this column (1AP) generated mainly due to decay of a fraction of DOC used as leaching solution. Around 80 % of the DOC gets consumed in this column and the leachates collected at the end of this column contained only 20 % DOC, which is used as leaching solution for column 2AP. As compared to column 1AP, in column 1BP—where the sediments are leached with water—do not show much As mobilization. This suggests that (i) DOC can mobilize significant quantity of As from sediments and, (ii) internal organic carbon (OC present in the sediments) is either in insufficient quantity or of refractory nature and do not play an important role in As mobilization. An external DOC source is thus responsible for As mobilization. Also, the concentration of OC present in sediments of both profiles is too low to induce the stronger reducing conditions on basin scale (Table 12.4).

12 Surface Generated Organic Matter

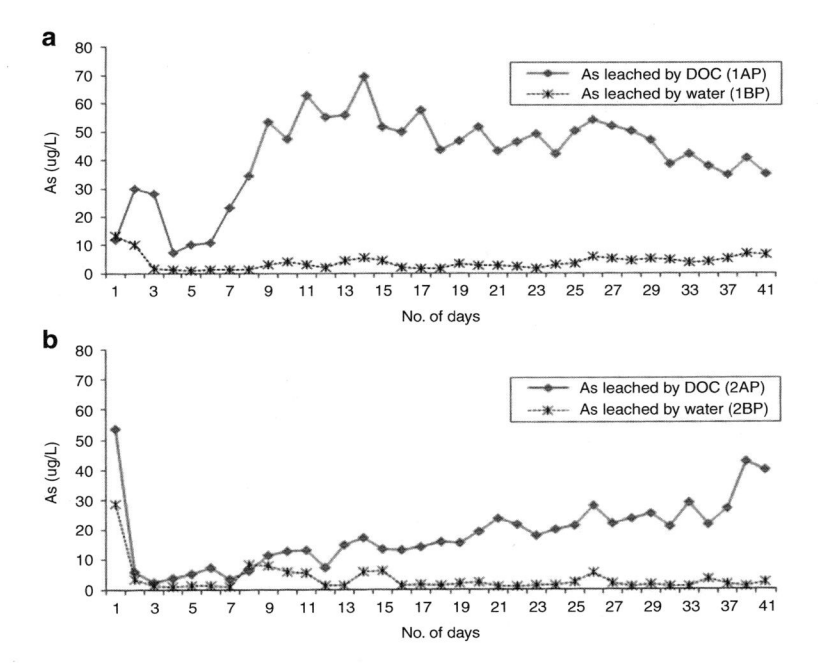

Fig. 12.4 (a) Arsenic concentrations in the leachates collected from the columns 1AP and 1BP. (b) Arsenic concentrations in the leachates collected from the columns 2AP and 2BP

Table 12.4 Distribution of carbon along the sediment profile

Site	Depth zone	Total carbon (%)	Inorganic carbon (%)	Organic carbon (%)
Paddy	Zone 1	0.6–2.5	0.4–1.6	0.1–1.0
	(0–3.2 m)	(av. 1.6)	(av. 1.3)	(av. 0.3)
	Zone 2	0.6–1.9	0.6–1.5	0.0–0.4
	(3.2–9.0 m)	(av. 1.1)	(av. 0.9)	(av. 0.1)
Pond	Zone 1	1.0–1.7	0.4–1.5	0.1–0.7
	(0–2.6 m)	(av. 1.4)	(av. 1.2)	(av. 0.2)
	Zone 2	0.8–2.6	0.3–1.6	0.0–1.2
	(2.6–6.1 m)	(av. 1.6)	(av. 1.3)	(av. 0.4)
	Zone 3	1.6–1.9	1.4–1.6	0.2–0.6
	(6.1–9.0 m)	(av. 1.8)	(av. 1.5)	(av. 0.3)

The leachates collected at the end of column 2A show a continuous increase in As concentration till the end of the experiment (Fig. 12.4b). However, if we look closely and consider the As concentration in the solution which is leaching column 2AP (i.e. leachates of column 1AP), it will be realized that actually column 2AP is showing a considerable As adsorption. On an average $\approx 55\%$ of As which is released from column 1AP and entered into column 2AP along with leaching solution is getting

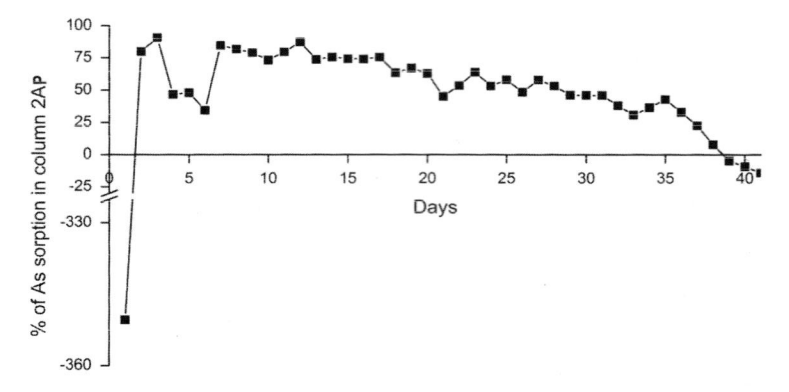

Fig. 12.5 Percentage of As sorption in column 2AP

adsorbed in this column (Fig. 12.5). However, as the experiment proceeded, the percentage adsorption of As in column 2AP decreased with time and from 39th day onwards the As adsorbed earlier in column 2AP started to remobilize. Most of the DOC of leaching solution consumed in column 1AP and very little DOC (20 %) could reach column 2AP to leach the sediments of Zone 2. Under such circumstances, it is reasonable to assume that the conditions in column 2A change from oxic in the start to mildly reducing as the experiment proceeds (i.e. strong reducing conditions could not develop). Under oxic condition Fe-oxide coating forms on the surface of sand grain which can trap As very effectively.

Thus, in general, column 2AP acted as sorption media for the As leached from column 1AP. However, with progressive occupation of available binding sites and development of reducing condition, adsorption of As became lesser and lesser and at the end of the experiment the As absorbed earlier again started to remobilize. Thus, it can be assumed that in future, if stronger reducing conditions are developed, more As will be mobilized and will reach the groundwater table.

(ii) *Pond Profile*

The column experiment run over 42 days shows that significant quantity of As mobilized from column 1AJ, notably during the first 2 weeks of the experiment (Fig. 12.6a). Studies have shown that in the sediments of Zone 1 up to 60 % of total As is in loosely bound form, associated with easily mobilizable and organically bound phases (Farooq et al. 2012). Consequently, it is reasonable to assume that As mobilized from this column is derived mainly from these phases. Gradually decreasing As concentration in leachates again suggests the consumption of easily mobilized As with time.

In the sediments of Zone 2 a significant fraction of As is attached to Mn oxides and amorphous Fe oxide phases, which are highly redox sensitive (Farooq et al. 2012). The lagging of the maximum As concentration in column 2AJ behind that in 1A is due to the development of progressively stronger reducing conditions in column 2A over the duration of experiment, which leads to a higher mobilization of As

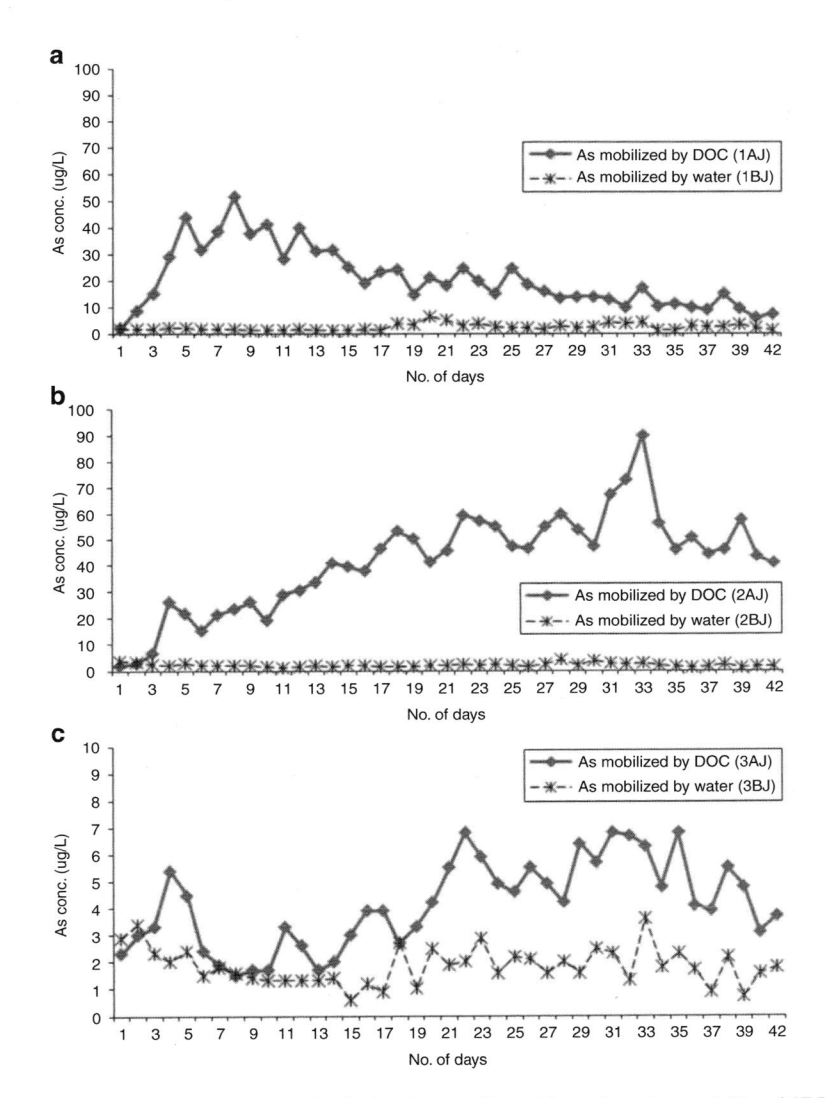

Fig. 12.6 (**a**) Arsenic concentrations in the leachates collected from the columns 1AJ and 1BJ. (**b**) Arsenic concentrations in the leachates collected from the columns 2AJ and 2BJ. (**c**) Arsenic concentrations in the leachates collected from the columns 3AJ and 3BJ

from Zone 2 sediments (Fig. 12.6b). Further, a weaker relationship of As between Fe for first few days of experiment and a stronger correlation at the later stage of experiment hints the redox driven As mobilization.

The As content in the sediment of third column (derived from Zone 3 of the core) seems not to be affected much by the percolating solution neither in column 3AJ nor in 3BJ (Fig. 12.6c). The solution that is entering not only in column 3BJ, but also in column 3AJ contains much too low amounts of DOC. The average DOC concentration in solution entering and exiting column 3AJ are 39.7 mg/L and 11.0 mg/L, respectively, which means effectively 28.7 mg/L of DOC is getting consumed in this column. Such a low DOC content may not mobilize significant quantity of arsenic. Even if, there may be some As mobilization in the upper part of this column, it must have been attenuated in the lower part of the same column as enough binding sites are available. Also, Zone 3 sediments contain very high clay content which may further increase the possibility of As re-adsorption. Thus, the water percolating through the jute decomposing pond does not pose an immediate threat of groundwater contamination but over the time if the DOC production in ponds continued Zone 2 will keep in shifting downward.

12.5 Conclusions

In Bengal delta, huge quantities of organic material remain available on the surface due to traditional practice of paddy cultivation and retting of jute plants to obtain jute fibre. Paddy fields are irrigated by As contaminated water, this causes accumulation of As in paddy fields. The decay of organic matter in paddy field triggers a series of reactions that help in mobilizing the As accumulated in the top sediments. The column experiment with paddy field sediments clearly demonstrate that the As mobilized from the top sediments ultimately reaches groundwater table and contaminates shallow groundwater resources. However, in case of jute decomposing ponds the As mobilized from upper horizon (0–2.6 m, Zone 1), subsequently gets fixed in the horizon below (2.6–6.1 m, Zone 2), and the lowest horizon (6.1–9 m, Zone 3) remains unaffected. Though, significant As mobilization takes place from upper horizon but still water reaching the groundwater table contains As lesser than the WHO prescribed limit of As in drinking water. Thus, it does not pose any imminent threat to the groundwater resources. However, the column experiment further demonstrates that if the DOC production in ponds continues for a longer time, the As fixation horizon will move further downward and ultimately will intersect the groundwater table.

In general, due to the availability of large quantities of organic matter on the surface of Bengal delta, groundwater resources are either getting contaminated (as in case of paddy fields) or posing a future threat (as in case of jute decomposing ponds). To protect the groundwater resources, it is advisable to reduce the availability of organic carbon by changing the cultivation practices and jute processing technique.

Acknowledgements The authors gratefully acknowledge support from German Academic Exchange Programme (DAAD) through research fellowship. Indian Institute of Technology Bombay (India), and Institute of Mineralogy and Geochemistry (IMG), Karlsruhe Institute of Technology (Germany) are thanked for providing laboratory facility to carry out this research work. Elsevier is thanked for permitting the use of few figures and excerpts from the work published by the author. Kaynat Tabassum and Hemant Kr. Singh are thanked for their valuable suggestions and never ending support.

References

Acharyya SK, Chakraborty P, Lahiri S, Raymahashay BC, Guha S, Bhowmik A (1999) Arsenic poisoning in the Ganges delta. Nature 401:545

Acharyya SK, Shah BA, Ashyiya ID, Pandey Y (2005) Arsenic contamination in groundwater from parts of Ambagarh-Chowki block, Chhattisgarh, India: source and release mechanism. Environ Geol 49:148–158

Ahamed S, Kumar Sengupta M, Mukherjee A, Amir Hossain M, Das B et al (2006) Arsenic groundwater contamination and its health effects in the state of Uttar Pradesh (UP) in upper and middle Ganga plain, India: a severe danger. Sci Total Environ 370:310–322

Ahsan DA, DelValls TA, Blasco J (2009) Distribution of arsenic and trace metals in the floodplain agricultural soil of Bangladesh. Bull Environ Contam Toxicol 82:11–15

Bagla P, Kaiser J (1996) India's spreading health crisis draws global arsenic experts. Science 274:174–175

Bhattacharjee S, Chakravarty S, Maity S, Dureja V, Gupta KK (2005) Metal contents in the groundwater of Sahebgunj district, Jharkhand, India, with special reference to arsenic. Chemosphere 58:1203–1217

Bhattacharya P, Chatterjee D, Jacks G (1997) Occurrence of arsenic-contaminated groundwater in alluvial aquifers from delta plains, eastern India: options for safe drinking water supply. Int J Water Resour Dev 13:79–92

Bhattacharya P, Jacks G, Ahmed KM, Routh J, Khan AA (2002) Arsenic in groundwater of the Bengal Delta Plain aquifers in Bangladesh. Bull Environ Contam Toxicol 69:538–545

Chakraborti D, Biswas BK, Roy Chowdhury T, Basu GK, Mandal BK et al (1999) Arsenic groundwater contamination and sufferings of people in Rajnandangaon, Madhya Pradesh. India Curr Sci 77:502–504

Chakraborti D, Mukherjee SC, Pati S, Sengupta MK, Rahman MM et al (2003) Arsenic groundwater contamination in Middle Ganga Plain, Bihar, India: a future danger? Environ Health Perspect 111:1194–1201

Chandrasekharam D (2008) Groundwater for sustainable development: problems, perspectives and challenges. Taylor and Francis, London

Chandrasekharam D, Karmakar J, Berner Z, Stuben D (2001) Arsenic contamination in groundwater, Murshidabad district, West Bengal. Proc Water Rock Interact 12:1051–1054

Das D, Samanta G, Mandal B, Roy Chowdhury T, Chanda C et al (1996) Arsenic in groundwater in six districts of West Bengal, India. Environ Geochem Health 18:5–15

Dittmar J, Voegelin A, Roberts LC, Hug SJ, Saha GC et al (2007) Spatial distribution and temporal variability of arsenic in irrigated rice fields in Bangladesh. 2. Paddy soil. Environ Sci Technol 41:5967–5972

Farooq SH, Chandrasekharam D, Berner Z, Norra S, Stüben D (2010) Influence of traditional agricultural practices on mobilization of arsenic from sediments to groundwater in Bengal delta. Water Res 44:5575–5588

Farooq SH, Chandrasekharam D, Abbt-Braun G, Berner Z, Norra S, Stueben D (2012) Dissolved organic carbon from the traditional jute processing technique and its potential influence on arsenic enrichment in the Bengal Delta. Appl Geochem 27:292–303

Islam FS, Gault AG, Boothman C, Polya DA, Chamok JM et al (2004) Role of metal-reducing bacteria in arsenic release from Bengal delta sediments. Nature 430:68–71

Mallick S, Rajagopal NR (1996) Groundwater development in the arsenic-affected alluvial belt of West Bengal – some questions. Curr Sci 70:956–958

Mathan KK, Natesan R (1988) Root distribution characteristics of paddy (IR 60) under mechanical impedance in wetlands. J Agron Crop Sci 161:300–304

McArthur JM, Ravenscroft P, Safiulla S, Thirlwall MF (2001) Arsenic in groundwater: testing pollution mechanisms for sedimentary aquifers in Bangladesh. Water Resour Res 37:109–117

McArthur JM, Banerjee DM, Hudson-Edwards KA, Mishra R, Purohit R et al (2004) Natural organic matter in sedimentary basins and its relation to arsenic in anoxic ground water: the example of West Bengal and its worldwide implications. Appl Geochem 19:1255–1293

Nickson R, McArthur J, Burgess W, Matin Ahmed K, Ravenscroft P, Rahman M (1998) Arsenic poisoning of Bangladesh groundwater. Nature 395:338

Nickson RT, McArthur JM, Ravenscroft P, Burgess WG, Ahmed KM (2000) Mechanism of arsenic release to groundwater, Bangladesh and West Bengal. Appl Geochem 15:403–413

Norra S, Berner ZA, Agarwala P, Wagner F, Chandrasekharam D, Stuben D (2005) Impact of irrigation with As rich groundwater on soil and crops: a geochemical case study in West Bengal Delta Plain, India. Appl Geochem 20:1890–1906

Ona-Nguema G, Morin G, Juillot F, Calas G, Brown GE (2005) EXAFS analysis of arsenite adsorption onto two-line ferrihydrite, hematite, goethite, and lepidocrocite. Environ Sci Technol 39:9147–9155

Otte ML, Kearns CC, Doyle MO (1995) Accumulation of arsenic and zinc in the rhizosphere of wetland plants. Bull Environ Contam Toxicol 55:154–161

Paul AB, Kar D (2004) Ground water arsenic in Assam: a report from Brahmaputra and Barak valley. Environ Ecol 22:588–589

Ravenscroft P, McArthur JM, Hoque BA (2001) Geochemical and palaeohydrological controls on pollution of groundwater by arsenic. Arsen Expos Health Effects IV:53–77

Rowland HAL, Polya DA, Lloyd JR, Pancost RD (2006) Characterisation of organic matter in a shallow, reducing, arsenic-rich aquifer, West Bengal. Org Geochem 37:1101–1114

Smedley PL, Kinniburgh DG (2002) A review of the source, behaviour and distribution of arsenic in natural waters. Appl Geochem 17:517–568

Stüben D, Berner Z, Chandrasekharam D, Karmakar J (2003) Arsenic enrichment in groundwater of West Bengal, India: geochemical evidence for mobilization of As under reducing conditions. Appl Geochem 18:1417–1434

Sun G (2004) Arsenic contamination and arsenicosis in China. Toxicol Appl Pharmacol 198:268–271

van Geen A, Zheng Y, Cheng Z, Aziz Z, Horneman A et al (2006) A transect of groundwater and sediment properties in Araihazar, Bangladesh: further evidence of decoupling between As and Fe mobilization. Chem Geol 228:85–96

van Geen A, Radloff K, Aziz Z, Cheng Z, Huq MR et al (2008) Comparison of arsenic concentrations in simultaneously-collected groundwater and aquifer particles from Bangladesh, India, Vietnam, and Nepal. Appl Geochem 23:3244–3251

WHO (2006) Guidelines for drinking-water quality, Recommendations. World Health Organization, Geneva

WHO (2011) Guidelines for drinking-water quality, 4th edn. World Health Organization, Geneva

Chapter 13
A Comparative Study on the Arsenic Levels in Groundwaters of Gangetic Alluvium and Coastal Aquifers in India

S. Chidambaram, R. Thilagavathi, C. Thivya, M.V. Prasanna, N. Ganesh, and U. Karmegam

13.1 Introduction

The elevated levels of arsenic in water threaten human health in large areas globally and its origin is intensively discussed by researchers worldwide. There remains considerable debate about the key control of arsenic concentrations in water. No specific sources of arsenic could be identified in Gangetic plain of Uttar Pradesh, India but several potential minor sources have been identified both in the Himalayan belt as well as, in peninsular India. Gangetic plain was formed due to the accumulation of bulk sediments from the Himalayan hill range, whereas the input of peninsular India is minor. The southern belt of the Himalayas is subjected to high erosion and intense rainfall during the Holocene time. The possibility of erosion, oxidation and transportation of arsenic-bearing products in suspension and solution in the Gangetic plain is high. Studies in Gangetic alluvium and few other reports in India are listed in Table 13.1. The current status of knowledge of various workers (Bhattacharya et al. 1997; Nickson et al. 2000; Ahmed et al. 2004; Singh et al. 2010; Kumar et al. 2010a, b) indicates predominantly Geogenic source (arsenic sediments deposited in the Ganga plains by Himalayan Rivers) and its release in groundwater through natural processes. As per the study, from the data it is seen that arsenic present in the groundwater sample of Tamil Nadu within the safe limit of 0.05 mg/L (50 µg/L) as fixed by the Bureau of Indian Standards for the drinking water. However the highest amount of arsenic 8.62, 16.67 and 11.5 µg/L have been noticed in the districts of

---

S. Chidambaram (✉) • R. Thilagavathi • C. Thivya • N. Ganesh • U. Karmegam
Department of Earth Sciences, Annamalai University,
Annamalai Nagar, Chidambaram 608002, Tamil Nadu, India
e-mail: chidambaram_s@rediffmail.com

M.V. Prasanna
Department of Applied Geology, School of Engineering and Science, Curtin University, CDT 250, 98009 Miri, Sarawak, Malaysia

Table 13.1 Few studies on arsenic in groundwaters of Gangetic alluvium and that of Indian aquifers

S. no.	Study details	Author	Year
1.	Geochemistries of arsenic, antimony, mercury and related elements in sediments of Puget sound	Crecerius et al.	1975
2.	Stable isotope study of the Ganga (Ganges) river system	Ramesh and Sarin	1992
3.	Occurrence of arsenic contamination of groundwater in alluvial aquifers from Delta Plain, eastern India: Option for safe drinking supply	Bhattacharya et al.	1997
4.	Arsenic levels in drinking water and the prevalence of skin lesions in West Bengal, India	Mazumder et al.	1998
5.	Arsenic poisoning of groundwater in Bangladesh	Nickson et al.	1998
6.	Groundwater arsenic in the Holocene alluvial aquifers of Bengal Delta Plains: Petrological, geochemical and isotope geochemical studies	Bhattacharya et al.	1999
7.	Contamination of drinking water by arsenic in Bangladesh: A public health emergency	Smith et al.	2000
8.	A report on isotope hydrology of groundwater in Bangladesh: Implications for characterization and mitigation of arsenic in groundwater	Aggarwal et al.	2000
9.	Arsenic in groundwater and health problems in Bangladesh	Karim	2000
10.	Groundwater arsenic-contamination status at four geo-morphological areas in Bangladesh (special reference to arsenic in biological samples and agricultural crop)	Chowdhury	2001
11.	Arsenic in groundwater in the Bengal Delta Plain: Slow poisoning in Bangladesh	Mukherjee and Bhattacharya	2001
12.	Arsenic contamination of ground and pond water and water purification system using pond water in Bangladesh	Yokota et al.	2001
13.	Arsenic in groundwater: Testing pollution mechanisms for sedimentary aquifers in Bangladesh	McArthur et al.	2001
14.	Arsenic contamination of groundwater in Bangladesh	Kinniburgh and Smedley	2001
15.	Arsenic mobility and groundwater extraction in Bangladesh	Harvey et al.	2002
16.	Arsenic groundwater contamination in Middle Ganga Plain, Bihar, India: A future danger?	Chakraborti et al.	2003
17.	Arsenic enrichment in groundwater of West Bengal, India: Geochemical evidence for mobilization of As under reducing conditions	Stuben et al.	2003
18.	Spatial variability of arsenic in 6000 tube wells in a 25 km^2 area of Bangladesh	Van Geen et al.	2003
19.	Antimony isotope variations in natural systems and implications for their use as geochemical tracers	Rouxel et al.	2003

(continued)

Table 13.1 (continued)

S. no.	Study details	Author	Year
20.	Arsenic levels in groundwater from quaternary alluvium in the Ganga Plain and the Bengal basin, Indian subcontinent: Insights into influence of stratigraphy	Acharyya	2005
21.	Risk of arsenic contamination in groundwater affecting Ganga alluvial Plain, India?	Acharyya and Shah	2004
22.	Arsenic enrichment in groundwater of the alluvial aquifers in Bangladesh: An overview	Ahmed et al.	2004
23.	Groundwater arsenic contamination and its health effects in the Ganga–Meghna–Brahmaputra plain	Chakraborti et al.	2004
24.	Mechanism and source of arsenic contamination in groundwater in Bangladesh	Takaaki and Takahashi	2005
25.	Groundwater arsenic in the Central Gangetic Plain in Ballia District of Uttar Pradesh, India	Tripathi et al.	2006
26.	Groundwater dynamics and arsenic mobilization in Bangladesh assessed using noble gases and tritium	Klump et al.	2006
27.	Deeper groundwater flow and chemistry in the arsenic affected western Bengal basin, West Bengal, India	Mukherjee	2006
28.	Comparison of Antimony Behavior with that of Arsenic under Various Soil Redox Conditions	Mitsunobu et al.	2006
29.	Response to comments on "Limited temporal variability of Arsenic concentration in 20 wells monitored for 3 years in Araihazar, Bangladesh"	Cheng et al.	2006
30.	Ground water Arsenic contamination and its health effects: Case studies from India and SE Asia	Ramanathan et al.	2007
31.	Regional-scale stable isotopic signatures of recharge and deep groundwater in the arsenic affected areas of West Bengal, India	Mukherjee et al.	2007
32.	Hydrological and geochemical constraints on the mechanism of formation of Arsenic contaminated groundwater in Sonargaon, Bangladesh	Itai et al.	2008
33.	A study of arsenic, iron and other dissolved ion variations in the groundwater of Bishnupur District, Manipur, India	Oinam et al.	2011
34.	Tracing the factors responsible for arsenic enrichment in groundwater of the middle Gangetic Plain, India: A source identification perspective	Kumar et al.	2010b
35.	Arsenic enrichment in groundwater in the middle Gangetic Plain of Ghazipur District in Uttar Pradesh	Kumar et al.	2010a
36.	Impact of groundwater abstraction and of the organic matter on release and distribution of arsenic in aquifers of the Bengal Delta Plain, India	Neidhardt	2012a
37.	Arsenic in groundwater of West Bengal: Implications from a field study	Neidhardt	2012b
38.	Surface complexation modeling of temporal variability of arsenic in groundwater: Estimating the role of competing ions in the mobilization processes	Biswas et al.	2013

(continued)

Table 13.1 (continued)

S. no.	Study details	Author	Year
39.	Reconstructing the sedimentation history of the Bengal Delta Plain by means of geochemical and stable isotopic data	Neidhardt et al.	2013b
40.	Influences of groundwater extraction on the distribution of dissolved As in shallow aquifers of West Bengal, India	Neidhardt et al.	2013a
41.	Role of competing ions in the mobilization of arsenic in groundwater of Bengal Basin: Insight from surface complexation modeling	Biswas et al.	2014c
42.	Spatial, vertical and temporal variation of arsenic in the shallow aquifers of Bengal Basin: Controlling geochemical processes	Biswas et al.	2014b
43.	Characterizing the controlling geochemical processes for spatial, vertical and temporal variation of arsenic in shallow groundwater of the Bengal basin	Biswas et al.	2014a

Kancheepuram, Perambalur and Virudhunagar respectively. In Theni and Erode district the lowest amount of Arsenic 1.00 μg/L have been noticed. Even for the geological sources, there are two explanations—one theory relies on the oxidation (addition of oxygen) and the other on reduction (loss of oxygen). These studies indicate the identification sources of As and their geochemical process, evolution of As, groundwater contamination and their effects on human people. It also focuses on human health, living aspects and prevention of polluting agents of arsenic to the groundwater. The present study indicates the origin and occurrence to identifying the more concentrated areas located with arsenic enrichment and mobilization. This paper attempted to obtain a comprehensive idea about the arsenic in Gangetic alluvium and a coastal hard rock aquifer in south India.

13.2 Study Area

13.2.1 Gangetic Alluvium

The Indo-Gangetic plain occurs on the convex side of the Himalayan and Burmese mountain arc and this plain parallels the Himalayan front for over 2000 km. The sediments of Indo-Gangetic plain consist of alluvial material in layers of sands and clays, with a depth of 3,000 m in the northwest (Punjab plains) to over 6,700 m in the central part (U.P. and Bihar). These sediment layers lie over the older formations of the Indian shield, and have deposited in the fore deep of Himalayas and this fore deep itself includes a frontal belt and a shelf zone which were created during the different phases of mountain building activities in the Himalayas. The basement to the fore deep is an extension of the Indian shield as depicted by three fault control ridges below the sediments, namely (i) the Delhi-Hardwar ridge which is a continuation of the north-northwest Aravalli range, (ii) the Faizabad ridge which is a northeast extension of the Bundelkhand massif, and (iii) the Mungher-Saharsa ridge which is an extension of the Satpura belt. The area is drained by many major rivers, served by a large number of streams. Since the alluvium conceals the geology of its floor, it is of interest mainly on account of its tectonic and hydrological parameters, particularly since the entire area is highly populated.

13.2.2 Southern Coast

Kalpakkam region located in northern part of Tamil Nadu State, India between 12° 37′ and 12° 25′ North latitudes, 80° 00′ and 80° 12′ East longitudes (Fig. 13.1). It comprises both hard rock and sedimentary formation covering a total area of 385

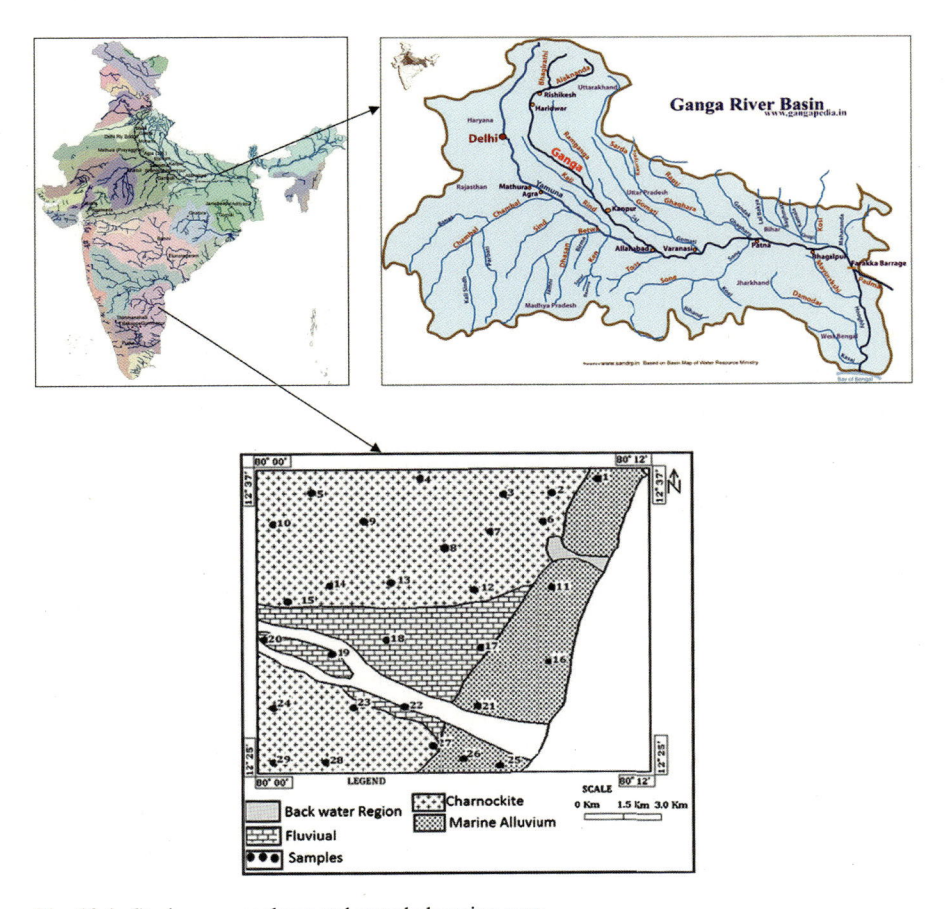

Fig. 13.1 Study area, geology and sample location map

sq.km. The study area lies between the back water from Bay of Bengal in the east and the landfill sites of Mahabalipuram town in NE part. River Palar, a major river course which drains this region, originates from Western Ghats in Karnataka state, and discharges in Bay of Bengal near Pudupattinam. The study area reveals that there are two major lithologies, i.e., charnockite (hard rock) and alluvium formation (fluvial and marine). The Palar River flows from the northwest to southeastern part of the study area fringed by this Quaternary alluvium (flood plain). The Quaternary formation is found along the river course of the Palar and it spatially divides the charnockites into northern and southern parts (Fig. 13.1). Marine alluvial formation is found parallel to the coast. The subsurface lithology of the study area shows different types of formations which include sand and clayey sand with weathered charnockite in west. Archaean basement at the bottom, which is made up of charnockites and is overlain by recent alluvium (Singh et al. 2004; Singh and Saxena 2004).

The depth of hard rock varies from 12.9 to 46.0 m below the ground surface. The bedrock is shallow on the northern and western sides and deeper in the central region. The weathered/fractured charnockite and alluvium form the major aquifer system. The thickness of alluvium is more on the southern and eastern sides, and lenses of clays were encountered in alluvial formations (Karmegam et al. 2010). Sand formations vary from 3 to 12 m in thickness and constitute a shallow unconfined aquifer. The rock mass consists of quartz, feldspar, biotite and pyroxene. Fractures in the charnockite rock mass consist of fracture filling material, which is greenish/blackish green in colour and is predominantly composed of clay minerals biotite, sericite and chlorite, altered feldspar and quartz. Dug wells tap the alluvium with depths ranging between 6 and 12 m bgl (below ground level) and that of fissured crystalline rock it ranges in depth between 6 and 17 m bgl. The depth to water level ranged from 3.50 to 8.34 m bgl during summer and 1.32–7.53 m bgl post monsoon.

13.3 Distribution of Arsenic

Arsenic was released later to groundwater mainly by reductive dissolution of hydrated iron-oxide and corresponding oxidation of sediment organic matter. Strong reducing nature of groundwater in the Bengal Basin and parts of affected middle Ganga floodplains is indicated by high concentration of dissolved iron (maximum 9–35 mg/l). The maximum concentration of arsenic in groundwater of Bihar state is shown in Fig. 13.2. It is worth to notice that the higher concentration of arsenic is observed in southern Gangetic alluvium compared to northern plain. The classification of hydrogeochemical facies of groundwaters indicates that the dominant hydrochemical facies of the study area is Ca-Mg -HCO_3.

The low concentration of HCO_3 between 10 and 20 m depth corresponds to the most important anion species, which competes with As for adsorption sites at mineral surface (e.g. Fe/Mn oxyhydroxides, clay minerals and weathered mica), consequently releasing As into the groundwater (Charlet et al. 2007). Groundwater

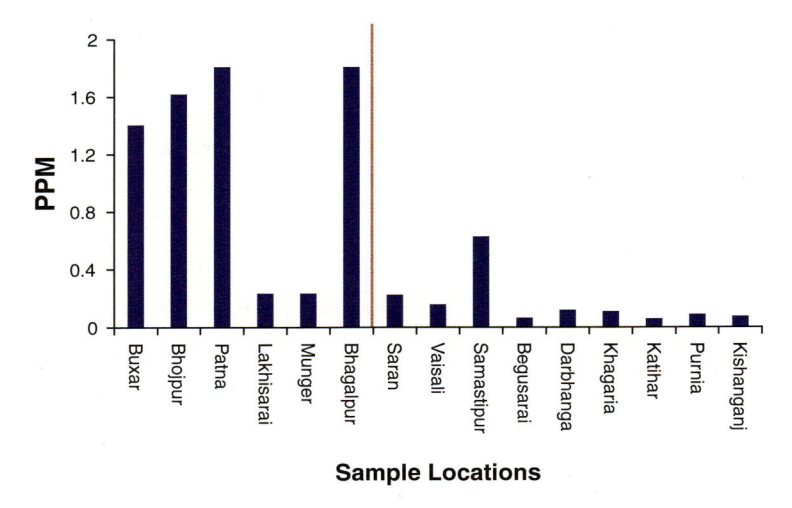

Fig. 13.2 Maximum arsenic concentration in groundwater in affected districts of Bihar state. Arsenic concentration is expressed in mg/l. The districts on the left side of the *red line* are on the southern bank of River Ganga and remaining districts are from the northern plain

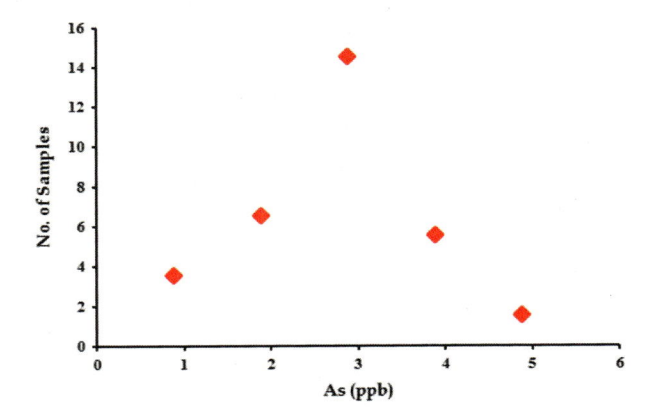

Fig. 13.3 Frequency chart for ranges as (µg/l)

being virtually stagnant under these settings, released arsenic accumulates and contaminates groundwater. The distribution of arsenic in the southern coastal aquifer shows that concentration in groundwater shows that it ranges from 0.84 to 8.66 ppb (Fig. 13.3). Higher number of samples fall in following order 4–6 ppb (14), followed by 2–4 ppb (7), by 6–8 ppb (5), by 0–2 ppb category and finally 8–10 ppb category. Though all the samples fall within the permissible limit, there is a fear for increase of concentration in near future. Na-Mg-Cl-HCO$_3$ (Type I) and Mg-Na-Cl-HCO3 (Type II) are the dominant water type of the study area. Majority of the samples from the charnockite shows the type II and higher As is mostly distributed in type I.

13.4 Depth

Hydrogeochemical comparison of groundwater chemistry in geologically divergent sub-basins near the western margin (Nadia, Bhagirathi sub-basin, West Bengal) was done by Mukherjee et al. (2008) to understand the controls of As occurrence in groundwater.

Results show that groundwater in the western site is mostly Ca^- and HCO_3 facies, while that in the eastern sub-basin consists of six different facies. Depth profiles of major solutes (Na, Ca and HCO_3) show different trends in the sub-basins; however, redox-sensitive solutes in both areas tend to behave similarly, strongly decreasing in concentration with depth, which suggests similarity in the redox processes along vertical flow paths.

Depth information of 68 tube wells in Uttar Pradesh indicates that 85 % of tube wells are from shallow depth (10–42 m) 51 tube wells have arsenic 40 µg/l within the depth of 10–54 m and 17 tube wells have arsenic 50 µg/l within the depth of 30–50 m. However, most of the arsenic contaminated tube wells occur within the depth of 25–45 m. Maximum value of arsenic (180 µg/l) corresponds to a depth of 42 m.

In Middle Ganga plain of Bihar, the area is underlain by a multi-layer sequence of sand (aquifer) alternating with aquitards like sandy-clay and clay, down to depth of 300 m. In an affected village Bariswan, an aquifer specific groundwater analysis revealed a rapid decline in arsenic load with depth, from 0.095 mg/l at 19 m to 0.006 mg/l at 194 m below ground. The upper part of the shallow aquifer (within ~50 m below ground) is affected by groundwater arsenic contamination. The low-land Terai belt in Nepal, where groundwater is an important source for agriculture and drinking, recorded high load of arsenic (0.02–2.6 mg/l) in shallow tube wells, where cases of arsenicosis have also been reported.

The shallow ground water of the southern coastal hard rock aquifer reveals the presence of oxidizing conditions and infers that they are frequently flushed by the rain water/river water along the flood plains during the monsoonal period, which may also dilute or reduce the concentrations. Moreover the hard rock aquifer has a high residence time and variation in temperature will also induce the release of arsenic (Molerio León and Toujague de la Rosa 2004). The control of temperature is more clearly because of its effect on the solubility patterns and the oxidation reduction potential of the system, in particular, to the development of aerobic or anaerobic conditions in the sampling station. The range of metal concentrations of the groundwater in the region are provided in the Table 13.2.

13.5 Variation in Source of Arsenic

Groundwater from the upland terraces in the Bengal Basin and in the Central Ganga Alluvial Plain, made up of the Pleistocene sediments are free of arsenic contamination. These sediments are weakly oxidized in nature and associated groundwater is

Table 13.2 Maximum, minimum and average values of heavy minerals in groundwater samples (all values in µg/l)

	Fe	Mn	Cu	Ni	Cr	Pb	Cd	Zn	As	Be	Al	Se	Mo	Sb	Ba
Maximum	7799.74	708.28	265.5	236.06	50.62	205.76	8.48	90023.16	8.66	0.98	6196.12	5900	25.78	1.98	8119.66
Minimum	146.92	6.94	2.06	3.74	12.16	1.32	0.14	415.08	0.84	0.02	62.62	0.18	0.52	0.14	88.44
Average	2300.90	372.54	124.69	80.84	25.60	90.57	2.71	22139.19	4.71	0.38	1837.67	389.11	5.53	0.63	2240.25

mildly reducing in general with low concentration of iron (<1 mg/l), and thus incapable to release arsenic. These sediments are also flushed free of arsenic, released if any, by groundwater flow due to high hydraulic head, because of their initial low-stand setting and later upland terraced position. Arsenic mobilization is also controlled by redox-dependent reactions (Mukherjee et al. 2008). Saturation index calculations for groundwater in West Bengal is generally near equilibrium with respect to carbonate phases such as calcite ($CaCO_3$) and dolomite [$MgCa(CO_3)$], suggesting that the high HCO_3^- values in Bengal Basin groundwater cannot be totally attributed to carbonate dissolution and may have a significant input from natural organic matter (NOM) oxidation (Ahmed et al. 2004). Arsenic is found along the landfill sites and in the south western part of the hard rock aquifer in the southern coastal region. A mechanism which could operate on a local scale is ion exchange of the adsorbed arsenic species by phosphate acquired in the aquifer from fertilizer application of ferrihydrite, goethite (Cornell and Schwertmann 1996) in charnockite and clay minerals (Reymond et al. 2001; Pfeifer et al. 2001) or decay of organic matter (Raymahashay and Khare 2003). Arsenic is not contained in primary sulfide minerals in the study area, but is adsorbed onto secondary alteration solids such as Fe-oxy-hydroxides. The presence of apatite in the charnockites of this type is reported by few authors (Chidambaram 2000; Ramanathan 1956). The apatites are also believed to be a source of arsenic (Stewart 1963).

13.6 Statistics

Liu et al. (2003) applied factor analysis to assess the quality of ground water in Yun-Lin county in Taiwan, and stated that the arsenic pollutant factors included As, total organic carbon, and alkalinity, which were all significantly positively correlated to each other. Correlation matrix and factor analysis were used to identify various factors influencing the gradual As enrichment in the middle Gangetic Plain. The poor relationship between As and Fe indicates that the As release into the groundwater depends on several processes such as mineral weathering, O_2 consumption, and NO_3 reduction and is de-coupled from Fe cycling (Kumar et al. 2010b). A negative correlation between arsenic and iron content in three blocks of Assam is observed with correlation coefficient of $r=0.168$ found to be non-significant (Mridulchetia et al. 2008) and poor correlation between As and Fe is observed in Manipur valley (Chakraborti et al. 2008). The depth profile of As revealed that low concentrations of NO_3 are associated with high concentration of As and that As depleted with increasing depth.

Results of correlation analyses indicate that arsenic contamination is strongly associated with high concentrations of Fe, PO_4 and NH_4 but relatively low Mn concentrations. Further, the enrichment of arsenic is more prevalent in the proximity of the Ganges River, indicating that fluvial input is the main source of arsenic (Kumar et al. 2010a, b). There was an inverse correlation between arsenic and Eh (redox potential), i.e., the value of Eh increases with decreasing arsenic concentration, which means concentration of arsenic increases with increasing redox status of any environment. Further, significant positive correlation between As and Fe, NH$_4$ and PO$_4$ were also noticed along with negative correlation between As and Mn, which substantiate the strong reducing character of this environment. In the post-monsoon season, Factors 1, 2, 3, 4 and 5 represent NO$_3$, SO$_4$, Cl and Na; HCO$_3$, Ca and Mg; pH and Fe; Cl and As; and F and PO$_4$ respectively. Factor 4 indicating redox sensitive Fe enrichment in the alkaline regime also reveals an association of As with Cl indicating the anthropogenic contributions, mainly through herbicide application in the agriculture fields. Moreover, Ravenscroft et al. (2001) studied the limiting depth of active groundwater circulation with anthropogenic influence on Cl and As concentrations, which may be controlled by the subdued topography and occurrence of a silty clay layer in shallow well in Gangetic basin.

The factor analysis and correlation matrix of physico-chemical parameters reveals the association of As with pH (pre-monsoon) and Cl (post-monsoon). Ghazipur district experiences frequent floods caused by the south-west monsoon generating a channelized natural water system, which accelerates the sediment load, weathering and other hydrogeochemical processes, consequently triggering the accumulation of As in the study area (Kumar et al. 2010a). Factor 1 exhibits high loading, i.e., strong geochemical associations between PO$_4$, Fe, As and As (III) and inverse association with depth, ORP and Mn in both seasons. Such loading for the most important factor i.e., component 1 is a strong indicator of a reducing environment which seems to be the main reason for arsenic enrichment in the groundwater (Kumar et al. 2010a, b).

In flood plain alluvium samples of Kalpakkam have the positive correlation Ca, Mg, Cl, NO$_3$, Cu, Ni, Cr, Pb and pH (Table 13.3). It is possible that the high concentration of nitrate in groundwaters result from excessive application of manure and inorganic fertilizer (Sharifi and Sinegani 2012). Chlorine (Cl), which is significantly higher in flood plain deposits, shows good positive relationship with arsenic. The land use pattern and its utilization (e.g. fertilizers, saltwater intrusion) can influence Cl concentrations, as Cl is quite mobile in the environment subject to leaching (Julie Wilson et al. 2008). Factor 3 represented by As, Ca, PO$_4$ and K indicates the significant contribution of fertilizers for the potential release of arsenic into the coastal groundwater.

13.7 Conclusion

This comparative study reveals that the upper part of the shallow aquifer is affected by groundwater arsenic contamination in middle Gangetic plain, whereas southern coastal shallow aquifer is frequently flushed by the rain water/river water along the flood plains which reduce the concentration. The major hydrochemical facies of groundwaters in gangetic alluvium is Ca$^-$-Mg$^-$-HCO$_3$, where the low concentration of HCO$_3$ competes with As for adsorption sites at mineral surface, consequently releasing As into the groundwater. In southern coastal aquifer, majority of the

Table 13.3 The correlation table of the heavy metal concentration in the groundwaters of the southern coastal aquifer

Parameters	pH	EC	Fe	Mn	Cu	Ni	Cr	Pb	Cd	Zn	As	Be	Al	Se	Mo	Sb	Ba
pH	1																
EC	0.33	1															
Fe	0.13	−0.15	1														
Mn	−0.1	−0.15	0.6	1													
Cu	−0.18	−0.2	0.16	0.38	1												
Ni	0.04	−0.2	0.95	0.53	0.24	1											
Cr	0.14	−0.18	0.97	0.56	0.27	0.97	1										
Pb	0.05	−0.25	0.92	0.55	0.35	0.97	0.98	1									
Cd	0.09	−0.16	0.93	0.6	0.35	0.96	0.95	0.94	1								
Zn	0.09	−0.18	0.97	0.56	0.15	0.99	0.97	0.95	0.96	1							
As	0.13	−0.14	0.42	0.2	0.49	0.46	0.53	0.55	0.39	0.38	1						
Be	0.18	−0.04	0.79	0.44	0.22	0.86	0.83	0.82	0.86	0.83	0.43	1					
Al	0.06	−0.2	0.94	0.55	0.2	0.99	0.96	0.96	0.97	0.99	0.4	0.87	1				
Se	0.24	0.15	−0.27	−0.25	−0.1	−0.24	−0.22	−0.21	−0.22	−0.23	−0.21	−0.3	−0.23	1			
Mo	0.33	0.13	−0.07	−0.16	0.06	−0.04	0.01	0.02	−0.04	−0.05	0.18	−0.04	−0.06	0.89	1		
Sb	0.07	−0.11	0.43	0.22	0.33	0.51	0.49	0.52	0.48	0.48	0.6	0.55	0.47	−0.02	0.36	1	
Ba	0.08	−0.18	0.96	0.54	0.16	0.99	0.97	0.96	0.96	1	0.39	0.85	0.99	−0.22	−0.05	0.45	1

samples from the charnockite shows Mg-Na-Cl-HCO$_3$ and higher As is mostly distributed in Na-Mg-Cl-HCO$_3$. The enrichment of arsenic is more prevalent in the proximity of the Ganges River, indicating that fluvial input is the main source of arsenic; whereas in Kalpakkam region, As is found along the landfill sites and in the south western part of the hard rock aquifer. Results of correlation analyses indicate that arsenic contamination is strongly associated with high concentrations of Fe, PO$_4$ and NH$_4$ but relatively low Mn concentrations in Gangetic plain. In coastal alluvium aquifer shows higher factor loading of As with Ca, Mg, Cl, NO$_3$ and pH which represents anthropogenic factor of fertilizer and their influence of landfill leachate.

References

Acharyya SK (2005) Arsenic levels in groundwater from quaternary alluvium in the Ganga Plain and the Bengal basin, Indian subcontinent: insights into influence of stratigraphy. Gondwana Res 8(1):55–66

Acharyya SK, Shah BA (2004) Risk of arsenic contamination in groundwater affecting Ganga alluvial plain, India? Environ Health Perspect 112:A19–A20

Aggarwal PK, Basu AR, Poreda RJ, Kulkarni KM, Froehlich K, Tarafdar SA, Ali M, Ahmed N, Hussain A, Rahman M, Ahmed SR (2000) A report on isotope hydrology of groundwater in Bangladesh: implications for characterization and mitigation of arsenic in groundwater. IAEA – TC Project (BGD/8/016)

Ahmed KM, Bhattacharya P, Hasan M, Akhter SH, Alam M, Bhuyian H, Imam MB, Khan AA, Sracek O (2004) Arsenic enrichment in groundwater of the alluvial aquifers in Bangladesh: an overview. Appl Geochem 19:181–200

Bhattacharya P, Chatterjee D, Jacks G (1997) Occurrence of arsenic contaminated groundwater in alluvial aquifers from Delta plains, eastern India: options for safe drinking water supply. Water Resour Dev 13(1):79–92

Bhattacharya P, Ahmed KM, Hasan MA (1999) Groundwater arsenic in the Holocene alluvial aquifers of Bengal Delta Plains: petrological, geochemical and isotope geochemical studies. In: Allauddin M (ed) Proceeding of international conference on arsenic in Bangladesh groundwater: world's greatest arsenic Calamity. Wagner College, New York

Biswas A, Gustafsson J, Neidhardt PH, Halder D, Kundu AK, Chatterjee D, Berner Z, Bhattacharya P (2013) Surface complexation modeling of temporal variability of arsenic in groundwater: estimating the role of competing ions in the mobilization processes. ICOBTE Abstract: 0220–000455

Biswas A, Neidhardt H, Kundu AK, Halder D, Chatterjee D, Berner Z, Jacks G, Bhattacharya P (2014a) Characterizing the controlling geochemical processes for spatial, vertical and temporal variation of arsenic in shallow groundwater of the Bengal Basin. Paper no. 283–12 2014 GSA annual meeting in Vancouver

Biswas A, Neidhardt H, Kundu AK, Halder D, Chatterjee D, Berner Z, Jacks G, Bhattacharya P (2014b) Spatial, vertical and temporal variation of arsenic in shallow aquifers of the Bengal Basin: controlling geochemical processes. Chem Geol 387:157–169

Biswas A, Gustafsson JP, Neidhardt H, Halder D, Kundu AK, Chatterjee D, Berner Z, Bhattacharya P (2014c) Role of competing ions on the mobilization of arsenic in groundwater of sedimentary aquifers: insight from surface complexation modeling. Water Res 15(55):30–39. doi: 10.1016/j.watres.2014.02.002

Chakraborti D, Mukherjee SC, Pati S, Sengupta MK, Rahman MM, Chowdhury UK, Lodh D, Chanda CR, Chakraborty AK, Basu GK (2003) Arsenic groundwater contamination in Middle Ganga Plain, Bihar, India: a future danger. Environ Health Perspect 111:1194–1201

Chakraborti D, Sengupta MK, Rahman MM, Ahamed S, Chowdhury UK, Mukherjee SC, Pati S, Saha KC, Dutta RN, Zaman QQ (2004) Groundwater arsenic contamination and its health effects in the Ganga–Meghna–Brahmaputra plain. J Environ Monit 6:74–83

Chakraborti D, Singh EJ, Das B, Shah BA, Hossain MA, Nayak B, Ahamed S, Singh NR (2008) Groundwater arsenic contamination in Manipur, one of the seven North Eastern Hill state in India: a future danger. Environ Geol 56:381–390

Charlet L, Chakraborty S, Appelo CAJ, Roman-Ross G, Nath B, Ansari AA, Lanson M, Chatterjee D, Mallik SB (2007) Chemodynamics of an arsenic "hotspot" in a West Bengal aquifer: a field and reactive transport modeling study. Appl Geochem 22:1273–1292

Cheng ZQ, van Geen A, Seddique AA, Ahmed KM (2006) Response to comments on limited temporal variability of arsenic concentration in 20 wells monitored for 3 years in Araihazar, Bangladesh. Environ Sci Technol 40:1718–1720

Chidambaram S (2000) Hydrogeochemical studies of groundwater in periyar district, Tamilnadu, India. Unpublished Ph.D. thesis, Department of Geology, Annamalai University, Chidambaram, Tamil Nadu

Chowdhury UK (2001) Groundwater arsenic-contamination status at four geo-morphological areas in Bangladesh (special reference to arsenic in biological samples and agricultural crop). Ph.D. dissertation, Jadavpur University, Kolkata

Cornell RM, Schwertmann U (1996) The iron oxides structure, properties, reactions, occurrence and uses. VCH Verlagsgesellschaft, Weinheim

Crecerius EA, Bothner MH, Carpenter R (1975) Geochemistries of arsenic, antimony, mercury, and related elements in sediments of Puget Sound. Environ Sci Technol 9:325–333

Harvey CF, Swartz CH, Badruzzaman AB, Keon-Blute N, Yu W, Ali MA, Jay J, Beckie R, Niedan V, Brabander D, Oates PM, Ashfaque KN, Islam S, Hemond HF, Ahmed MF (2002) Arsenic mobility & groundwater extraction in Bangladesh. Science 298(5598):1602–1606

Itai T, Masuda H, Seddique AA, Mitamura M, Maruoka T, Li X, Kusakabe M, Dipak BK, Farooqi A, Yamanaka T, Nakaya S, Matsuda J, Ahmed KM (2008) Hydrological and geochemical constraints on the mechanism of formation of arsenic contaminated groundwater in Sonargaon, Bangladesh. Appl Geochem 23:3155–3176

Karim M (2000) Arsenic in groundwater and health problems in Bangladesh. Water Res 34:304–310

Karmegam U, Chidambaram S, Prasanna MV, Sasidhar P, Manikandan S, Johnsonbabu G, Dheivanayaki V, Paramaguru P, Manivannan R, Srinivasamoorthy K, Anandhan P (2010) A study on the mixing proportion in groundwater samples by using Piper diagram and Phreeqc model. Chin J Geochem 30:490–495

Kinniburgh DG, Smedley PL (2001) Arsenic contamination of groundwater in Bangladesh. BGS technical report WC/00/19, British Geological Survey, Keyworth, UK

Klump S, Kipfer R, Cirpka OA, Harvey CF, Brennwald MS, Ashfaque KN, Badruzzaman AB, Hug SJ, Imboden DM (2006) Groundwater dynamics and arsenic mobilization in Bangladesh assessed using noble gases and tritium. Environ Sci Technol 40(1):243–50

Kumar M, Kumar P, Ramanathan AL, Bhattacharya P, Thunvik R, Singh UK, Tsujimura M, Sracek O (2010a) Arsenic enrichment in groundwater in the middle Gangetic plain of Ghazipur district in Uttar Pradesh. India J Geochem Explor 105:83–94

Kumar P, Kumar M, Ramanathan AL, Tsujimura M (2010b) Tracing the factors responsible for arsenic enrichment in groundwater of the middle Gangetic Plain, India: a source identification perspective. Environ Geochem Health 32:129–146

Liu CW, Lin KH, Kuo YM (2003) Application of factor analysis in the assessment of groundwater quality in a blackfoot disease area in Taiwan. Sci Total Environ 313:77–89

Mazumder GDN, Haque R, Ghosh N, De BK, Santra A, Chakraborti D, Smith AH (1998) Arsenic levels in drinking water and the prevalence of skin lesions in West Bengal, India. Int J Epidemiol 27:871–877

McArthur JM, Ravenscroft P, Safiulla S, Thirlwall MF (2001) Arsenic in groundwater testing pollution mechanisms for sedimentary aquifers in Bangladesh. Water Resour Res 37:109–117

Mitsunobu S, Harada T, Takahashi Y (2006) Comparison of antimony behavior with that of arsenic under various soil redox conditions. Environ Sci Technol 40:7270–7276

Molerio León LF, Toujague de la Rosa R (2004) Arsenic in a hard rock aquifer: multivariate optimisation of the groundwater monitoring network. 32nd International Geology Congress, Florence

Mridulchetia S, Saurabh KR, Bora K, Kalita H, Saikia LB, Goswami DC, Srivastava RR, Sarma HP (2008) Groundwater arsenic contamination in three blocks of Golaghat district of Assam. J Indian Water Works Assoc 40:150–154

Mukherjee AB, Bhattacharya P (2001) Arsenic in groundwater in the Bengal Delta Plain: slow poisoning in Bangladesh. Environ Rev 9(3):189–220

Mukherjee A (2006) Deeper groundwater flow and chemistry in the arsenic affected western Bengal basin, West Bengal, India. Ph.D. thesis, University of Kentucky, Lexington

Mukherje A, Fryar AE, Rowe HD (2007) Regional-scale stable isotopic signatures of recharge and deep groundwater in the arsenic affected areas of West Bengal, India. J Hydrol 334:151–161

Mukherjee A, von Brömssen M, Scanlon BR, Bhattacharya P, Fryar AE, Hasan MA, Ahmed KM, Chatterjee D, Jacks G, Sracek O (2008) Hydrogeochemical comparison and effects of overlapping redox zones on groundwater arsenic near the Western (Bhagirathi sub-basin, India) and Eastern (Meghna sub-basin, Bangladesh) margins of the Bengal Basin. J Contam Hydrol 99(1–4):31–48. doi:10.1016/j.jconhyd.2007.10.005

Neidhardt H (2012a) Impact of groundwater abstraction and of the organic matter on release and distribution of arsenic in aquifers of the Bengal Delta Plain, India. Universität, Karlsruhe

Neidhardt H (2012b) Arsenic in groundwater of West Bengal: implications from a field study. KIT Scientific Publishing, Karlsruhe

Neidhardt H, Berner Z, Freikowski D, Biswas A, Winter J, Chatterjee D, Norra S (2013a) Influences of groundwater extraction on the distribution of dissolved As in shallow aquifers of West Bengal, India. J Hazard Mater 262: 941–950

Neidhardt H, Biswas A, Freikowski D, Majumder S, Chatterjee D (2013b) Reconstructing the sedimentation history of the Bengal Delta Plain by means of geochemical and stable isotopic data. Appl Geochem 36:70–82

Nickson R, McArthur JM, Burgess WS, Ahmed KM, Ravenscroft P, Rahman M (1998) Arsenic poisoning of groundwater in Bangladesh. Nature 395:338

Nickson R, McArthur JM, Ravenscroft P, Burges WG, Ahmed KM (2000) Mechanism of arsenic release to groundwater, Bangladesh and West Bengal. Appl Geochem 15:403–413

Oinam JD, Ramanathan AL, Linda A, Singh G (2011) A study of arsenic, iron and other dissolved ion variations in the groundwater of Bishnupur District, Manipur, India. Environ Earth Sci 62(6):1183–1195

Pfeifer HR, Giradet A, Lavanchy JC, Reymond D, Schlegel C, Schmidt V (2001) Pathways of natural arsenic from rocks and acid soils to groundwaters and plants in Southern Switzerland. In: Weber J et al (eds) Biochemical processes and cycling of elements in the environment. Polish Society of Humic Substances, Wroclaw, pp 391–392

Ramanathan S (1956) Ultrabasic rock of Salem and Dodkanya and their relationship with Charnocite. Unpublished Ph.D. thesis, University of Madras

Ramanathan AL, Prasad MBK, Chidambaram S (2007) Ground water Arsenic contamination and its health effects-case studies from India and SE Asia. Indian J Geochem 22(2):371–384

Ramesh R, Sarin MM (1992) Stable isotope study of Ganga (Ganges) river system. J Hydrol 139:49–62

Ravenscroft P, McArthur JM, Hoque B (2001) Geochemical & palaeo-hydrological controls on pollution of groundwater by arsenic. In: Chappell WR, Abernathy CO, Calderon R (eds) Arsenic exposure and health effects IV. Elsevier Science Ltd., Oxford

Raymahashay BC, Khare AS (2003) The arsenic cycle in late quaternary fluvial sediments: mineralogical considerations. Special section: late Cenozoic fluvial deposits. Curr Sci 84(8):25

Reymond D, Pfeifer HR, Hesterberg D, Weiqing Chou J (2001) X-ray adsorption spectroscopy of some arsenic-contaminated soils. In: Weber J et al (eds) Biochemical processes and cycling of elements in the environment. Polish Society of Humic Substances, Wroclaw, pp 393–394

Rouxel O, Ludden J, Fouquet Y (2003) Antimony isotope variations in natural systems and implications for their use as geochemical tracers. Chem Geol 200:25–40

Sharifi Z, Sinegani AAS (2012) Arsenic and other irrigation water quality indicators of groundwater in an agricultural area of Qorveh Plain, Kurdistan, Iran. Am Eurasian J Agric Environ Sci 12(4):548–555

Singh VS, Saxena VK (2004) Assessment of utilization ground water resources in a coastal shallow aquifer. In: Proceeding of the 2nd Asia Pacific Association of hydrology & water resources conference, vol 2. Singapore

Singh VS, Saxena VK, Prakash BA, Mondal NC, Jain SC (2004) Augmentation of ground water resources in saline ingress coastal deltaic area. NGRI-Technical report no. NGRI-2004-GW-422

Singh M, Singh AK, Swati, Srivastava N, Singh S, Chowdhary AK (2010) Arsenic mobility in fluvial environment of the Ganga Plain, northern India. Environ Earth Sci 59:1703–1715

Smith AH, Lingas EO, Rahman M (2000) Contamination of drinking-water by arsenic in Bangladesh: a public health emergency. Bull World Health Organ 78(9):1093–1103. Technical report

Stewart FH (1963) In: Fleischer M (ed) Marine evaporates 1–52 in data of geochemistry, 6th edn. Geological Survey, Washington, DC

Stuben D, Berner Z, Chandrasekharam D, Karmakar J (2003) Arsenic enrichment in groundwater of West Bengal, India: geochemical evidence for mobilization of as under reducing conditions. Appl Geochem 18:1417–1434

Takaaki I, Takahashi Y (2005) Mechanism and source of arsenic contamination in groundwater in Bangladesh W1-14-05 (Workshop 1) Development of phyto-technology for decreasing heavy metal in food. MARCO Symposium

Tripathi P, Ranjan RK, Ramanathan AL, Bhattacharya P (2006) Groundwater arsenic in the Central Gangetic Plain in Ballia District of Uttar Pradesh, India: a future concern. In: Ramanathan AL et al (eds) International conference on groundwater for sustainable development: problems, perspectives and challenges (IGC-2006), vol 1–4, Abstract

Van Geen A, Zheng Y, Versteeg R, Stute M, Horneman A, Dhar R, Steckler M, Gelman A, Small C, Ahsan H, Graziano JH, Hussain I, Ahmed KM (2003) Spatial variability of arsenic in 6000 tube wells in a 25 km^2 area of Bangladesh. Water Resour Res 39(5):1140–1156

Wilson J, Schreier H, Brown S (2008) Arsenic in groundwater in the Surrey-Langley area. Institute for Resources & Environment, The University of British Columbia, New Westminster Fraser Health Authority Environmental Health Services Abbotsford, B.C. and Ministry of Environment Lower Mainland Region Surrey, B.C.

Yokota H, Tanabe K, Sezaki M, Akiyoshi Y, Miyata T, Kawahara K, Tsushima S, Hironaka H, Takafuji H, Rahman M, Ahmad SKA, Sayed MHSU, Faruquee MH (2001) Arsenic contamination of ground and pond water and water purification system using pond water in Bangladesh. Eng Geol 60:323–331

Section IV
Arsenic in Food Chain, Health and Its Remediation

Chapter 14
Groundwater Arsenic Contamination in Bengal Delta and Its Health Effects

Mohammad Mahmudur Rahman, Khitish Chandra Saha,
Subhas Chandra Mukherjee, Shyamapada Pati, Rathindra Nath Dutta,
Shibtosh Roy, Quazi Quamruzzaman, Mahmuder Rahman,
and Dipankar Chakraborti

14.1 Introduction

Arsenic contamination of groundwater has been detected in more than 70 countries and has become a major public health concern worldwide (Bundschuh et al. 2010). Arsenic contamination in groundwater of Southeast Asian regions received significant interest in recent years. In this region, countries affected with As in groundwater include Bangladesh, several states of India, Nepal, Myanmar, Pakistan, Vietnam, Lao People's Democratic Republic, Cambodia, several provinces of China

---

M.M. Rahman
School of Environmental Studies (SOES), Jadavpur University,
Kolkata 700 032, West Bengal, India

Centre for Environmental Risk Assessment and Remediation (CERAR), University of South Australia, Mawson Lakes Campus, Mawson Lakes, SA 5095, Australia

K.C. Saha • D. Chakraborti (✉)
School of Environmental Studies (SOES), Jadavpur University,
Kolkata 700 032, West Bengal, India
e-mail: dcsoesju@gmail.com

S.C. Mukherjee
Department of Neurology, Medical College, Kolkata, West Bengal, India

S. Pati
Department of Obstetrics and Gynaecology, Institute of Post Graduate Medical Education and Research, SSKM Hospital, Kolkata, West Bengal, India

R.N. Dutta
Department of Dermatology, Institute of Post Graduate Medical Education and Research, SSKM Hospital, Kolkata, West Bengal, India

S. Roy • Q. Quamruzzaman • M. Rahman
Dhaka Community Hospital, Dhaka, Bangladesh

(Inner Mongolia, Shanxi, Xinjiang, Ningxia, Jilin, Shandong, Qinghai, Sichuan, Anhui, Heilongjiang, Henan, Gansu, Jiangsu, Yunnan and Hunan) and lowlands of Sumatra in Indonesia (Rahman et al. 2009; Yu et al. 2007).

Several epidemiological studies showed that exposure of inorganic As via drinking water and food causes skin disorders, cardiovascular disease, neurological complications, reproductive disorders, respiratory effects, diabetes mellitus as well as various types of cancers including skin, lung, bladder and kidney (NRC 1999, 2001; IPCS 2001; IARC 2004; Chakraborti 2011). According to the World Bank Policy Report (2005), about 0.7 million people had been thus far affected by As-related diseases in Southeast Asian regions. Based on our research surveys, it was predicted that large parts of Ganga-Brahmaputra basin in India (Uttar Pradesh, Bihar, Jharkhand, West Bengal, Arunachal Pradesh and Assam), six out of seven North Eastern hill states in India (except Mizoram) and Padma-Meghna-Brahmaputra plain in Bangladesh are adversely contaminated with As in groundwater (Chakraborti et al. 2004).

This article reports As contamination of groundwater in Bengal delta of West Bengal, India and Bangladesh to understand the extent and severity of the problem and consequences on human health including dermal, neurologic and obstetric effects based on our analytical, environmental, epidemiological and clinical surveys for more than two decades.

14.2 Arsenic Contamination in Groundwater

14.2.1 West Bengal

In 1983, K.C. Saha identified the first case of arsenicosis in West Bengal while examining patients in an outdoor department of the School of Tropical Medicine, Calcutta (Chakraborti et al. 2002). The first published article of As contamination of groundwater and its health effects in the lower Gangetic plain (West Bengal) reported that 16 patients in three families were diagnosed with different types of arsenical skin lesions such as hyper-pigmentation, hyper-keratosis, dorsal keratosis, oedema, ascites, wasting, weakness, pain and burning sensation in toes and fingers from a village in 24-Parganas district (Garai et al. 1984). Since 1988, School of Environmental Studies (SOES), Jadavpur University has been conducting analytical, epidemiological and clinical surveys in As-contaminated areas of West Bengal to understand the magnitude of As contamination. According to our preliminary survey, 3,417 villages in 107 blocks in nine districts are contaminated (Chakraborti et al. 2009). We identified increasing number of contaminated villages and affected people with every additional survey in As-affected districts. The magnitude of groundwater As contamination in West Bengal received scientific attention only after 1995 (International Conference on Arsenic 1995).

We collected and analyzed 150 hand tube-wells water samples from 7,823 villages of 241 blocks in 19 districts of West Bengal (Chakraborti et al. 2009).

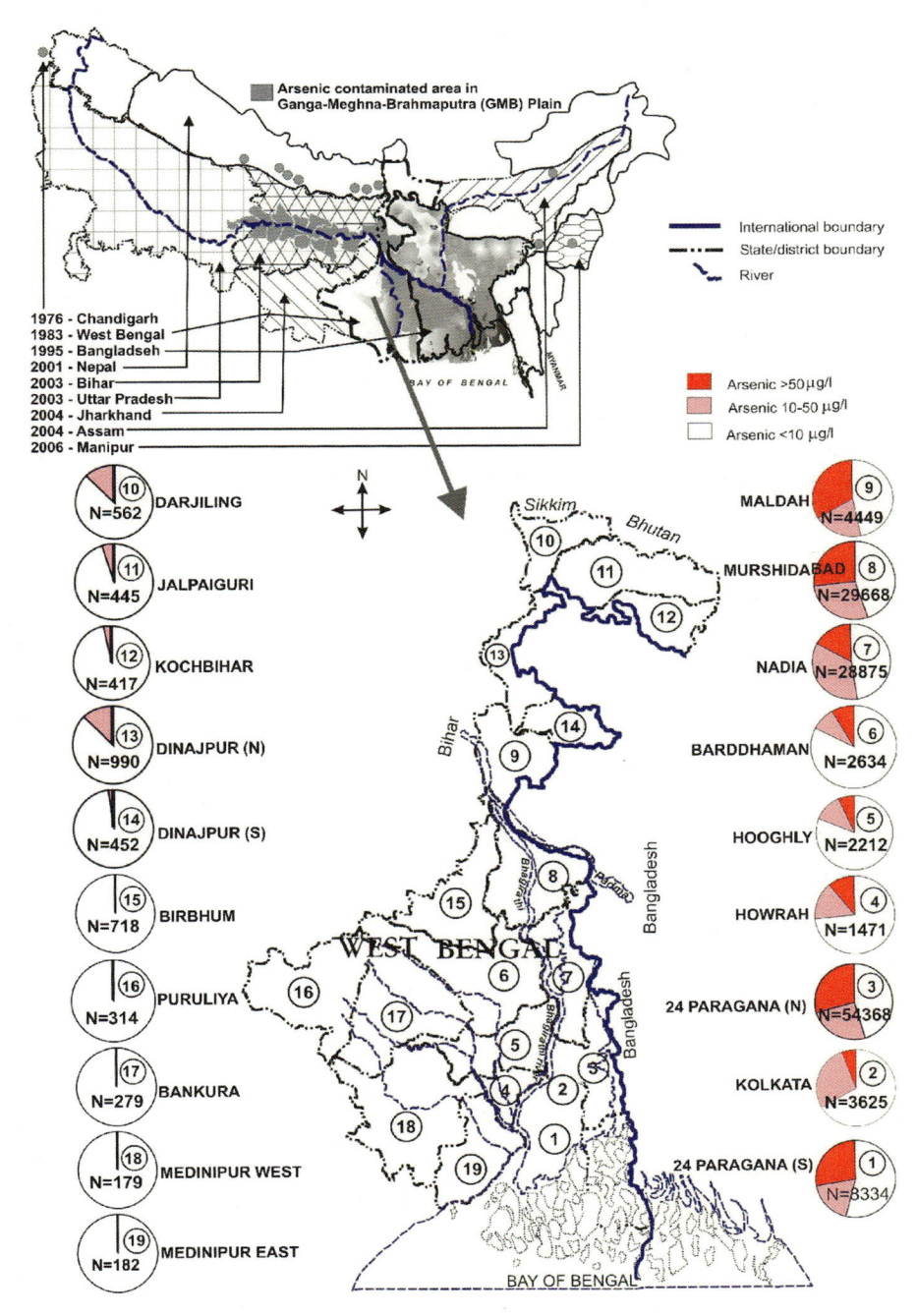

14 Groundwater Arsenic Contamination in Bengal Delta and Its Health Effects

Fig. 14.1 Status of As contamination in groundwater of all districts in West Bengal, India (Chakraborti et al. 2009)

Table 14.1 Distribution of As concentration in groundwater of West Bengal, India

Status	Districts	Area in km²	Population	Total no. of blocks	No. of blocks surveyed	No. of blocks with As >10 µg/l	No. of blocks with As >50 µg/l
Highly contaminated	North-24-PGS	4,094	8,934,286	22	22	22	21
	South-24-PGS	9,960	6,906,689	29	17	12	11
	Murshidabad	5,324	5,866,569	26	26	25	24
	Nadia	3,927	4,604,827	17	17	17	17
	Maldah	3,733	3,290,468	15	14	13	9
	Haora	1,467	4,273,099	14	12	12	7
	Hugli	3,147	5,041,976	18	17	16	11
	Kolkata	185	4572876[a]	141[b]	100[b]	65[b]	30[b]
	Bardhaman	7,024	6,895,514	31	24	12	7
	Sub total	38,861	50,386,304	172	149	129	107
Less contaminated	Cooch Behar	3,387	2,479,155	12	5	4	1
	Darjiling	3,149	1,609,172	12	4	3	0
	Dinajpur (N)	3,140	2,441,794	9	7	6	2
	Dinajpur (S)	2,219	1,503,178	8	6	2	1
	Jalpaiguri	6,227	3,401,173	13	7	4	0
	Sub total	18,122	11,434,472	54	29	19	4
Non-contaminated	Bankura	6,882	3,192,695	22	17	0	0
	Birbhum	4,545	3,015,422	19	11	0	0
	Purulia	6,259	2,536,516	20	14	0	0
	Medinipur (E)	14,081	9,610,788	25	10	0	0
	Medinipur (W)			29	8	0	0
	Sub total	31,767	18,355,421	115	60	0	0
	Grand total	88,750	80,176,197	341	241	148	111

Chakraborti et al. (2009)

Note: [a]Night time population = 4.6 million; [b]Number of wards

The current As contamination status in groundwater in West Bengal is presented in Fig. 14.1. The district-wise distribution of As concentrations in 19 districts of West Bengal is given in Table 14.1. The data revealed that 48.1 % of samples had As >10 µg/l and 23.8 % had As >50 µg/l. About 3.3 % samples had As >300 µg/l and 0.13 % (187 samples) has As >1,000 µg/l. The highest As concentration was 3,700 µg/l observed in Ramnagar village, South 24-Parganas district, West Bengal.

We categorized entire West Bengal into three zones based on tube-well As concentrations such as highly affected, mildly affected and un-affected (Chakraborti et al. 2009). Nine districts (Murshidabad, Malda, Nadia, North 24-Parganas, South 24-Parganas, Bardhaman, Hoara, Hoogly and Kolkata), in which As concentration >300 µg/l was detected in some tube-wells, are considered as highly affected (Chakraborti et al. 2009). A total of 135,555 hand tube-wells water samples were analyzed from these highly affected districts, 49.7 % had As >10 µg/l and 24.7 %

No. of samples analyzed	Number of samples								Max. As (µg/l)
	≤3	4–10	11–50	51–100	101–200	201–300	301–500	501–1,000	>1,000	
54,368	22,221	3,129	13,001	6,403	5,531	2,249	1,308	477	49	2,830
8,333	4,407	427	1,141	743	741	327	305	212	30	3,700
29,668	11,471	2,244	8,042	3,267	2,366	941	884	382	71	3,003
28,794	11,431	2,613	9,810	2,265	1,520	630	360	152	13	3,200
4,449	1,754	373	810	488	559	183	163	97	22	1,904
1,471	889	226	192	87	41	22	12	1	1	1,333
2,212	1,469	346	251	77	52	14	2	1		600
3,626	2,224	855	345	85	75	27	10	5		800
2,634	2,091	79	244	86	89	27	11	6	1	2,230
135,555	57,957	10,292	33,836	13,501	10,974	4,420	3,055	1,333	187	
474	403	57	13	1						54
562	502	50	10							19
990	817	57	112	4						68
452	398	47	6	1						51
445	355	74	16							27
2,923	2,475	285	157	6						
279	279									<3
718	718									<3
314	314									<3
182	182									<3
179	179									<3
1,672	1,672									
140,150	62,104	10,577	33,993	13,507	10,974	4,420	3,055	1,333	187	

had As >50 µg/l (Chakraborti et al. 2009). About 3.4 % samples had As >300 µg/l (Chakraborti et al. 2009).

We also analyzed 2,923 hand tube-well water samples from five districts (North Dinajpur, East Dinajpur, Jalpaiguri, Darjeeling and Cooch Bihar) situated in the northern part of West Bengal. Arsenic concentration in groundwater of these districts was generally <50 µg/l (only a few sample >50 µg/l but none <100 µg/l), are considered mildly affected. The results of As in water samples showed only six (0.2 %) samples had As >50 µg/l and 163 (5.7 %) samples had As >10 µg/l (Chakraborti et al. 2009). We analyzed 1,672 hand tube-wells water samples from five districts (Birbhum, Bakura, Purulia, Mednipur East and Mednipur West) situated in the western and southwestern parts of West Bengal. The results do not show As concentrations in groundwater >3 µg/l (Chakraborti et al. 2009). The groundwater of these districts is considered unaffected or safe (Chakraborti et al. 2009).

State government of West Bengal with the support of UNICEF analyzed 132,262 hand tube-well water samples from eight As-contaminated districts by the silver diethyldithiocarbamate (SDDC) method (Nickson et al. 2007). Their result shows that 25.5 % of samples were found to contain As >50 µg/l and 58 % had >10 µg/l. Our data reveals that groundwater of 107 blocks in eight districts (excluding Kolkata) had As >50 µg/l whereas Nickson et al. (2007) reported As >50 µg/l in 79 blocks of West Bengal. Both studies showed that one fourth of the analyzed samples had As >50 µg/l.

14.2.1.1 Arsenic Concentration in Deeper Tube-Wells (>100 m)

We analyzed 5,338 hand tube-wells for As at depth range 100–651 m from four highly affected districts (North 24-Parganas, Nadia, Murshidabad and South 24-Parganas) of West Bengal (Chakraborti et al. 2009). About 28.6 % of 5,338 tube-wells >100 m had As >10 µg/l. Bureau of Indian Standard (BIS) recommended permissible limit of As in drinking water of 10 µg/l (BIS 2009). Still some tube-wells >350 m had As between 10 and 50 µg/l. Therefore, we cannot say tube-wells >100 m would be safe.

14.2.1.2 Estimation of the Number of As-Contaminated Hand Tube-Wells

Field survey also included the number of people using a hand tube-well. We have information of users for 37,833 tube-wells from eight highly affected districts. We excluded Kolkata district from this calculation, because the principal source of drinking water is treated Ganges river water supplied by the city corporation. The estimated (average) number of people using each tube-well is 34 (Chakraborti et al. 2009). Based on this estimate and the population in each district, we estimated 1.34 million tube-wells in these As-affected districts out of which 0.53 million tube-wells had As >10 µg/l and 0.28 million tube-wells had As >50 µg/l (Table 14.2). Extrapolating these data, the total number of tube-wells throughout West Bengal (excluding Kolkata) is expected to be 2.2 million and out of which 0.54 million tube-wells had As >10 µg/l and 0.28 million tube-wells had As >50 µg/l (Chakraborti et al. 2009).

14.2.1.3 Estimation of Population Exposed to As-Contaminated Drinking Water

We estimated the number of people who could be drinking As-contaminated water by multiplying average number of a tube-well user ($n=34$) and number of As-contaminated tube-wells in different concentration ranges. We report here only the As-affected blocks

Table 14.2 Estimated number of tube-wells at different As concentration levels in groundwater of West Bengal, India

Districts	Population	Expected No. of tube-wells	Number of tube-wells contaminated with As								
			>10	>50	>100	>125	>200	>250	>300	>500	>1,000
North-24-PGS	8,934,286	260,627	139,175	76,885	46,131	37,009	19,547	13,031	8,861	2,606	261
South-24-PGS	6,906,689	201,479	84,621	57,018	39,087	32,841	21,155	16,521	13,298	5,843	806
Murshidabad	5,866,569	171,137	92,072	45,694	26,868	21,563	13,178	9,926	7,701	2,567	342
Nadia	4,604,827	134,330	68,777	22,970	12,358	9,672	5,239	3,493	2,284	672	1
Maldah	3,290,468	95,988	50,106	32,636	22,077	18,430	10,079	7,775	6,143	2,592	480
Howrah	4,273,099	124,653	30,166	13,836	6,482	5,235	2,992	2,119	1,122	249	125
Hugli	5,041,976	147,082	26,328	9,707	4,560	3,383	1,030	441	147	147	
Bardhaman	6,895,514	201,153	35,403	16,696	10,058	6,839	3,218	2,012	1,207	402	2
Cooch Behar	2,479,155	72,321	2,097	145							
Darjiling	1,609,172	46,942	845								
Dinajpur (N)	2,441,794	71,231	8,334	285							
Dinajpur (S)	1,503,178	43,850	658	88							
Jalpaiguri	3,401,173	99,217	3,572								
Bankura	3,192,695	93,136									
Birbhum	3,015,422	87,964									
Purulia	2,536,516	73,994									
Medinipur (E&W)	9,610,788	280,361									
Grand total	80,176,197	2,205,465	542,154	275,960	167,621	134,972	76,438	55,318	40,763	15,078	2,017

Chakraborti et al. (2009)

in eight As-affected districts that we surveyed (Table 14.3). It is apparent that about 9.5 and 4.2 million people could be drinking As-contaminated water with levels >10 µg/l and >50 µg/l, respectively, and an estimated 0.53 million drinking water containing As >300 µg/l, the concentration predicted to cause overt arsenical skin lesions (Chakraborti et al. 2009). Throughout West Bengal, 26 million people could be potentially exposed to As-contaminated water at levels >10 µg/l (Chakraborti et al. 2009). This was calculated by multiplying the population of the affected blocks of each district by percentage of hand tube-well water samples >10 µg/l.

14.2.2 Bangladesh

Arsenic contamination of groundwater in Bangladesh was first discovered in 1992 through reports of arsenical skin lesions in few Bangladeshi married women living in Gobindapur village of West Bengal, India (Dhar et al. 1997). The magnitude of groundwater As contamination in Bangladesh surfaced only after the International Conference on As held in Dhaka, Bangladesh (International Conference on Arsenic 1998). Since 1996, extensive field survey had been conducted in collaboration with the Dhaka Community Hospital (DCH) to determine the magnitude of As contamination in groundwater in all 64 districts of Bangladesh. In 1997, we reported groundwater As calamity in Bangladesh based on the preliminary survey results of 3,106 hand tube-well water samples from 28 districts and predicted that more than 50 million people could be at risks from As poisoning (Dhar et al. 1997). We further reported that 66 % and 51 % of 8,065 hand tube-wells water samples from 60 districts of Bangladesh contained As >10 µg/l and >50 µg/l, respectively (Dhar et al. 1998). We also reported that As was detected in groundwater of 52 districts >10 µg/l and in 41 districts >50 µg/l (Dhar et al. 1998). In 2000, we reported that 59 % of the 10,991 hand tube-well water samples collected from districts where we identified patients with arsenical skin lesions had As >50 µg/l (Chowdhury et al. 2000).

To date, we analyzed 52,202 hand tube-well water samples from all 64 districts of Bangladesh by the FI-HG-AAS method (Chakraborti et al. 2010). We surveyed 3,600 villages (out of 68,000 villages) in 338 out of 490 thanas. The survey results indicated that groundwater from 2,000 villages of 50 districts and 2,500 villages of 59 districts were found to contain As >10 and >50 µg/l (Chakraborti et al. 2010). Figure 14.2 shows groundwater As contamination status in all the districts of Bangladesh. Table 14.4 shows the district-wise distribution of hand tube-well water samples against different As concentration range. From the overall water analysis of 64 districts of Bangladesh, it appears that 42.1 % of the samples had As >10 µg/l and 27.5 % >50 µg/l (Chakraborti et al. 2010). The results from 50 As-affected districts of Bangladesh where we detected As >50 µg/l showed that out of 46,321 hand tube-wells, 48.1 % and 30.9 % were found to have As >10 µg/l and >50 µg/l, respectively (Chakraborti et al. 2010).

Table 14.3 Estimated number of population exposed to As-contaminated water in eight contaminated districts of West Bengal, India

Districts	Total Population	Population exposed								
		>10	>50	>100	>125	>200	>250	>300	>500	>1,000
North-24-Parganas	4,290,233	1,921,371	959,377	568,240	458,595	244,314	161,979	106,534	32,898	3,980
South-24-Parganas	2,577,369	854,916	524,922	340,894	273,083	168,136	129,625	90,057	24,257	3,583
Murshidabad	5,249,116	2,568,707	1,208,863	692,033	494,444	292,359	220,726	152,854	48,437	8,075
Nadia	3,855,122	2,075,328	589,810	281,406	216,897	111,853	77,031	47,579	13,066	742
Maldah	2,751,151	971,975	571,224	366,803	308,535	168,581	127,124	104,399	47,975	8,439
Howrah	2,437,846	498,312	201,938	82,256	67,303	38,667	27,713	14,609	2,772	1,801
Hugli	3,272,749	379,311	82,722	32,669	25,237	9,156	4,259	2,820	301	0
Bardhaman	1,860,326	258,808	112,741	68,240	47,098	20,407	11,803	7,593	3,016	456
Total	**26,293,912**	**9,528,728**	**4,251,597**	**2,432,541**	**1,891,192**	**1,053,473**	**760,260**	**526,445**	**172,722**	**27,076**

Chakraborti et al. (2009)

Note: (a) Total population of those areas (villages and GPs) in those blocks from where we collected water samples for As analysis and found As concentration >10 μg/l were considered in this estimation and (b) assuming 34.3 persons using a contaminated tube-well

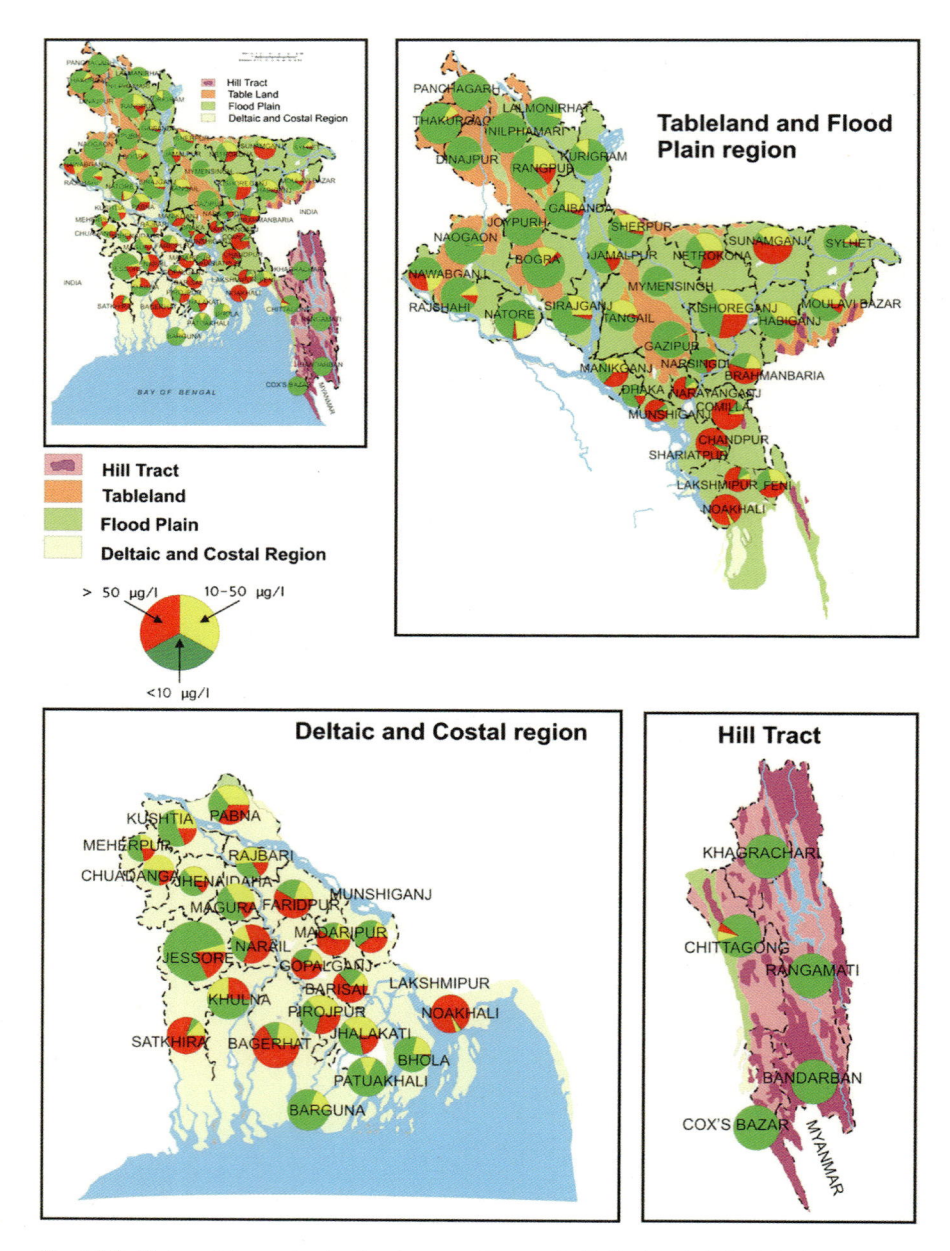

Fig. 14.2 Status of As contamination in groundwater of all districts in Bangladesh along with geo-morphological regions (Chakraborti et al. 2010)

Table 14.4 Distribution of As concentrations in tube-wells in all 64 districts of Bangladesh

Division	District	Area (km²)	Population	Total no. of thana	No. of surveyed thana	No. of thana >50 µg/l thana	No. of samples analyzed	Number of samples								Maximum As (µg/l)
								<10	10–50	51–99	100–299	300–499	500–699	700–1,000	>1,000	
Rajshahi	Bogra	2898.25	3,013,056	11	11	4	767	607	125	17	16	1	–	–	1	1,040
	Dinajpur	3437.98	2,642,850	13	13	–	2,641	2,612	28	1	–	–	–	–	–	77
	Gaibanda	2179.27	2,138,181	7	7	5	1,233	863	308	40	17	4	1	–	–	512
	Joypurhat	965.44	846,696	5	5	–	398	388	10	–	–	–	–	–	–	32
	Kurigram	2296.10	1,792,073	9	7	–	539	467	72	–	–	–	–	–	–	50
	Lalmanirhat	1241.46	1,109,343	5	5	–	464	434	30	–	–	–	–	–	–	50
	Naogaon	3435.67	2,391,355	11	10	–	537	527	10	–	–	–	–	–	–	22
	Natore	1896.05	1,521,336	6	3	2	117	91	22	4	–	–	–	–	–	63
	Nawabganj	1702.56	1,425,322	5	5	4	1,902	920	434	173	273	57	23	12	10	1,600
	Nilphamari	1580.85	1,571,690	6	6	–	523	505	18	–	–	–	–	–	–	50
	Pabna	2371.50	2,176,270	9	9	7	5,117	1,595	1,807	807	691	124	52	25	16	2,108
	Panchagarh	1404.63	836,196	5	5	–	462	458	4	–	–	–	–	–	–	15
	Rajshahi	2407.01	2,286,874	13	10	7	2,698	2,197	266	105	121	8	1	–	–	524
	Rangpur	2370.45	2,542,441	8	8	2	464	285	114	19	20	15	9	2	–	939
	Sirajganj	2497.92	2,693,814	9	5	3	278	187	79	8	4	–	–	–	–	216
	Thakurgaon	1809.52	1,214,376	5	5	2	461	416	38	6	1	–	–	–	–	130

(continued)

Table 14.4 (continued)

Division	District	Area (km²)	Population	Total no. of thana	No. of surveyed thana	No. of thana >50 μg/l thana	No. of samples analyzed	Number of samples								Maximum As (μg/l)
								<10	10–50	51–99	100–299	300–499	500–699	700–1,000	>1,000	
Khulna	Bagherhat	3959.11	1,549,031	9	3	3	371	90	72	34	79	67	17	12	–	958
	Chuadanga	1177.40	1,007,130	4	4	4	457	124	223	73	11	11	13	2	–	841
	Jessore	2570.42	2,471,554	8	6	6	5,465	4,227	248	317	457	105	76	32	3	1,120
	Jhenaidaha	1949.62	1,579,490	6	3	3	388	185	142	24	27	6	4	–	–	592
	Khulna	4394.46	2,378,971	14	12	9	1,000	518	233	55	138	39	6	5	6	3,143
	Kushtia	1621.15	1,740,155	6	6	5	2,065	1,082	557	154	168	42	33	13	16	2,190
	Meherpur	716.08	591,436	3	3	3	1,024	526	271	95	95	27	5	4	1	1,230
	Magura	1048.61	824,311	4	4	3	496	243	168	44	33	5	–	–	3	1,050
	Narail	990.23	698,447	3	3	3	371	96	56	35	164	20	–	–	–	375
	Satkhira	3858.33	1,864,704	7	5	5	532	32	73	56	236	133	1	1	–	750
Barisal	Barguna	1831.31	848,554	5	2	–	43	35	8	–	–	–	–	–	–	15
	Barisal	2790.51	2,355,967	10	6	6	803	179	113	40	227	106	75	47	16	1,770
	Bhola	3737.21	1,703,117	7	3	–	74	57	17	–	–	–	–	–	–	50
	Jhalakati	758.06	694,231	4	3	2	42	17	16	3	5	1	–	–	–	310
	Patuakhali	3220.15	1,460,781	7	1	–	15	13	2	–	–	–	–	–	–	10
	Pirojpur	1307.61	1,111,068	6	4	4	124	42	41	8	24	8	–	1	–	731
Sylhet	Habiganj	2636.58	1,757,665	8	2	2	103	59	37	6	1	–	–	–	–	100
	Moulavi Bazar	2799.39	1,612,374	6	5	2	152	72	65	12	3	–	–	–	–	133
	Sunamganj	3669.58	2,013,738	10	2	2	89	6	34	29	17	3	–	–	–	302
	Sylhet	3490.40	2,555,566	11	5	1	391	331	44	14	2	–	–	–	–	177

Dhaka	Dhakaª	1459.56	8,511,228	27(6)	6	3	574	449	29	26	63	7	–	–	–	352
	Faridpur	2072.72	1,756,470	8	5	5	707	243	171	67	142	40	24	12	8	1,630
	Gazipur	1741.53	2,031,891	5	4	1	3,386	3,312	33	8	16	16	1	–	–	533
	Gopalganj	1489.92	1,165,273	5	5	5	384	86	74	55	146	19	2	2	–	920
	Jamalpur	2031.98	2,107,209	7	2	1	144	89	30	7	6	2	2	6	–	1,172
	Kishoreganj	2731.21	2,594,954	13	12	11	1,328	527	429	238	133	1	–	–	2	365
	Madaripur	1144.96	1,146,349	4	4	4	2,309	453	480	336	622	316	76	21	–	1,200
	Manikganj	1383.06	1,285,080	7	3	3	282	79	101	44	55	2	1	–	5	586
	Munshiganj	954.96	1,293,972	6	5	5	151	10	6	12	80	36	7	–	–	529
	Mymensingh	4363.48	4,489,726	12	9	6	1,825	1,705	101	12	6	1	–	–	–	330
	Narayanganj	687.76	2,173,948	5	3	2	412	54	42	34	147	68	36	26	5	1,750
	Narshingdi	1140.76	1,895,984	6	5	3	336	252	16	7	23	24	10	4	–	1,000
	Netrokona	2747.91	1,988,188	10	10	10	533	201	180	84	49	13	6	–	–	580
	Rajbari	1118.80	951,906	4	3	2	174	79	72	5	15	1	–	2	–	714
	Shariatpur	1181.53	1,082,300	6	3	3	152	63	29	20	26	12	2	–	–	580
	Sherpur	1363.76	1,279,542	5	5	3	303	191	100	7	5	–	–	–	–	275
	Tangail	3375.00	3,290,696	11	11	4	597	443	131	21	2	–	–	–	–	224

(continued)

Table 14.4 (continued)

Division	District	Area (km²)	Population	Total no. of thana	No. of surveyed thana	No. of thana >50 μg/l thana	No. of samples analyzed	Number of samples								Maximum As (μg/l)
								<10	10–50	51–99	100–299	300–499	500–699	700–1,000	>1,000	
Chittagong	Bandarban	4479.03	298,120	7	2	–	41	41	–	–	–	–	–	–	–	–
	Brahmanbaria	1927.21	2,398,254	8	2	2	47	12	9	9	17	–	–	–	–	210
	Chandpur	1704.06	2,271,229	8	7	7	1,165	50	36	30	675	294	54	21	5	1,318
	Chittagong	5282.98	6,612,140	26	14	2	366	319	26	13	8	–	–	–	–	275
	Comilla	3085.17	4,595,557	13	6	6	545	113	26	29	128	123	75	25	26	1,769
	Cox's Bazar	2491.86	1,773,709	7	2	–	58	58	–	–	–	–	–	–	–	–
	Feni	928.34	1,240,384	5	2	2	186	58	53	40	28	5	1	1	–	1,000
	Khagrachari	2699.55	525,654	8	3	–	39	39	–	–	–	–	–	–	–	–
	Lakshmipur	1455.96	1,489,901	4	4	4	2,662	304	235	339	852	421	246	177	88	2,030
	Noakhali	3600.99	2,577,244	6	4	4	843	5	36	92	413	80	79	48	90	4,730
	Rangamati	6116.13	508,182	10	2	–	47	47	–	–	–	–	–	–	–	–
	Total	147,570	124,355,263	508 (486)	339	197	52,202	29,768	8,230	3,714	6,487	2,263	938	501	301	
	Percentage							57.0	15.8	7.1	12.4	4.3	1.8	1.0	0.6	

Chakraborti et al. (2010)

Note: [a]Dhaka city has 21 Thanas and we consider Dhaka city as a Thana

Based on groundwater surveying efforts, we classified all districts of Bangladesh into five categories: highly affected (As >300 µg/l), moderately affected (As = 100–300 µg/l), mildly affected (As = 50–100 µg/l), very mildly affected (As = 10–50 µg/l) and unaffected/As safe (As <10 µg/l). The groundwater of four districts (Rangamati, Khagrachari, Bandarban and Cox's Bazar) is unaffected/safe (As <10 µg/l). Ten districts (Dinajpur, Joypurhat, Kurigram, Lalmonirhat, Naogaon, Nilphamari, Panchagarh, Barguna, Bhola and Patuakhali) are considered as very mildly affected (As levels were between 10 and 50 µg/l) and Natore is termed as mildly affected (As concentrations between 50 and 100 µg/l). Nine districts (Sirajganj, Thakurgaon, Habiganj, Moulavibazar, Sylhet, Sherpur, Tangail, Brahmanbaria and Chittagong) are regarded as moderately affected (As concentrations between 100 and 300 µg/l). The remaining 40 districts where As >300 µg/l was detected in tube-well water samples are considered as highly affected. The comparative results of As concentration of hand tube-wells water samples from Bangladesh and West Bengal is presented in Fig. 14.3.

British Geological Survey (BGS) and the Department of Public Health Engineering (DPHE) reported that 46 % and 27 % of the 3,534 analyzed tube-wells exceeded 10 and 50 µg/l of As, respectively (BGS-DPHE 2001). Based on their study, BGS and DPHE (2001) reported that 57 million and 35 million people may be drinking As-contaminated water >10 µg/l and >50 µg/l, respectively. BGS-DPHE (2001) and our data show that 27 % of the analyzed tube-wells had As >50 µg/l although we analyzed greater number of samples from Bangladesh. van Geen et al. (2007)

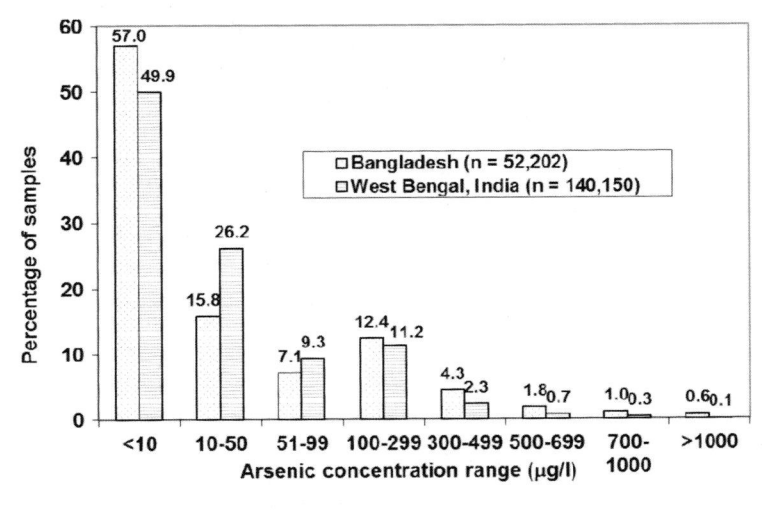

Fig. 14.3 Comparison of As concentrations in hand tube-wells of Bangladesh and West Bengal (India)

reported half of the 6,500 tube-wells sampled in Araihazar thana of Bangladesh had As >10 µg/l, and one quarter contained As >50 µg/l. The As database (tested by using various field testing kits) of National Arsenic Mitigation Information Centre (NAMIC) showed that 1.4 out of 4.8 million (30 %) tube-wells analyzed contained As >50 µg/l (Johnston and Sarker 2007). Nationwide about 20 % of shallow tube-wells are contaminated, and there are more than 8,000 villages where 80 % of the tube-wells are contaminated (As >50 µg/l) (www.unicef.org/bangladesh/Arsenic.pdf).

14.2.2.1 Arsenic Concentration in Deeper Tube-Wells (>100 m)

To ensure the supply of As-safe water to the villagers in the affected areas of Bangladesh, thousands of hand tube-wells >100 m were installed. To assess As contamination in these tube-wells, we analyzed 1,349 hand tube-wells (depth from 102 to 415 m) from four principal geomorphological regions of Bangladesh. Arsenic concentrations >50 µg/l were not present in tube-wells >350 m in the flood plain and deltaic regions including coastal regions. BGS and DPHE (2001) study reported that only 1 and 5 % of the 327 deep tube-wells (>150 m) were contaminated with As >50 µg/l and >10 µg/l, respectively. BGS-DPHE (1999) data also showed that 4 % of the 909 sampled wells >200 m are contaminated. Consequently, it is not possible to conclude that tube-wells >200 m will be free of As contamination. Out of 1,349 deep tube-wells 68 % had As <10 µg/l, 32 % had As >10 µg/l and 11 % had As >50 µg/l.

DPHE-DFID-JICA (2006) reported that there are substantial knowledge gaps regarding deep aquifers, including geology, water quality and hydraulic properties. As a result, it is very difficult to make any recommendations on the use of deep aquifers without a complete compilation of all these information. Recent study indicated that most of the Bengal Basin is highly vulnerable to downward migration of high-As groundwater caused by increased withdrawals of deeper groundwater for irrigation, and the use of low-As deep groundwater for irrigation should therefore be discouraged, particularly in the areas that are vulnerable to As contamination (Radloff et al. 2011). The study also recommended that deep community wells now in use throughout Bengal delta clearly need to be tested periodically to prevent renewed exposure to As (Radloff et al. 2011).

14.2.2.2 Estimation of the Number of As-Contaminated Hand Tube-Wells

During our field surveys, we also collected number of users for each hand tube-well. Based on this information, it was estimated an average of 24 people used a tube-well; however this estimate may vary significantly throughout Bangladesh as well as within each district. The total population in 50 As-affected districts of Bangladesh was roughly 102 million (BBS 2001). We estimated approximately 4.9 million hand tube-wells existed in these 50 districts (Table 14.5). Extrapolation of this estimate predicts approximately 5.2 million hand tube-wells across Bangladesh,

14 Groundwater Arsenic Contamination in Bengal Delta and Its Health Effects

Table 14.5 Estimated number of tube-wells at different As concentration levels in 50 contaminated districts of Bangladesh

Districts	Population	Expected no. of tube-wells	>10 μg/l	>50 μg/l	>100 μg/l	>125 μg/l	>200 μg/l	>250 μg/l	>300 μg/l	>500 μg/l	>1,000 μg/l
Thakurgaon	1,141,000	47,542	4,612	713							
Nawabganj	1,366,000	56,917	28,914	16,790	11,440	10,214	6,474	4,494	3,074	1,423	296
Rajshahi	2,255,000	93,958	17,476	8,174	4,510	2,924	870	487	282		
Natore	1,500,000	62,500	13,875	2,125							
Sirajganj	2,568,000	107,000	34,989	4,601		1,534	1,151				
Bogra	3,053,000	127,208	26,459	5,852	2,799	1,960	1,470	653	165		165
Gaibandha	2,218,000	92,417	27,725	4,621	1,664	1,261	816	519	370	74	
Rangpur	2,475,000	103,125	39,806	14,438	10,209	8,801	7,898	6,544	5,775	2,475	
Pabna	2,266,000	94,417	64,958	31,630	16,712	13,616	8,078	5,744	3,966	1,700	293
Kushtia	1,691,000	70,458	32,552	13,035	7,891	6,237	3,763	3,136	2,818	1,691	543
Meherpur	555,000	23,125	11,239	5,111	2,983	2,392	1,435	1,048	833	224	21
Chuadanga	921,000	38,375	27,975	9,248	3,108	2,553	2,306	2,141	2,187	1,266	
Jhenaidaha	1,540,000	64,167	32,789	9,882	7,443	4,113	2,632	1,974	1,989	770	
Jessore	2,387,000	99,458	22,378	18,002	12,333	9,858	5,706	4,883	3,879	2,089	50
Magura	815,000	33,958	17,319	5,807	2,819	2,533	890	685	543	204	204
Satkhira	1,780,000	74,167	69,643	59,556	51,694	47,524	34,131	27,794	18,838	297	
Khulna	2,417,000	100,708	50,656	26,386	20,545	16,906	9,706	7,097	6,043	1,813	604
Narail	705,000	29,375	21,855	18,154	15,451	13,331	7,823	3,752	1,674		
Rajbari	941,000	39,208	26,779	6,469	5,058	2,852	713	713	823	549	
Faridpur	1,678,000	69,917	47,124	31,113	23,912	19,187	12,657	10,648	9,019	4,475	769
Gopalganj	1,169,000	48,708	37,798	28,397	21,432	18,654	12,880	7,106	2,874	487	
Madaripur	1,185,000	49,375	39,698	29,428	24,302	19,935	15,045	11,604	8,937	2,173	104

(continued)

Table 14.5 (continued)

Districts	Population	Expected no. of tube-wells	>10 µg/l	>50 µg/l	>100 µg/l	>125 µg/l	>200 µg/l	>250 µg/l	>300 µg/l	>500 µg/l	>1,000 µg/l
Shariatpur	1,054,000	43,917	25,691	17,347	11,550	11,043	7,705	5,136	4,040	571	
Barisal	2,481,000	103,375	80,322	65,747	60,578	78,397	67,021	199,083	31,426	17,781	1,964
Pirojpur	1,178,000	49,083	31,953	14,332	11,584	9,516	5,009	3,506	2,209		
Bagherhat	161,100	6,713	6,021	4,477	3,739	3,000	2,323	2,030	2,061	624	
Jhalakathi	745,000	31,042	18,470	6,643	4,439						
Brahmanbaria	2,458,000	102,417	76,198	56,636	37,075	16,387	2,048				
Chandpur	2,309,000	96,208	92,071	89,089	86,588	83,210	61,881	41,651	30,883	7,216	414
Comilla	4,751,000	197,958	177,569	167,473	155,793	133,667	114,780	102,067	107,293	45,728	9,304
Feni	1,258,000	52,417	36,063	21,124	9,854	6,929	2,410	2,410	1,939	577	
Lakshmipur	1,502,000	62,583	55,449	49,879	41,931	38,395	30,435	25,973	21,904	12,016	2,065
Noakhali	2,547,000	106,125	105,488	100,925	89,357	73,621	44,563	39,165	37,356	27,274	11,249
Habiganj	1,727,000	71,958	30,726	5,613							
Sunamganj	1,941,000	80,875	75,376	44,481	18,197	15,164	6,318	5,055	2,750		
Sylhet	2,464,000	102,667	15,708	4,209	513	263					
Moulavibazar	1,556,000	64,833	34,102	6,354	1,232	398					
Gazipur	1,899,000	79,125	1,741	950	791	701	514	421	396	24	
Manikganj	1,293,000	53,875	38,736	19,503	11,044	8,408	2,336	1,246	593	189	
Munshiganj	1,309,000	54,542	50,942	48,760	44,397	16,666	10,908	8,181	15,544	2,509	
Mymensingh	4,450,000	185,417	7,231	556	315	611	102		167		
Narayanganj	2,013,000	83,875	72,887	64,332	57,371	40,804	29,121	24,238	27,427	13,588	1,007
Narsinghdi	1,864,000	77,667	19,417	15,533	13,980	13,180	10,356	9,414	8,776	3,184	
Netrokona	1,938,000	80,750	50,307	23,014	10,255	8,922	3,176	2,873	2,826	888	
Sherpur	1,279,000	53,292	19,665	2,078	853	879	352	176			

District											
Kishoreganj	2,574,000	107,250	64,672	30,030	10,832	8,531	3,250	1,381	75		
Tangail	3,371,000	140,458	36,238	5,337	421	255	255				
Jamalpur	2,111,000	87,958	33,600	15,217	10,995	1,079	1,079	1,079	7,301	6,069	1,143
Dhaka	7,751,000	322,958	70,405	53,934	39,401	41,747	23,773	17,395	3,876		
Chittagong	6,292,000	262,167	33,557	14,944	5,506	6,081	4,054	676			
Dinajpur	2,642,850	110,119	958								
Joypurhat	846,696	35,279	600								
Kurigram	1,792,073	74,670	6,048								
Lalmanirhat	1,109,343	46,223	2,588								
Naogaon	2,391,355	99,640	1,295								
Nilphamari	1,571,690	65,487	1,834								
Panchagarh	836,196	34,842	150								
Barguna	848,554	35,356	2,546								
Bhola	1,703,117	70,963	6,671								
Total	116,643,974	4,860,166	2,079,914	1,298,047	984,989	824,238	570,211	594,179	382,931	159,945	30,194

Chakraborti et al. (2010)

based on a total population of 124 million (BBS 2001). Using current population estimate of 153 million, the number of tube-wells is predicted to be 6.4 million. BGS and DPHE (1999) reported that about four million drinking water tube-wells exist in Bangladesh. BGS and DPHE (2001) further reported that there are roughly 6–11 million tube-wells nationwide. UNICEF estimated that the number of tube-wells in Bangladesh was 9.7 million (as of 2007), based on the As mitigation project in Bangladesh (www.unicef.org/bangladesh/arsenic.pdf).

14.2.2.3 Estimation of Population Exposed to As-Contaminated Drinking Water

Based on our data (water analyses and number of tube-well users), we have estimated population exposed to As-contaminated water in the As-affected districts of Bangladesh (Table 14.6). For this estimation, only population of the surveyed thanas was considered and did not consider the population of thanas where we found As <10 µg/l. It is estimated that about 36.6 million from 59 districts and 22.7 million from 50 districts could be exposed to As >10 and >50 µg/l, respectively. Up to 6.8 million people could be exposed to As >300 µg/l. Based on As surveys in Bangladesh and several states in India, we believe that exposure to 300 µg/l of As-contaminated water for 2 years or more may produce arsenical skin lesions (Chakraborti et al. 2004). The manifestation of arsenical skin lesions depends on several factors including: (i) concentration of As in drinking water, (ii) volume of water consumed (including cooking water), (iii) duration of water consumption and (iv) health and nutritional status of the individual. Based on our survey (1996–2002), ~80 million people in As-affected areas of Bangladesh were potentially at risks of As-contaminated groundwater >10 µg/l.

It is practically difficult to assess population size currently exposed to As-contaminated water at various concentration levels in Bangladesh. This is mainly because: (i) tube-wells are continuously installed or re-bored at different depths throughout Bangladesh and we need to know the contamination status of all tube-wells, (ii) information on temporal variation of As in tube-wells is lacking and (iii) alternative safe water sources are installed regularly in affected areas and those realizing the danger of As toxicity are using water from these sources. It was also reported that previously safe tube-wells (<50 µg/l) were being contaminated (>50 µg/l) over the course of time (Chakraborti et al. 2001; Sengupta et al. 2004). Arsenic content in many hand tube-wells has increased by as much as 5–20 fold (Chakraborti et al. 2001). Others have also reported temporal increase in groundwater As contamination depending on sub-surface geology (Burren 1998; Rosenboom 2004; BGS-DPHE 1999). Due to these reasons, it is hard to get the exact number of population exposed to As-contaminated water.

Table 14.6 Estimated number of population exposed to As-contaminated water in 50 contaminated districts of Bangladesh

District	Population*	>10 µg/l	>50 µg/l	>100 µg/l	>125 µg/l	>200 µg/l	>250 µg/l	>300 µg/l	>500 µg/l	>1,000 µg/l
Thakurgaon	385,525	37,396	5,783	771						
Nawabganj	1,425,322	724,064	420,470	286,490	255,787	162,129	112,546	76,967	35,633	7,412
Rajshahi	1,916,279	356,428	166,716	91,981	59,640	17,750	9,940	5,749		
Natore	633,840	140,712	21,551							
Sirajganj	1,328,774	434,509	57,137		19,051	14,288				
Bogra	2,183,277	454,122	100,431	48,032	33,632	25,224	11,211	2,838		2,838
Gaibandha	2,138,181	641,454	106,909	38,487	29,173	18,876	12,012	8,553	1,710	
Rangpur	2,542,441	981,382	355,942	251,702	216,970	194,716	161,337	142,377	61,019	
Pabna	2,176,270	1,497,266	729,050	385,200	313,834	186,191	132,391	91,403	39,173	6,746
Kushtia	1,740,155	803,952	321,929	194,897	154,049	92,946	77,455	69,606	41,764	13,399
Meherpur	591,436	287,438	130,707	76,295	61,183	36,710	26,804	21,292	5,736	532
Chuadanga	1,007,130	734,198	242,718	81,578	66,998	60,514	56,192	57,406	33,235	
Jhenaidaha	942,843	481,793	145,198	109,370	60,439	38,681	29,011	29,228	11,314	
Jessore	2,065,300	464,693	373,819	256,097	204,707	118,495	101,404	80,547	43,371	1,033
Magura	824,311	420,399	140,957	68,418	61,491	21,605	16,619	13,189	4,946	4,946
Satkhira	1,329,327	1,248,238	1,067,450	926,541	851,802	611,749	498,175	337,649	5,317	
Khulna	2,049,403	1,030,850	536,944	418,078	344,045	197,507	144,414	122,964	36,889	12,296
Narail	698,447	519,645	431,640	367,383	316,958	185,999	89,204	39,811		
Rajbari	599,245	409,284	98,875	77,303	43,581	10,895	10,895	12,584	8,389	
Faridpur	1,391,731	938,027	619,320	475,972	381,926	251,951	211,959	179,533	89,071	15,309
Gopalganj	1,165,273	904,252	679,354	512,720	446,275	308,142	170,009	68,751	11,653	
Madaripur	1,146,349	921,665	683,224	564,226	462,827	349,312	269,414	207,489	50,439	2,407
Shariatpur	662,321	387,458	261,617	174,190	166,549	116,197	77,464	60,934	8,610	
Barisal	1,361,534	1,057,912	865,936	797,859	1,032,551	882,717	2,622,093	413,906	234,184	25,869

(continued)

Table 14.6 (continued)

District	Population*	>10 µg/l	>50 µg/l	>100 µg/l	>125 µg/l	>200 µg/l	>250 µg/l	>300 µg/l	>500 µg/l	>1,000 µg/l
Pirojpur	824,702	536,881	240,813	194,630	159,891	84,153	58,907	37,112		
Bagherhat	565,194	506,977	376,984	314,813	252,566	195,585	170,944	173,515	52,563	
Jhalakathi	563,867	335,501	120,668	80,633						
Brahmanbaria	896,585	667,059	495,812	324,564	143,454	17,932				
Chandpur	2,146,121	2,053,838	1,987,308	1,931,509	1,856,160	1,380,378	929,101	688,905	160,959	9,228
Comilla	2,755,361	2,471,559	2,331,035	2,168,469	1,860,501	1,597,604	1,420,654	1,493,406	636,488	129,502
Feni	639,727	440,132	257,810	120,269	84,562	29,413	29,413	23,670	7,037	
Lakshmipur	1,489,901	1,320,052	1,187,451	998,234	914,067	724,545	618,323	521,465	286,061	49,167
Noakhali	2,021,416	2,009,288	1,922,367	1,702,032	1,402,303	848,820	745,999	711,538	519,504	214,270
Habiganj	542,130	231,490	42,286							
Sunamganj	701,776	654,055	385,977	157,900	131,583	54,826	43,861	23,860		
Sylhet	1,442,021	220,629	59,123	7,210	3,688					
Moulavibazar	1,378,654	725,172	135,108	26,194	8,458					
Gazipur	321,454	7,072	3,857	3,215	2,848	2,089	1,709	1,607	96	
Manikganj	681,551	490,035	246,721	139,718	106,369	29,547	15,758	7,497	2,385	
Munshiganj	1,155,864	1,079,577	1,033,342	940,873	353,181	231,173	173,380	329,421	53,170	
Mymensingh	3,566,006	139,074	10,698	6,062	11,743	1,957		3,209		
Narayanganj	1,188,533	1,032,835	911,605	812,957	578,205	412,651	343,464	388,650	192,542	14,262
Narsinghdi	1,651,444	412,861	330,289	297,260	280,245	220,193	200,175	186,613	67,709	
Netrokona	1,988,188	1,238,641	566,634	252,500	219,669	78,187	70,741	69,587	21,870	
Sherpur	1,279,542	472,151	49,902	20,473	21,115	8,446	4,223			
Kishoreganj	2,472,754	1,491,071	692,371	249,748	196,696	74,932	31,846	1,730		
Tangail	2,259,533	582,960	85,862	6,779	4,101	4,101	0			

Jamalpur	884,733	337,968	153,059	110,592	10,856	10,856	10,856	73,433	61,047	11,502
Dhaka	1,091,142	237,869	182,221	133,119	141,045	80,317	58,769	13,094		
Chittagong	6,612,140	846,354	376,892	138,855	153,374	102,250	17,042			
Dinajpur	789,476	6,868								
Joypurhat	472,497	8,032								
Kurigram	1,792,073	145,158								
Lalmanirhat	193,185	10,818								
Naogaon	710,321	9,234								
Nilphamari	1,355,991	37,968								
Panchagarh	315,631	1,357								
Barguna	421,782	30,368								
Bhola	863,393	81,159								
Total	80,339,402	36,636,313	22,749,872	17,342,196	14,480,145	10,092,548	9,785,712	6,791,089	2,783,886	520,719

Chakraborti et al. (2010)

Note: *Total population of Thanas we surveyed. We did not consider the population of Thanas where we found As <10 µg/l

14.3 Arsenic Concentration in Biological Samples and Sub-clinically Affected Population

Total As concentrations in hair, nail, blood and the total or metabolites of As in urine was used to identify As exposure in exposed subjects (IPCS 2001). Usually As accumulates in keratin-rich tissues—such as palm and sole, skin, hair and nails whereas hair and nails may be used as an indicator of past As exposure (IPCS 2001). Inorganic As is rapidly cleared from human blood. So, blood As reflects recent exposure and is very time dependent (IPCS 2001). Since As is rapidly metabolized and excreted through urine, total As, inorganic As and sum of As metabolites (i.e., inorganic As, MMA and DMA) in urine have all been used as biomarkers of recent As exposure (IPCS 2001).

We analyzed 8,400 hair, 8,665 nails, 11,000 urine and 230 skin-scale samples from As affected villages in West Bengal and 4,536 hair, 4,471 nail, 1,586 urine and 705 skin-scale samples from As-affected villages in Bangladesh. About 32 % of these biological samples were collected from people with arsenical skin lesions and the rest from non-patients who lived in the As-affected villages (skin-scale is only from those having keratosis). We realize that many villagers may not be affected by arsenical skin lesions but have elevated levels of As in their hair and nails, and thus may be sub-clinically affected. The elevated As concentrations in urine demonstrate that most of the villagers were drinking contaminated water. Table 14.7 shows the concentration of As in biological samples including hair, nail, urine and skin scales from West Bengal and Bangladesh compared with control population (As <3 μg/l). Mean As concentrations in biological samples of Bangladesh are considerably higher than those of West Bengal. The results also revealed that people of Bangladesh were exposed to high level of As than West Bengal. Mean As concentrations in hair, nail and urine of exposed population were much higher compared to the control population of both West Bengal and Bangladesh.

14.4 Evaluation of Clinical Signs and Symptoms in Bengal Delta

14.4.1 Arsenical Skin Lesions

Arsenic skin manifestations were diagnosed by the experienced dermatologists on the basis of characteristics detailed in Rahman et al. (2001) and Mukherjee et al. (2005). Salient dermatological features are diffuse melanosis (blackening of body), spotted melanosis (pigmentation), diffuse keratosis (thickening of skin in palm and sole), spotted keratosis (rough, dry and papular skin lesions) and dorsal keratosis. Melanosis usually appears earlier than keratosis. Melanosis is diffused in palms, soles and later over whole body if As exposure continues. Early melanosis in palms is often overlooked. Melanosis may be spotted like raindrop pigmentation often

Table 14.7 Arsenic in hair, nail, urine (metabolites) and skin-scale collected from As-contaminated subjects of West Bengal (India) and Bangladesh

	Exposed group								Control group					
	West Bengal				Bangladesh				West Bengal			Bangladesh		
Parameters	As in hair[a] (μg/kg)	As in nail[b] (μg/kg)	As in urine[c] (μg/kg)	As in skin scale[d] (μg/kg)	As in hair[a] (μg/kg)	As in nail[b] (μg/kg)	As in urine[c] (μg/kg)	As in skin scale[d] (μg/kg)	As in hair (μg/kg)	As in nail (μg/kg)	As in urine (μg/kg)	As in hair (μg/kg)	As in nail (μg/kg)	As in urine (μg/kg)
No. of observations	8,400	8,665	11,000	230	4,536	4,471	1,586	705	75	75	75	62	62	62
Mean	1,480	4,560	180	6,820	3,390	8,570	280	5,730	341	748	16	410	830	31
Median	1,320	3,870	115	4,460	2,340	6,400	115.78	4,800	338	743	15	210	90	6
Minimum	180	380	10	1,280	280	260	24	600	217	540	10	850	1,580	94
Maximum	20,340	44,890	3,147	15,510	28,060	79,490	3,086	53,390	499	1,066	41	180	680	20
Standard deviation	1,550	3,980	268	4,750	3,330	7,630	410	9,790	–	–	–	–	–	–
% of samples having As above normal/toxic levels[e]	62	84	89	–	83	93	95	–	–	–	–	–	–	–

Note: [a]Normal level of As in hair ranges 80–250 μg/kg; 1,000 μg/kg is an indication of toxicity (Arnold et al. 1990)
[b]Normal level of As in nail ranges 430–1,080 μg/kg (Ioanid et al. 1961)
[c]Normal excretion of As in urine ranges from 5 to 40 μg per 1.5 l (per day) (Farmer and Johnson 1990)
[d]There is no normal value for skin-scale in literature
[e]For hair, this is the % of samples above the toxic level of 1,000 μg/kg

noticed over chest, back and limbs. Leucomelanosis (pigmentation and depigmentation side by side) usually appears in later stage (after discontinuing consumption of contaminated water). Guttate melanosis (like black mole) may sometimes be found. Buccal mucus membrane melanosis on the tongue, gums and lips (diffuse, patchy and spotted) may also be observed. Starting from palms and soles (palmo-plantar keratosis), in advanced cases, keratosis is often distributed over dorsal of hands and feet and even over other body parts. Early keratosis is better felt than noticed and may be missed unless palms and soles are carefully palpated. Large nodules are often pre-runners of cancer. Bowen's disease is a premalignant condition called intraepidermal carcinoma in situ, which may be noted as a complication in chronic arsenicosis. Skin ulcer is a late feature of cutaneous arsenicosis that may turn into malignancy and skin cancers are late complications.

We examined 96,000 people (including children) randomly selected from As-contaminated areas of West Bengal and identified 9,356 (9.7 %) people having or had arsenical skin symptoms. We surveyed all nine highly As-contaminated districts of West Bengal and arsenicosis patients were identified in seven districts. A total 602 villages were surveyed and arsenicosis patients were identified in 488 villages of West Bengal. From Murshidabad, we screened 25,274 people for arsenical skin lesions and registered 3,320 males, 1,371 females and 122 children (total 4,813) with various types of arsenical symptoms.

Arsenicosis patients were found in 229 villages of Bangladesh. People with arsenical skin lesions were identified in 69 thanas out of 77 surveyed. We have screened 18,991 people including children and 3,762 (19.8 %) people were registered with arsenical skin lesions. Out of 18,991 people, we had registered 1,906 males, 1,558 females and 298 children, with arsenical skin lesions. Although the percentage of people including children showing arsenical skin lesions in As affected villages is 19.8 %, this does not mean that such a high prevalence of arsenicosis patients would be found at the same rate all over As affected districts of Bangladesh. The reason for such high percentage of arsenicosis patients is due to the fact that we had usually surveyed in those villages where people were drinking As-contaminated water and we had the information of arsenicosis patients.

Figure 14.4 tabulates the type of skin involvement of 3,762 and 9,356 arsenicosis patients including children of Bangladesh and West Bengal, respectively. We registered patients with various arsenical skin symptoms such as melanosis, leucomelanosis, keratosis, hyperkeratosis, dorsal keratosis, non-petting edema, Bowen's, gangrene and cancer. We have also found some symptoms in arsenicosis patients that are not even mentioned in the literature. A few examples are: 80 % of the patients having arsenical skin lesions reported severe itching sensation when they are exposed to sunlight; and arsenical skin lesions usually do not appear on the face (except in some cases of diffuse melanosis). Figure 14.5a–g show some arsenicosis patients from affected districts of West Bengal and Bangladesh including different types of melanosis and keratosis.

Squamous cell carcinoma, basal cell carcinoma, Bowen's diseases, carcinoma affecting lung, liver, uterus, bladder, genitourinary tract or other sites often appear in an advance stage in cases that have been neglected for many years. Persistent

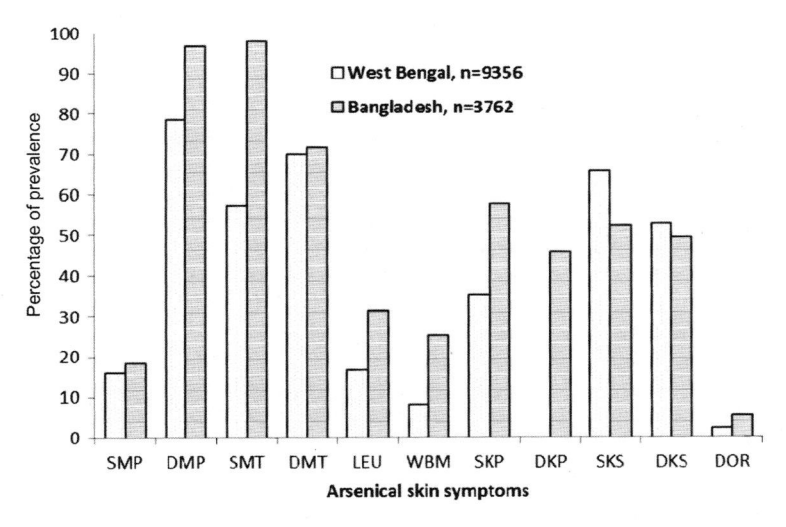

Fig. 14.4 Comparison of arsenical skin symptoms (in percentage) in arsenicosis patients observed in Bangladesh and West Bengal (India). *SMP* spotted melanosis on palm, *DMP* diffuse melanosis on palm, *SMT* spotted melanosis on trunk, *DMT* diffuse melanosis on trunk, *LEU* leuco – melanosis, *WBM* whole - body melanosis, *SKP* spotted keratosis on palm, *DKP* dorsal keratosis on palm, *SKS* spotted keratosis on sole, *DKS* diffuse keratosis on sole, *DOR* dorsal keratosis

enlargement of nodule or ulcer of skin is the cause of gangrene or malignancy. Gangrene patients with amputation and cancer are not rare in the affected villages. Figure 14.5h–k show some arsenicosis patients from the affected districts of West Bengal and Bangladesh with Bowen's, Gangrene and carcinoma. In addition to above symptoms, we observed some common problems in arsenicosis patients such as burning sensation on whole body, weakness and respiratory problems.

Our study efforts of more than two decades had only identified a small percentage of the total population in As affected villages who had arsenical skin lesions. It appeared that only 15–20 % of the total number of people suffering from arsenicosis really come to our camp for clinical examination. This is due to following reasons: (i) affected people sometimes being isolated; (ii) young girls and women from conservative families do not want to be examined; (iii) people are sometimes frustrated and feel that there is no cure of this disease; and (iv) normally during the day most males were working in the field, school and college going children were also not available in the village.

14.4.2 Arsenic Affected Children

Infants and children are considered to be more susceptible to adverse effects of toxic substances (NRC 1999). From our field experience in West Bengal and Bangladesh, we observed that children (<11 years of age) usually do not show arsenical skin

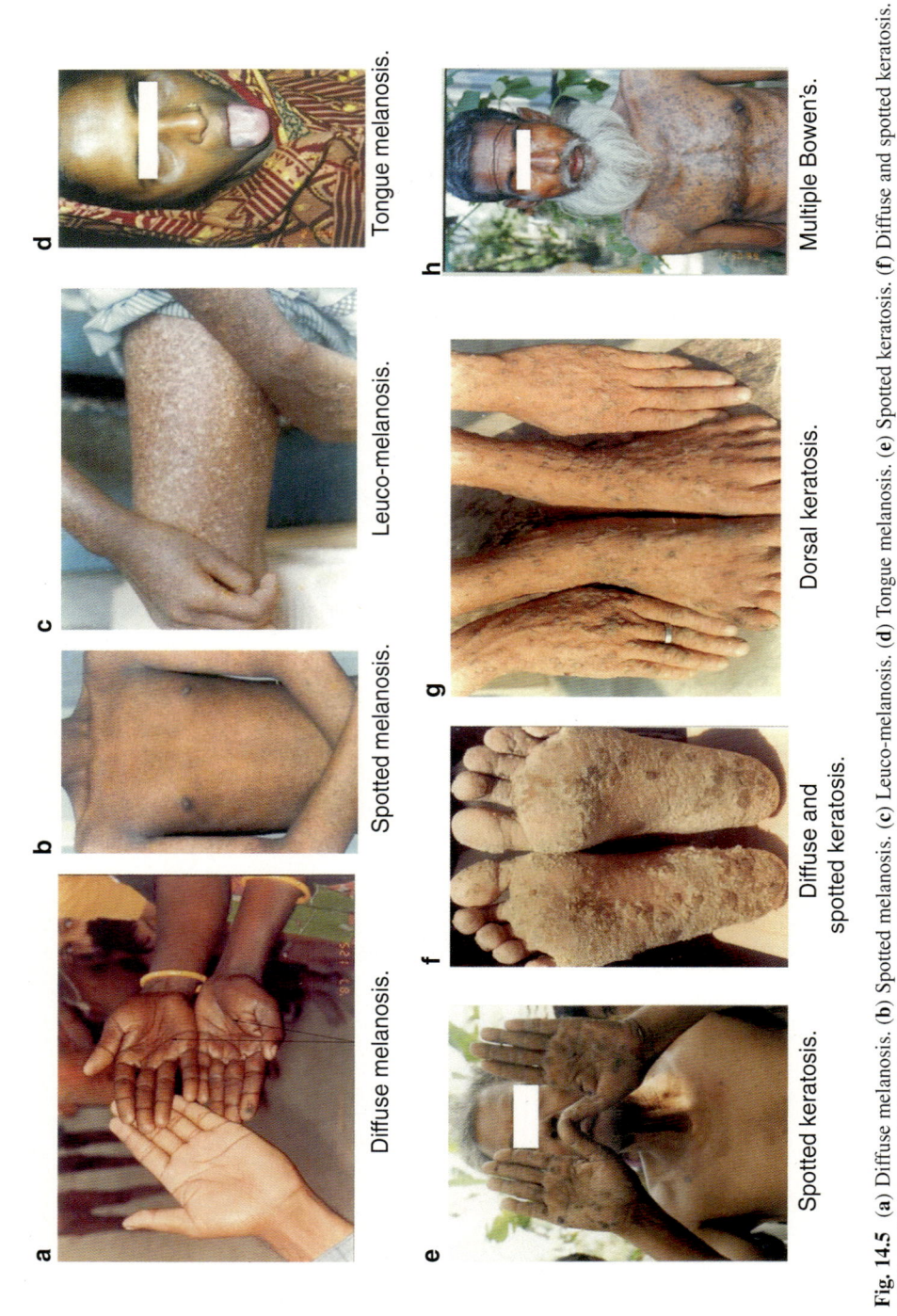

Fig. 14.5 (**a**) Diffuse melanosis. (**b**) Spotted melanosis. (**c**) Leuco-melanosis. (**d**) Tongue melanosis. (**e**) Spotted keratosis. (**f**) Diffuse and spotted keratosis. (**g**) Dorsal keratosis. (**h**) Multiple Bowen's. (**i**) Gangrene. (**j**) Leg amputation due to gangrene. (**k**) Squamous cell carcinoma. (**l**) A child arsenicosis patient from Bangladesh. (**m**) A child arsenicosis patient from West Bengal

i. Gangrene.

j. Leg amputation due to gangrene.

k. Squamous cell carcinoma.

l. A child arsenicosis patient from Bangladesh.

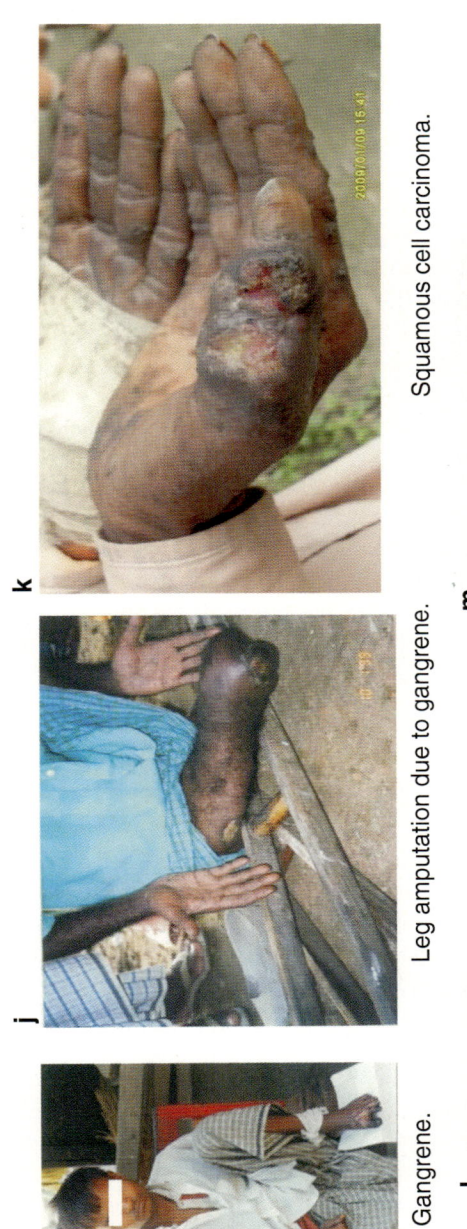

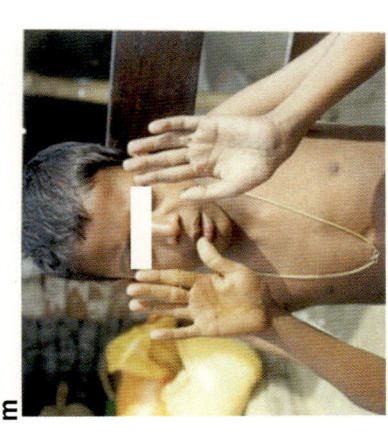

m. A child arsenicosis patient from West Bengal.

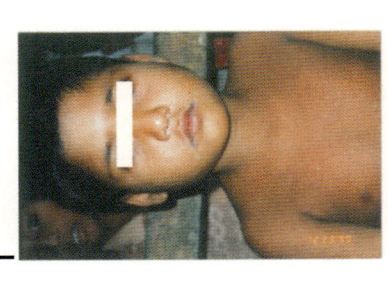

Fig 14.5 (continued)

lesions although their biological samples such as hair, nail and urine contain high levels of As. However, we noticed exceptions in cases when (i) As content in water consumed is very high ($\geq 1,000$ µg/l) and (ii) As content in drinking water is not so high (around 500 µg/l) but their nutrition status is poor (Chowdhury et al. 2000). Figure 14.5l–m show two children with arsenical skin lesions from West Bengal and Bangladesh, respectively.

In West Bengal, 778 (5.6 %) of the children out of 14,000 screened were detected with arsenical skin lesions (Chakraborti et al. 2009). We examined 4,864 children <11 years of age from affected villages of Bangladesh. Out of these, we have registered 298 children (6.1 %) with arsenical skin lesions. It was observed that children of As-contaminated areas of Bangladesh are more affected than those in West Bengal (Rahman et al. 2001). Analyses of 1,600 hair and nail samples from children in As-affected areas of Bangladesh and West Bengal demonstrated that about 90 % of the samples had As above normal levels (Chakraborti et al. 2004). The results showed that children in the As-affected areas of Bangladesh and West Bengal have a higher body burden, but less dermatological symptoms.

The major arsenical skin lesions observed in children are diffuse melanosis and spotted melanosis. We have not observed children suffering from severe (+++) stage of melanosis and keratosis. We have not observed any child <11 years of age with non-pitting edema, gangrene, Bowen's or cancer. We have registered children patients from 24 out of 31 districts surveyed in Bangladesh. We have further observed that children recover from diffuse melanosis (blackening of colour) and light spotted melanosis (+) quickly if they use safe water and with better diet and with vitamins recovery is enhanced. Mild keratosis (+) also disappears but the children having moderate to high spotted melanosis (++) and spotted keratosis (++) even after drinking safe water and nutritious food, do not recover completely.

Watanabe et al. (2007) reported the effects of chronic As exposure through consumption of contaminated water among 241 children (age range 4–15 years) living in two villages in Bangladesh. The As levels of the tube-well waters ranged from below detection limit to 535 µg/l (Watanabe et al. 2007). Guha Mazumder et al. (1998) reported pigmentation and keratosis among children (age <9 years) exposed to As-contaminated water >50 µg/l from South 24-Parganas district of West Bengal. Twelve (1.9 %) out of 613 boys and 9 (1.7 %) out of 536 girls had pigmentation due to high level of As exposure. Keratosis was found in one girl and three boys (Guha Mazumder et al. 1998). An epidemiological study of As in drinking water and the prevalence of respiratory effects in 7,683 participants of all ages in West Bengal, India showed that about 2 % of the children in the age groups ≤ 9 and 10–19 years manifested cough, shortness of breath and abnormal chest sounds (crepitations and/or bronchi) (Guha Mazumder et al. 2000). In Eurani village, Comilla district, Bangladesh four children below 16 years of age with arsenical skin lesions due to As-contaminated groundwater were identified as having sensory neuropathy, a more common finding in adults from the same village with arsenical skin lesions (Ahamed et al. 2006).

14.4.3 Neurological Involvements

We examined several groups of As-exposed subjects from three highly As-contaminated districts (Murshidabad, Naida and Bardhaman) of West Bengal for neurologic involvements due to As toxicity (Mukherjee et al. 2003, 2005). The presenting features of As induced peripheral neurological involvements from three groups of West Bengal is given in Table 14.8. Basu et al. (1996) reported a sensory predominant distal polyneuropathy in eight arsenicosis patients exposed to As-contaminated water (range: 200–2,000 µg/l) from West Bengal. Guha Mazumder et al. (1997) found abnormal electromyographic in 10 subjects and altered nerve conduction velocity and electromyographic in 11 subjects.

From Bangladesh, altogether 166 subjects (104 females and 62 males) were examined for arsenical neuropathy involvements (age ranged from 9 to 80 years) from Eruaini village of Bangladesh (Ahamed et al. 2006). The majority of the examined subjects presented with sensory features of distal paresthesias (57.2 %), limb pains (18.7 %), and distal hypesthesias (46.9 %) that outnumbered motor features of distal limb weakness or atrophy. The presenting features of As-induced neurological involvements were considerably higher in patients of Bardhaman district compared to other groups in West Bengal and Bangladesh. Hafeman et al. (2005) examined the association between As exposure and peripheral neuropathy in 137 Bangladeshi subjects, chronically exposed to As in drinking water and reported that increased As exposure, as measured by both cumulative and urinary measures, was associated with evidence of subclinical sensory neuropathy (Hafeman et al. 2005). Wasserman et al. (2004) reported children's intellectual function decrease with increasing As exposure, based on their study from the Araihazar thana of Bangladesh.

14.4.4 Obstetric Outcome

The obstetric outcomes due to As toxicity in the As-exposed women were examined from both West Bengal and Bangladesh (Mukherjee et al. 2005; Ahamed et al. 2006). The respondents of the As-exposed group of women from West Bengal were divided into: Group A, with As concentration in drinking water ranging 284–400 µg/l and Group B, with As concentration ranging 401–1,474 µg/l. Table 14.9 shows the results of these two groups compared with those of control group (Group C) from Midnipore district, West Bengal. Arsenical skin lesions were present in four out of six subjects of Group A and 9 out of 11 subjects of Group B. It was noted that the rate of spontaneous abortion increased with increase in As levels in water, i.e., 2 out of 21 pregnancies of Group A compared to 8 out of 44 pregnancies of Group B. In comparison to control Group C, both Groups A and B showed increasing trends in spontaneous abortion, preterm birth, and low birth weight rates. But no significant change was observed in preterm birth, low birth weight, and neonatal death rates between Groups A and B.

Table 14.8 Presenting features of As-induced peripheral neuropathy in West Bengal (India) and Bangladesh

Parameters	Bangladesh (Ahamed et al. 2006)		West Bengal (Mukherjee et al. 2003, 2005)					
	Eruani, Comilla (n=166)		Group A Murshidabad and Nadia (n=413)		Group B Bardhaman (n=38)		Group C Murshidabad (n=249)	
	No. of patients	Percentage	No. of patients	Percentage	No. of patients	Percentage	No. of patients	Percentage
Distal paresthesias	95	57.2	76	18.4	28	73.7	104	41.7
Limb pains	31	18.7	46	11.1	9	23.7	30	12.0
Hyperpathia/allodynia	13	7.8	17	4.1	15	39.5	15	6.0
Distal hypesthesias	78	46.9	133	32.2	21	55.3	86	34.5
Calf tenderness	19	11.5	15	3.6	7	18.4	12	4.8
Distal limb weakness/atrophy	14	8.4	30	7.3	4	10.5	38	15.3
Diminished or absent tendon reflexes	21	12.6	33	8.0	10	26.3	28	11.2

Table 14.9 Arsenic in drinking water and pregnancy outcome in West Bengal (India) and Bangladesh

Parameters	Bangladesh (Ahamed et al. 2006)						West Bengal (Mukherjee et al. 2005)					
	Exposed group A	Per 1,000 live births	Exposed group B	Per 1,000 live births	Non-exposed group C	Per 1,000 live births	Exposed group A	Per 1,000 live births	Exposed group B	Per 1,000 live births	Non-exposed group C	Per 1,000 live births
Number of observation	4		18		18		6		11		7	
Arsenic in water (μg/l)	201–500		501–1,200		<3		200–400		401–1,474		<3	
Skin lesions	2		11		Nil		4	–	9	–	–	–
Number of pregnancies	10		56		47		21	–	44	–	18	–
Spontaneous abortion (per 1,000 pregnancies)	2	200	15	267	8	170	2	95	8	182	1	55
Stillbirth (per 1,000 total births)	2	250	7	170	2	51	3	158	2	55	1	59
Preterm birth	–	–	5	147	1	27	8	470	12	353	2	125
Low birth weight	1	166	8	235	2	54	6	352	12	353	2	125
Neonatal death	1	166	7	205	5	135	1	59	2	59	–	–
Range of As in hair (μg/kg)	867–3,067		567–4,684				453–2,564		536–3,546		288–732	
Range of As in nail (μg/kg)	1,010–6,440		1,001–12,914									
Range of As in urine (μg/1.5 L)	860–3,892		15–3,992									

Twenty-two women drinking As-contaminated water (201–1,200 µg/l) were studied for their pregnancy outcome from Eruani village of Comilla district where most of the villagers were exposed to As-contaminated hand tube-well water for the past 8 years (1997–2004). The pregnancy outcome of the group was compared with 18 control group, from Dhaka city, drinking As-safe water (<10 µg/l). Table 14.9 also shows the pregnancy outcome of the exposed and control group from Bangladesh. Although several studies (Rahman et al. 2007, 2011; von Ehrenstein et al. 2006; Milton et al. 2005; Ahmad et al. 2001) have reported potential reproductive effects of As, the evidence of As and its effect on pregnancy outcome is not considered very conclusive due to lack of various other confounding factors such as other potential exposure, duration of exposure, minimum concentration of As exposure necessary for adverse effect on pregnancy, personal factors, congenital malformation, repeated child birth, malnutrition and causes of spontaneous abortion.

14.4.5 Methylation Capacity of As-Exposed Adults and Children

Urinary As is generally considered as the most reliable indicator of recent exposure to inorganic As and is used as the main bio-marker of exposure. However, due to different toxicity of As compounds, speciation of As in urine is generally considered to be more convenient for health risk assessment than measuring total As concentration. Additionally, it can give valuable information about the metabolism of As species within the human body.

Mean As levels and range of total As metabolites in urine and the percentage of urinary As species of adults and children of exposed group from West Bengal and Bangladesh is given in Table 14.10. For urine speciation study, 54 subjects (30 adults and 24 children) were selected from two As affected blocks (Raninagar II and Bhagowangola II) in the Murshidabad district of West Bengal exposed to As through drinking water (average As concentration in drinking water was 446 µg/l and the range was 118–1,003 µg/l). The As species [arsenite (AsIII), arsenate (AsV), monomethylarsonic acid (MMA) and dimethyl arsenic acid (DMA)] in the urine samples were separated and quantified by using high performance liquid chromatography inductively coupled plasma mass spectrometry (HPLC-ICP-MS) method.

It was observed that In-As and its metabolites in the examined subjects contained on average 17.2 % In-As (range: 7.7–43.4 %), 11.5 % MMA (range: 3.8–22.5 %) and 71.25 % DMA (range: 50.5–85.5 %). The efficiency of first and second methylation processes were assessed by the ratios of MMA/In-As and DMA/MMA, respectively. Higher values of MMA/In-As and DMA/MMA indicated higher methylation capacity. The results showed that the values of MMA/In-As ratio for adults and children were 0.77 and 0.69, respectively. These results indicated that first reaction of the metabolic pathway was slightly active in adults than children. But a significant increase in the values of DMA/MMA ratio in children than adults of exposed group (10.3 vs. 5.3 respectively) indicated second methylation step as more active in children than adults.

Table 14.10 Comparative mean As (μg/l) and range of total As metabolites in urine (μg/l) and the percentage of urinary As species of adults and children of West Bengal (India) and Bangladesh

Area	Sex	Age (year)	Sum of As species (μg/l)	Inorganic As		% $(As^{III}+As^V)$	%MMA	%DMA
				% As^{III}	% As^V			
West Bengal (Rahman 2004)	Adults (n = 30)	27.7 (12–55)	614.3 (192.2–1447.6)	16.95 (7.90–32.30)	3.42 (ND–22.10)	20.37 (7.90–43.40)	13.72 (6.10–22.50)	65.88 (50.50–80.20)
	Children (n = 24)	7.9 (1.5–11)	347.2 (39.3–1180.8)	10.5 (4.8–18.6)	2.8 (ND–12.1)	13.2 (7.7–20.5)	8.7 (3.8–17.2)	77.9 (68.7–85.5)
Bangladesh (Chowdhury et al. 2003)	Adults (n = 24)	32.25 (13–70)	484.14 (89.9–1325.3)	7.33 (ND–22.85)	11.21 (ND–24.74)	18.54 (ND-26.81)	17.20 (10.30–29.51)	64.26 (50.35–79.38)
	Children (n = 18)	8.11 (3–11)	563.49 (129.3–1089.5)	4.34 (ND–17.31)	11.84 (1.99–27.59)	16.18 (8.92–34.19)	10.65 (5.32–18.81)	73.17 (58.46–84.44)

For Bangladesh study, 42 urine samples were collected from Datterhat (South) village of Madaripur district and an average As concentration in their drinking water was 376 μg/l (range 118–620 μg/l). Results indicate that average total urinary As-metabolites in children is higher than adults and total As excretion per kg body weight is also higher for children than adults (Chowdhury et al. 2003). Results show the values of MMA/In-As ratio for adults and children are 0.93 and 0.74 respectively (Chowdhury et al. 2003). These results indicate that first reaction of the metabolic pathway is more active in adults than children. But a significant increase in the values of the DMA/MMA ratio in children than adults of exposed group (8.15 vs. 4.11 respectively) indicates 2nd methylation step as more active in children than adults (Chowdhury et al. 2003). It has also been shown that the distribution of the values of DMA/MMA ratio to exposed group decrease with increasing age (2nd methylation process) (Chowdhury et al. 2003). Thus from these results we may infer that children retain less arsenic in their body than adults. This may also explain why children do not show skin lesions compared to adults when both are drinking same As-contaminated water.

14.5 How to Combat the Present Arsenic Crisis

Elimination of arsenic crisis in the Ganges basin requires concerted action that includes: (i) a moratorium on the installation of more tube-wells in contaminated areas until all the installed tube-wells are checked for As contamination. The local and national governments should frame and implement regulation of new tube-wells. Around 95 % of the people in Bangladesh and West Bengal, India depend on tube-wells for drinking water. If the mouths of all safe tube-wells are coloured green, and unsafe wells are coloured red, villagers can use green tube-wells for drinking and cooking purposes, and red tube-wells for bathing, cleaning and washing.

We have disturbing evidence from West Bengal, India that previously safe tube-wells now show As contamination (Chakraborti et al. 2001). The currently safe tube-wells require monitoring every 3–6 months to track this new development; (ii) traditional water management like dug-well, three-Kalsi system and rain water harvesting with controls of bacterial and other chemical contamination; (iii) public awareness of As calamity; (iv) recognition that, so far, there is no effective therapy. Safe water and optimal nutrition are the only proven measures; and (v) world-wide effort by the scientific community addressing the problem that has put 100 million people in Bangladesh and West Bengal (India) at risks of cancer, vascular disease and other health complications.

Although tube-wells provide drinking water free of microbial contamination, the merciless exploitation of groundwater for irrigation without effective watershed management to harness huge surface water resources and rain water is seen as a gross miscalculation. In Bangladesh and West Bengal, there are huge surface resources of sweet water in the rivers, wetlands, flooded river basins and oxbow lakes. Per capita available surface water in Bangladesh is about 11,000 cubic metres. These areas are known as the land of rivers and have approximately 2,000 mm annual rainfall. Watershed management and villager participation are needed to assure appropriate utilization of these water resources.

Acknowledgements The authors thank the field workers of the School of Environmental Studies (SOES), Jadavpur University for their extensive help in the field sampling in arsenic affected villages of West Bengal in India. We also thank the field workers and the Management of the Dhaka Community Hospital, Bangladesh for their active participation in the field survey in Bangladesh. Financial support from SOES is greatly acknowledged.

References

Ahamed S, Sengupta MK, Mukherjee SC, Pati S, Mukherjee A, Rahman MM, Hossain MA, Das B, Nayak B, Pal A, Zafar A, Kabir S, Banu SA, Morshed S, Islam T, Rahman MM, Quamruzzaman Q, Chakraborti D (2006) An eight-year study report on arsenic contamination in groundwater and health effects in Eruani village, Bangladesh and an approach for its mitigation. J Health Popul Nutr 24:129–141

Ahmad SA, Sayed MHSU, Barua S, Khan MH, Faruquee MH, Jalil A, Hadi SA, Talukder HK (2001) Arsenic in drinking water and pregnancy outcomes. Environ Health Perspect 109:629–631

Arnold HL, Odam RB, James WD (1990) Disease of the skin. In: Clinical dermatology. W. B. Saunders, Philadelphia

Basu D, Dasgupta J, Mukherjee A, Guha Mazumder DN (1996) Chronic neuropathy due to As intoxication from geo-chemical source – a five year follow up. JANEI 1:45–47

BBS (2001) Bangladesh Bureau of Statistics-Statistical yearbook of Bangladesh

BGS-DPHE (1999) Groundwater studies for arsenic contamination in Bangladesh. Final report. Mott MacDonald Ltd.

BGS-DPHE (2001) Arsenic contamination of groundwater in Bangladesh. BGS technical report WC/00/19, British Geological Survey, Keyworth

BIS (2009) Draft Indian standard. Drinking water specification (Second revision of IS 10500). Doc: FAD 25(2047) C. Last Date for Comments: 24/12/2009

Bundschuh J, Litter MI, Bhattacharya P (2010) Targeting arsenic-safe aquifers for drinking water supplies. Environ Geochem Health 32:307–315

Burren M (1998) Small scale variability of arsenic in groundwater in the district of Meherpur, Western Bangladesh. MSc. thesis, University College, London

Chakraborti D (2011) Arsenic occurrence in groundwater. In: Nriago JO (ed) Encyclopedia of environmental health. Elsevier, Amsterdam

Chakraborti D, Basu GK, Biswas GK, Chowdhury UK, Rahman MM, Paul K, Chowdhury TR, Ray SL (2001) Characterization of arsenic bearing sediments in Gangetic Delta of West Bengal, India. In: Chappell WR, Abernathy CO, Calderon RL (eds) Arsenic exposure and health effects. Elsevier Science, New York

Chakraborti D, Rahman MM, Paul K, Chowdhury UK, Sengupta MK, Lodh D, Chanda CR, Saha KC, Mukherjee SC (2002) Arsenic calamity in the Indian sub-continent: what lessons have been learned? Talanta 58:3–22

Chakraborti D, Sengupta MK, Rahman MM, Ahamed S, Chowdhury UK, Hossain MA, Mukherjee SC, Pati S, Saha KC, Dutta RN, Zaman QQ (2004) Groundwater arsenic contamination and its health effects in the Ganga-Meghna-Brahmaputra plain. J Environ Monit 6:74–83

Chakraborti D, Das B, Rahman MM, Chowdhury UK, Biswas BK, Goswami AB, Nayak B, Pal A, Sengupta MK, Ahamed S, Hossain MA, Basu G, Chowdhury TR, Das D (2009) Status of groundwater arsenic contamination in the state of West Bengal, India: a 20 years study report. Mol Nutr Food Res 53:542–551

Chakraborti D, Rahman MM, Das B, Murrill M, Dey S, Mukherjee SC, Dhar RK, Biswas BK, Chowdhury UK, Roy S, Sorif S, Selim M, Rahman M, Zaman QQ (2010) Status of groundwater arsenic contamination in Bangladesh: a 14-year study report. Water Res 44:5789–5802

Chowdhury UK, Biswas BK, Chowdhury TR, Samanta G, Mandal BK, Basu GK, Chanda CR, Lodh D, Saha KC, Mukherjee SC, Roy S, Kabir S, Quamruzzaman Q, Chakraborti D (2000) Groundwater arsenic contamination in Bangladesh and West Bengal, India. Environ Health Perspect 108:393–397

Chowdhury UK, Rahman MM, Sengupta MK, Lodh D, Chanda CR, Roy S, Zaman QQ, Tokunaga H, Ando M, Chakraborti D (2003) Pattern of excretion of arsenic compounds [arsenite, arsenate, MMA(V), DMA(V)] in urine of children compared to adults from an arsenic exposed area in Bangladesh. J Environ Sci Health 38:87–113

Dhar RK, Biswas BK, Samanta G, Mandal BK, Chakraborti D, Roy S, Jafar A, Islam A, Ara G, Kabir S, Khan AW, Ahmed SA, Hadi SA (1997) Groundwater arsenic calamity in Bangladesh. Curr Sci 73:48–59

Dhar RK, Biswas BK, Samanta G, Mandal BK, Roychowdhury T, Chanda CR, Basu G, Chakraborti D, Roy S, Kabir S, Zafar A, Faruk I, Islam KS, Choudhury M, Arif AI (1998) Groundwater arsenic contamination and sufferings of people in Bangladesh may be the biggest arsenic calamity in the world. In: International conference on arsenic pollution of groundwater in Bangladesh: cause, effects and remedies, Dhaka, 8–12 Feb 1998

DPHE-DFID-JICA (2006) Final report on "Development of deep aquifer database and preliminary deep aquifer map (first phase)". Department of Public Health Engineering with support from Arsenic Policy Support Unit, DFID Bangladesh and JICA, Bangladesh

Farmer JG, Johnson LR (1990) Assessment of occupational exposure to inorganic arsenic based on urinary concentrations and speciation of arsenic. Br J Ind Med 47:342–348

Garai R, Chakraborty AK, Dey SB, Saha KC (1984) Chronic arsenic poisoning from tubewell water. J Indian Med Assoc 82:34–35

Guha Mazumder DN, Dasgupta J, Santra A, Pal A, Ghose A, Sarkar S, Chattopadhaya N, Chakraborti D (1997) Non-cancer effects of chronic arsenicosis with special reference to liver damage. In: Abernathy CO, Calderon RL, Chappell WR (eds) Arsenic exposure and health effects. Chapman & Hall, London

Guha Mazumder DN, Haque R, Ghosh N, De BK, Santra A, Chakraborty D, Smith AH (1998) Arsenic levels in drinking water and the prevalence of skin lesions in West Bengal, India. Int J Epidemiol 27:871–877

Guha Mazumder DN, Haque R, Ghosh N, De BK, Santra A, Chakraborti D, Smith AH (2000) Arsenic in drinking water and the prevalence of respiratory effects in West Bengal, India. Int J Epidemiol 29:1047–1052

Hafeman DM, Ahsan H, Louis ED, Siddique AB, Slavkovich V, Cheng Z, van Geen A, Graziano JH (2005) Association between arsenic exposure and a measure of subclinical sensory neuropathy in Bangladesh. J Occup Environ Med 47:778–784

IARC (International Agency for Research on Cancer) (2004) IARC monographs on the evaluation of carcinogenic risks to humans. Some drinking water disinfectants and contaminants including arsenic. World Health Organization, Lyon

International Conference on Arsenic (1995) International conference on arsenic in groundwater: cause, effect and remedy. School of Environmental Studies, Jadavpur University, Calcutta, 6–8 Feb 1995

International Conference on Arsenic (1998) International conference on arsenic pollution of groundwater in Bangladesh: cause, effects and remedies. School of Environmental Studies, Jadavpur University, India and Dhaka Community Hospital, Dhaka, 8–12 Feb 1998

Ioanid N, Bors G, Popa I (1961) Beitage zur kenntnis des normalen arsengehaltes von nageln and des Gehaltes in den Faillen von Arsenpolyneuritits [in German]. Dstch Z Gesamte Gerichtl Med 52:90–94

IPCS (2001) Arsenic and arsenic compounds, IPCS environmental health criteria 224. International programme on chemical safety, World Health Organization, Geneva

Johnston RB, Sarker MH (2007) Arsenic mitigation in Bangladesh: national screening data and case studies in three upazilas. J Environ Sci Health 42:1889–1896

Milton AH, Smith W, Rahman B, Hasan Z, Kulsum U, Dear K, Rakibuddin M, Ali A (2005) Chronic arsenic exposure and adverse pregnancy outcomes in Bangladesh. Epidemiology 16:82–86

Mukherjee SC, Rahman MM, Chowdhury UK, Sengupta MK, Lodh D, Chanda CR, Saha KC, Chakraborti D (2003) Neuropathy in arsenic toxicity from groundwater arsenic contamination in West Bengal, India. J Environ Sci Health 38:165–183

Mukherjee SC, Saha KC, Pati S, Dutta RN, Rahman MM, Sengupta MK, Ahamed S, Lodh D, Das B, Hossian MA, Nayak B, Mukherjee A, Chakraborti D, Dutta SK, Palit SK, Kaies I, Barua AK, Asad KA (2005) Murshidabad—one of the nine groundwater arsenic-affected districts of West Bengal, India. Part II: dermatological, neurological, and obstetric findings. J Toxicol Clin Toxicol 43:835–848

Nickson R, Sengupta CS, Mitra P, Dave SN, Banerjee AK, Bhattacharya A, Basu S, Kakoti N, Moorthy NS, Wasuja M, Kumar M, Mishra DS, Ghosh A, Vaish DP, Srivastava AK, Tripathi RM, Singh SN, Prasad R, Bhattacharya S, Deverill P (2007) Current knowledge on the distribution of arsenic in groundwater in five states of India. J Environ Sci Health 42:1707–1718

NRC (National Research Council) (1999) Arsenic in drinking water. National Academy Press, Washington, DC

NRC (National Research Council) (2001) As in drinking water–2001 update. National Academy Press, Washington, DC

Radloff KA, Zheng Y, Michael HA, Stute M, Bostick BC, Mihajlov I, Bounds M, Huq MR, Choudhury I, Rahman MW, Schlosser P, Ahmed KM, van Geen A (2011) Arsenic migration to deep groundwater in Bangladesh influenced by adsorption and water demand. Nat Geosci. doi:10.1038/NGEO1283

Rahman (2004) Present status of groundwater arsenic contamination in Bangladesh and detailed study of Murshidabad, one of the affected neighboring districts in West Bengal-India. PhD thesis, Jadavpur University

Rahman MM, Chowdhury UK, Mukherjee SC, Mandal BK, Paul K, Lodh D, Biswas BK, Chanda CR, Basu GK, Saha KC, Roy S, Das R, Palit SK, Quamruzzaman Q, Chakraborti D (2001)

Chronic arsenic toxicity in Bangladesh and West Bengal, India – a review and commentary. Clin Toxicol 39:683–700

Rahman A, Vahter M, Ekstrom EC, Rahman M, Mostafa AHMG, Wahed MH, Yunus M, Persson LA (2007) Association of arsenic exposure during pregnancy with fetal loss and infant death: a cohort study from Bangladesh. Am J Epidemiol 165:1389–1396

Rahman MM, Naidu R, Bhattacharya P (2009) Arsenic contamination in groundwater in the Southeast Asia Region. Environ Geochem Health 31:9–21

Rahman A, Vahter M, Ekström EC, Persson LA (2011) Arsenic exposure in pregnancy increases the risk of lower respiratory tract infection and diarrhea during infancy in Bangladesh. Environ Health Perspect 119:719–724

Rosenboom JW (2004) Department of Public Health Engineering (Bangladesh), Department for International Development (UK), UNICEF. Arsenic in 15 Upazilas of Bangladesh: water supplies, health and behaviour – an analysis of available data

Sengupta MK, Ahamed S, Hossain MA, Rahman M, Lodh D, Das B, Dey B, Paul B, Rey PK, Chakraborti D (2004) Increasing time trends in hand tubewells and arsenic contamination in affected areas of West Bengal, India. In: Proceedings of the 5th international conference on arsenic: developing country perspectives on health, water and environmental issues, Dhaka, 15–17 Feb 2004

van Geen A, Cheng Z, Jia Q, Seddique AA, Rahman MW, Rahman MM, Ahmed KM (2007) Monitoring 51 community wells in Araihazar, Bangladesh, for up to 5 years: implications for arsenic mitigation. J Environ Sci Health 42:1729–1740

Von Ehrenstein OS, Guha Mazumder DN, Smith MH, Ghosh N, Yuan Y, Windham G, Ghosh A, Haque R, Lahri S, Kalman D, Das S, Smith AH (2006) Pregnancy outcomes, infant mortality and arsenic in drinking water in West Bengal, India. Am J Epidemiol 163:662–669

Wasserman GA, Liu X, Parvez F, Ahsan H, Factor-Litvak P, van Geen A, Slavkovich V, LoIacono NJ, Cheng Z, Hossain I, Momotaj H, Graziano JH (2004) Water arsenic exposure and children's intellectual function in Araihazar, Bangladesh. Environ Health Perspect 112:1329–1333

Watanabe C, Matsui T, Inaoka T, Kadono T, Miyazaki K, Bae MJ, Ono T, Ohtsuka R, Bokul ATMMH (2007) Dermatological and nutritional/growth effects among children living in arsenic-contaminated communities in rural Bangladesh. J Environ Sci Health 42:1835–1841

World Bank Policy Report (2005) Towards a more effective operational response: arsenic contamination of groundwater in South and East Asian countries, vol I and II. World Bank, Washington DC

Yu G, Sun D, Zheng Y (2007) Health effects of exposure to natural arsenic in groundwater and coal in China: an overview of occurrence. Environ Health Perspect 115:636–642

Chapter 15
Impact of Arsenic Contaminated Irrigation Water on Some Edible Crops in the Fluvial Plains of Bihar

N. Bose, A.K. Ghosh, R. Kumar, and A. Singh

15.1 Introduction

With increasing detection of arsenic contaminated aquifers upstream of the Bengal delta plain in last decade, groundwater contamination by arsenic is being perceived as the most severe health hazard over entire Gangetic Plain. The issue of mass arsenic poisoning was initially linked to direct consumption of arsenic contaminated groundwater through hand pumps, but the growing realization that such aquifers are also being increasingly tapped for obtaining irrigation water fuelled fears of arsenic ingestion by the affected population both directly through drinking water sources and indirectly through the water-soil-crop route in this highly fertile region. A significant Arsenic contamination has been reported in soil and major cereals cultivated in Gangetic plains of Bihar (Maner Block of Patna District) where source of irrigation water is mainly tube wells with very high arsenic content (Singh and Ghosh 2011).

Rice, being the staple diet in the Indian sub-continent, has drawn considerable attention of scientists. WHO (1992) permissible limit of arsenic is 10 µg/L for drinking water, FAO (1985) permissible limit of arsenic is 100 µg/L for irrigation water. The permissible limit of concentration of arsenic in rice grain as per WHO recommendation is 1.0 µg/g dry weight (Bhattacharya et al. 2009). It has been estimated that irrigating a rice field with groundwater containing 550 µg/L of arsenic with a water requirement of 1,000 mm results in an approximate addition of 5.5 kg of arsenic per ha per annum (Huq et al. 2006). As rice is cultivated in As-contaminated soils under anaerobic conditions, at which As is highly available for plant uptake (Meharg 2004; Carbonell-Barrachina et al. 1998), rice grain has been reported to

N. Bose (✉) • A.K. Ghosh • R. Kumar • A. Singh
Department of Environment and Water Management, A.N. College, Patna, Bihar, India
e-mail: nupur.bose@gmail.com

accumulate arsenic up to 2.0 µg/g (Meharg and Rahman 2003). In 2009, 18 districts in Bihar were declared as "Arsenic affected" on the basis of contaminated drinking water (hand pump) sources.

15.2 Materials and Methods

15.2.1 Sample Collection

The eastern Indian state of Bihar is located between latitudes 24°20′10″N and 27°31′15″N, and longitudes 82°19′50″E and 88°17′40″ E, and occupies 94,163 sq. km. in the Mid Ganga plains. Bihar lies mid-way between the humid West Bengal in the east and the sub humid Uttar Pradesh in the west. This fluvial plain is constituted of 16 river basins, including the major Ganga basin. With a total population of 103 million spread over its 38 districts, the average district population is 2.73 million. The state of Bihar has an agro-based economy (Department of Agriculture, Govt. of Bihar 2010), in which there is heavy and increasing dependence on ground water irrigation. Out of about 44,000 km^2 of gross irrigated area 24,000 km^2 or 54.5 % is being irrigated by the groundwater resources. Keeping in view the vast scope for accelerated development of irrigated agriculture based especially on groundwater wells, Bihar Ground Water Irrigation Scheme (BIGWIS) aimed at providing irrigation to further 9,280 km^2 of agricultural land of the State by installing 464,000 units of private shallow tube wells with pump sets dug over a period of 3 years ending 2011–2012 (Minor Irrigation Dept., Govt. of Bihar 2009).

Field tests were conducted in rural agricultural belts to determine the infiltration of arsenic in the irrigation water-soil-crop route. Water samples of 406 functioning bore wells were collected. Basic information on depth of borewells, periods of their use, and pH values of the sample collected at source were obtained. Upon laboratory confirmation of presence of arsenic in the water, soil and plant samples, pot experiments were undertaken in which ten locally produced rice varieties were grown under controlled laboratory conditions. In pot experiments, the young seedlings of rice were irrigated by different concentrations of arsenic spiked water of 50, 500, 1,000, 1,500, 2,000 and 5,000 µg/L.

15.2.2 Preparation of Soil Samples for Arsenic Analysis

The collected root soil samples from both field and laboratory grown plants were converted into powdery form. Each sample was then dried and ground to pass through a 0.5 mm sieve. Five grams of powdered and dried soil was taken in a 100-ml conical flask and 50 ml of 0.5 M NaHCO$_3$ (sodium bicarbonate) solution was added. Then the whole material was put in for 6 h. in a horizontal shaker, after which the suspension was filtered through Whatman filter paper No. 42. The filtrate was collected for arsenic analysis with UV spectrophotometer using SDDC method.

15.2.3 Plant Sample Digestion Procedure for Arsenic Analysis

The plant samples were dried, grounded and allowed to pass through a 0.5 mm sieve. After filtration, 0.5–1.0 g of filtered sample was transferred into digestion tubes. 6–8 ml of nitric acid was mixed in the sample. The samples were placed in the hot oven for digestion and the temperature was set at 40 °C for 8 h, at 60 °C for 3 h, at 80 °C for 3 h, at 120 °C for 2 h and finally at 140 °C for 4 h. The samples were then digested to reduce its quantity to reduce the total volume to about 1.0 ml. After digestion the crucibles were removed from hot oven, and allowed to cool. The leftover acid on the surface of the crucibles were washed by deionized water. The volume of the sample was maintained at 10–20 ml with deionized water. After dilution the samples were allowed to mix through vortex mixer. The samples were allowed to settle overnight before filtering through Whatman Filter No. 42. On the next day the samples were filtered again through Whatman Filter No. 42, and diluted up to 35 ml. for arsenic detection using SDDC Method on UV1 spectrophotometer (ImamulHuq and Alam 2005).

15.2.4 GIS Applications for Spatial Analysis

In order to prepare the primary database of arsenic contamination in the food chain in Bihar, coordinates of the sources of all types of field samples (irrigation water, rice and tomato cultivars, and soil) were recorded on GPS. Thereafter, the political division map of Bihar was geo-referenced, where each district could be individually assessed for the spatial spread of the stated contaminations. The entire database was entered in the mapping software (Geomatica 10). Layers of points of contaminated irrigation water samples, plant samples and soil samples were produced for final analysis. Isolines and Choropleths were prepared to highlight spatial variations in the intensity of arsenic contaminations in irrigation bore wells, and their impact on plants and root soil.

15.3 Results and Discussion

15.3.1 Analysis of Field Work Data

More than 80 % of the irrigation bore wells had depths ranging from 20 to 80 ft in the new floodplains. However, in a few exceptional areas, maximum bore well depths of 300 ft (Buxar district) were also found. In the older alluvial soil at the southern plateau edge of Chotanagpur Plateau, aquifers at depths more than 100 ft were common.

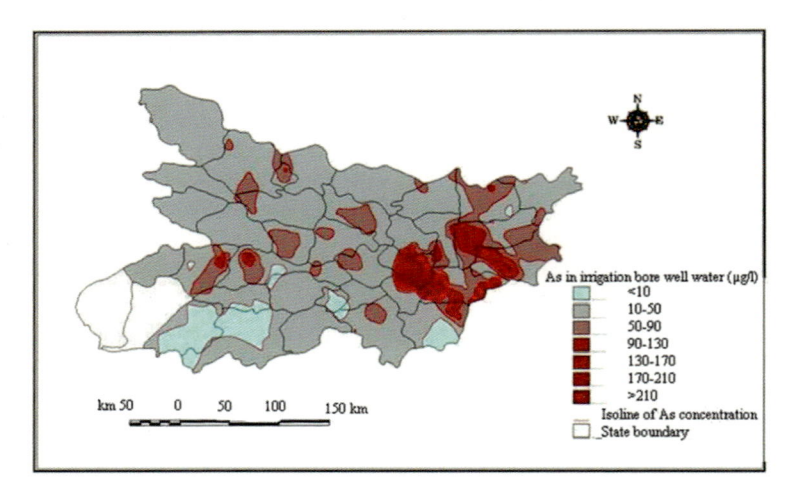

Fig. 15.1 Irrigation bore wells: Spatial patterns of As contamination, Bihar

Figure 15.1 reveals the incidences of high arsenic contaminated irrigation water from two zones in the fluvial plains. The hotspots of the eastern districts are frequent with Khagaria district with a mean district contamination value of 327.31 µg/l, registering the highest water contamination of 857 µg/l. An important fact to be noted was that the contaminations form a continuous belt in the Kosi river basin, while being fragmented along the Ganga. They also indicate the extension of the Bengal contaminations at the political borders. In contrast, western Bihar has more diffuse arsenic contaminations in irrigation bore wells, which is in contrast to the already documented 1,861 µg/l arsenic level in one of the drinking water hand pumps in Bhojpur.

The mean contamination values in irrigation water also exhibit similar trends in Fig. 15.2. Arsenic level of <10 µg/l along the southern boundary located on plateau edge were also found, thereby extending the area of geogenic arsenic contaminations from the Himalayan Terai in the north to the northern boundary of Chotanagpur plateau in the south.

European Committee Recommendation for Agricultural Soil (Bhattacharya et al. 2009) states the acceptable limit of arsenic in soil to be 20 mg/kg. Arsenic content of 100 µg/kg were detected in the root soil samples in Sheohar, Sitamarhi, Supal and Madhepura districts in the north Bihar plains; and in Buxar, Bhojpur, Nawada, Nalanda, Sheikpura and Lakhisarai districts in south Bihar Plains, as depicted in Fig. 15.3. Higher As concentrations in soil (>500 µg/kg) were found in Supaul, Buxar and Nawada. Four zones of different arsenic contaminated cropland soils were identified: Madhubani-Supaul-Madhepura belt and Champaran-Sheohar-Sitamarhi belt in north Bihar; Nawadah-Nalanda-Sheikpura-Lakhisarai belt in south Bihar; and Bhojpur-Buxar belt in western Bihar.

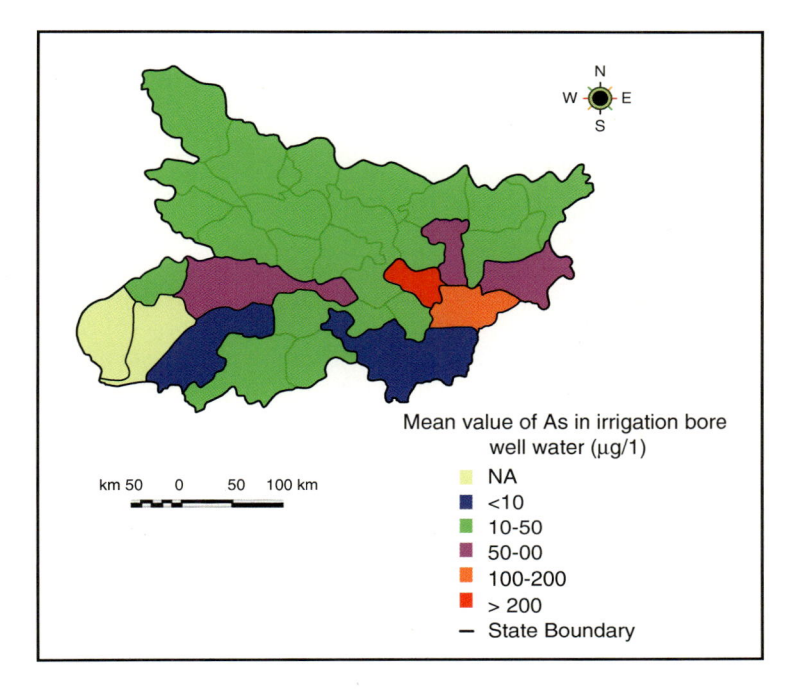

Fig. 15.2 District-wise mean As values in irrigation bore wells, Bihar

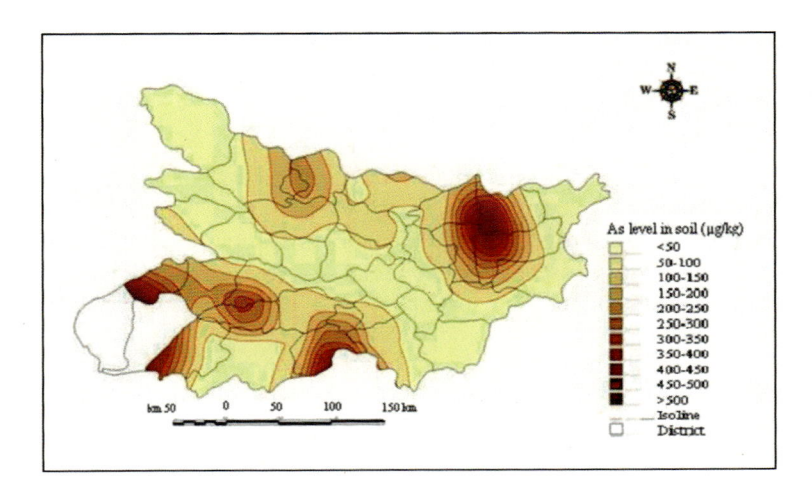

Fig. 15.3 Isolines of As values in root soil of plant samples, Bihar

In the second belt, arsenic contaminated irrigation water was being drawn from deeper wells of up to 170 ft. deep. This district also lies on the fringe of the Chotanagpur Plateau. These zones coincide with the upper reaches of the feeder channels of the Ganga, namely, the Kosi Basin and the Burhi-Gandak Basin in north Bihar; the Harohar Basin and the Sone-Kanhar and Kao-Gangi Basin.

Whole rice plants, rice grains and tomato fruit collected from the cropland irrigated from the water sample source were tested for arsenic content. The other crops grown were maize (maximum As uptake of 0.026 µg/g), sunflower and sugarcane crops, sugarcane juice having 52 µg/l of arsenic. In south Bihar, however, rice is the major *Kharif* crop, and accounted for much of the plant samples collected. All the rice grain samples collected tested positive for arsenic accumulations. Maximum plant uptake of arsenic was recorded from Khagaria district (0.052 µg/g) in north Bihar. In the south, arsenic uptake by rice plants in the range 0.030–0.040 µg/g predominated. In tomato fruits, 0.040–0.050 µg/g of As were detected in the eastern Bihar plains, thereby confirming this fruit also to be an accumulator of ground water arsenic.

15.3.2 Analysis of Pot Experiment Data

The following 10 local varieties of Oryza sativa L. were selected for pot experiments, after consultation with the Director of ICAR, Patna Centre:

- *Saroj SDV, Turantha ESDV, PNR 381 MDV, IET 20800, Rajendra Sweta, BPT 5204* and *Sugandha*—the seeds were procured from ICAR, Patna.
- *Sarna, Katarni* and *Sonam*—the seeds were purchased from local market on the basis of larger market demand of these three cultivars.

The pots were arranged in series of five pots each for the five rice varieties – each series having one pot labeled "Control", and the other 4 pots bearing the label of the particular amount of arsenic contamination in the irrigation solution. The solutions of concentrations 50 µg/l, 500 µg/l, 1,000 µg/l, 1,500 µg/l, 2,000 µg/l and 5,000 µg/l were prepared from 1 mg/l As stock solution in the laboratory. For each level of As solution, two pots were taken to guarantee adequate number of available plant specimens for final tests. The control plants were irrigated with plain water, while the four remaining pots of each variety were irrigated with As solutions of the aforementioned concentrations, respectively.

Two local cultivars of the tomato plant (*Lycopersicom esculentum*), *Gulshan* and *S 32*, were also tested for arsenic uptake under controlled laboratory conditions and periodically irrigated with As solutions of 50, 500, 1,000 and 2,000 µg/l.

In pot experiments, the responses of rates of arsenic uptake, plant growth and appearance for different varieties were varied. Fluctuations in growth rates occurred, as the length of the longest surviving plant in Control and each variety varied on each date. Growth observations in the early post-transplantation stage, mid-growth period and mature stage of the cultivars indicate that the irrigation inputs of As solution

Table 15.1 Arsenic accumulation in selected local rice cultivars

Provided As concentration (µg/l)	As accumulation in selected local rice cultivars (µg/g)			
	Turantha	PNR-381	Sarna	Katarni
Control	0.01	0.01	0.003	0.002
50	0.013	0.028	0.045	0.023
500	0.136	0.042	0.055	0.052
1,000	0.349	0.131	0.039	0.165
1,500	0.122	0.15	0.033	0.044
2,000	0.077	0.101	0.017	0.03
5,000	0.065	0.05	0.012	0.007

Table 15.2 Arsenic accumulation in selected local rice cultivars

Provided As concentration (µg/l)	As accumulation in selected local rice cultivars (µg/g)			
	Sonam	Sugandha	Saroj	IET
Control	0.008	0.053	0.005	UR
50	0.045	UR	UR	0.007
500	0.057	UR	UR	0.017
1,000	0.029	UR	UR	0.019
1,500	0.051	0.004	UR	–
2,000	0.05	UR	UR	0.026
5,000	0.06	UR	UR	–

UR under range

resulted in selective enhancement of plants' growth. In the early post transplant and mid-growth phases, growth was enhanced in *Saroj SDV, IET* and *Katarni* at 500 µg/l of As solution, in *Sugandha, Rajendra Shweta* and *Turantha ESDV* at 1,000 µg/l, in *Sarna* at 1,500 µg/l, in *Sonam* and *BPT* at 2,000 µg/l. In *PNR 381 MDV* maximum plant growth of up to 60 cm occurred during mid-growth phase at 5,000 µg/l of As solution, after which period the plant did not survive. Thereafter, with increasing concentrations of arsenic solution plant growths were retarded, with gradual chlorosis at leaf margins, indicating toxic effects of As accumulation.

It is recognized, and is also reflected in this study (Tables 15.1, 15.2 and 15.3), that rates of arsenic uptake between different rice varieties and in different countries vary greatly (Brammer and Ravenscroft 2009). Arsenic amount varied between 0.058 and 1.835 mg/kg in 13 different varieties in Bangladesh (Meharg and Rahman 2003), while another study showed variations of arsenic concentrations in the range of 0.032–0.046 mg/kg in aromatic rice from the countries of the Indian subcontinent (Duxbury and Zavala 2005).

At higher levels of As concentration there was absence of seed formation, indicative of "Straighthead disease" (Rahman et al. 2007), a physiological disorder of rice (*Oryza sativa* L.) characterized by sterility of the florates/spikelets leading to reduced grain yield. There is a direct correlation of increase of soil As concentration with the severity of straighthead in rice plants (Rahman et al. 2007).

Table 15.3 Arsenic accumulation in selected local rice cultivars

Provided As concentration (µg/l)	As accumulation in local rice cultivars (µg/g)	
	Rajendra Sweta	BPT
Control	UR	UR
50	0.017	UR
500	0.021	UR
1,000	0.023	UR
1,500	–	–
2,000	0.03	UR
5,000	–	–

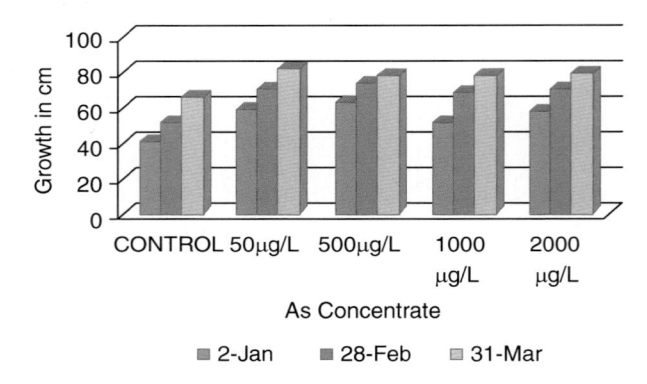

Fig. 15.4 Growth trends in *Gulshan*

In the pot experiments, three arsenic resistant rice cultivars, with "under range" readings in rice grains were identified: *Sugandha, Saroj* and *BPT 5204*. The other cultivars registered peaks of As uptake concentrations of 1,000–1,500 µg/l. Thereafter, there were abrupt decreases in As accumulation at different higher As solutions. The results are indicative of a possible inhibited As absorbing mechanism at higher concentrations. Previous pot studies (Jahiruddin et al. 2004) showed that higher As concentration in irrigation water (above 1.0 mg/l) resulted in lower yield of a local rice variety (BR-29).

In the tomato cultivars, prominent enhancements of growth in both the cultivars were noted commensurate with increased inputs of As concentrates as compared to the control. In the early stages, height of 64 cm was attained at 500 µg/l, in the later stages, the growth rates picked up at 50 µg/l in *Gulshan*. Figure 15.4 reveals the persistence of uniform high growth under As concentrate inputs of increasing levels.

In tomato cultivar *S 32* (Fig. 15.5), similar trends were noted, the magnitude of growth being higher in this case. Another feature was that among all the dates of plant growth measurements, the plant under 2,000 µg/L registered highest growth.

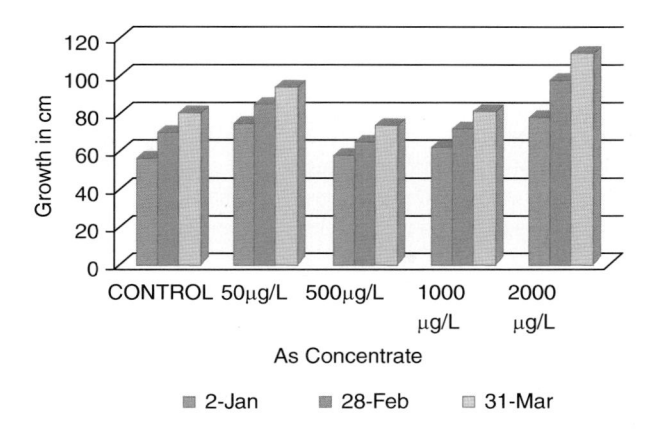

Fig. 15.5 Growth trends in *S* 32 stage of the plant

This is indicative of a nutrient absorption enhancement property of arsenic in irrigation water under 50 and 500 µg/l of as concentrate in *Gulshan* and under 2,000 µg/l in S 32 cultivars.

Lush growth was prominent in the growing period in all treated plants, although chlorosis had set in during the mature phase. In both the tomato cultivars, the flowering was not recorded even at mature stage of the plant. The possible reason for this is that due to presence of tall trees, sunlight infiltration was reduced, which could have interfered with the light exposure duration and inhibition of synthesis of Florigen hormone, thereby inhibiting flowering. However, under such conditions, appearance of flower and fruit occurred only in *Gulshan*, under 1,000 and 2,000 µg/l concentrate. This implies that inspite of the inhibitory effect of reduced sunlight on Florigen, high arsenic inputs triggered the growth process in these plants. Further, not much research has been conducted on the impact of arsenic contaminated irrigation water on the metabolism activity in edible plants, due to which the findings of this research could not be collated with plant chemistry at present.

15.4 Conclusion

It is now established that the major tributaries of the Ganga in Bihar, both of Himalayan and Peninsular origin, have hydro-geological environments that promote the release of arsenic in the aquifers along the zones where they debouch on to the Bihar Plains, thereby increasing the area of geogenic arsenic. Hence, the geology of arsenic hotspots in Bihar requires further investigation. The patterns of arsenic content of the soil may be correlated with the geomorphological diversities, particularly in south-west and south-east Bihar in future studies.

This is the first confirmatory study of arsenic infiltration in the food chain of Bihar. The development of cultivars that assimilate less As, or restrict As translocation to fruits/seeds would lead to reduced dietary exposure to As (Meharg and Hartley-Whitaker 2002), as found in the experiments on *Saroj, BPT* and *Sugandha* rice varieties.

The physiology of As uptake and retention by different varieties of rice grown in Bihar need to be investigated further. Studies are also required to understand the mechanism behind the enhanced plant growth of these rice varieties. Judicious use of groundwater for irrigation, water conservation for agriculture, further identification and adoption of arsenic resistant edible crop varieties, and constant monitoring and evaluation of arsenic release through the water-soil-crop route are immediate options to reduce dietary intake of arsenic through the water-soil-crop route.

References

Bhattacharya P, Samal AC, Majumdar J, Santra SC (2009) Transfer of arsenic from groundwater and paddy soil to rice plant (*Oryza sativa* L.): a micro level study in West Bengal, India. World J Agric Sci 5(4):425–431

Brammer H, Ravenscroft P (2009) Arsenic in groundwater: a threat to sustainable agriculture in South and South East Asia. Environ Int 35:647–654

Carbonell-Barrachina AA, Aarabi M, DeLaune RD, Gambrell RP, Patrick WH Jr (1998) The influence of arsenic chemical form and concentration on Spartina patens and Spartina alterniflora growth and tissue arsenic concentration. Plant Soil 198:33–43

Department of Agriculture, Govt. of Bihar, India (2010) website: http://krishi.bih.nic.in

Duxbury JM, Zavala YJ (2005) What are safe levels of arsenic in food and soils. CIMMYT/USGS. Symposium on the behaviour of arsenic in aquifers, soils and plants: implications for management, Dhaka, 16–18 Jan 2005. Centro Internacional de Mejoramiento de Maíz y Trigo and the U.S. Geological Survey

FAO (1985) Water quality guidelines for maximum crop production. Food and Agricultural Organization/UN. www.fao.org/docrep/T0551E.2006/9/13

Huq SMI, Joardar JC, Parvin S, Correll R, Naidu R (2006) Arsenic contamination in food-chain: transfer of arsenic into food materials through groundwater irrigation. J Health Popul Nutr 24(3):305–316

ImamulHuq SM, Alam MD (2005) A handbook on analyses of soil, plant and water. In: ImamulHuq SM, Alam MD (eds). Momin Offset Press, Dhaka

Jahiruddin MMR, Islam AL, Shah S, Islam M, Ghani A (2004) Effects of arsenic contamination on yield and arsenic accumulation in crops. In: Shah MAL et al (eds) Workshop on arsenic in the water–soil–crop systems, vol 147. BRRI, Gazipur, pp 39–52

Meharg AA (2004) Arsenic in rice – understanding a new disaster for South-East Asia. Trends Plant Sci 9:415–417

Meharg AA, Hartley-Whitaker J (2002) Arsenic uptake and metabolism in arsenic resistant and nonresistant plant species. New Phytol 154:29–43

Meharg AA, Rahman MM (2003) Arsenic contamination of Bangladesh paddy field soil: implication for rice contribution to arsenic consumption. Environ Sci Technol 37(2):224–234

Minor Irrigation Department, Govt. of Bihar (2009) http://biharirrigation.nic.in

Rahman MA, Hasegawa H, Rahman MM, Miah MAA, Tasmin A (2007) Straighthead disease of rice (*Oryza sativa* L.) induced by arsenic toxicity. Environ Exp Bot 62(1):54–59

Singh SK, Ghosh AK (2011) Entry of arsenic into food material – a case study. World Appl Sci J 13(2):385–390

WHO (1992) Guideline for drinking water quality, vol 1, 2nd edn, Recommendation. World Health Organization, Geneva, p 41

Chapter 16
A Greenhouse Pot Experiment to Study Arsenic Accumulation in Rice Varieties Selected from Gangetic Bengal, India

Piyal Bhattacharya, Alok C. Samal, and Subhas C. Santra

16.1 Introduction

It is predicted that around 100 million people living in the Ganga-Meghna-Brahmaputra plain are at the risk of serious arsenic toxicity through exposure of contaminated groundwater (Chakraborti et al. 2008). Groundwater arsenic contamination in the Gangetic Bengal has been termed as the largest mass poisoning in the history of human kind (Smith et al. 2000). Arsenic pollution has spread in fourteen out of total nineteen districts of Gangetic Bengal (Chakraborti et al. 2009). Application of arsenic-contaminated groundwater for irrigation in Gangetic Bengal has shown to influence accumulation of arsenic in rice, the major staple food in West Bengal (Meharg 2004; Signes-Pastor et al. 2008; Meharg et al. 2009; Bhattacharya et al. 2010a; Samal et al. 2011; Banerjee et al. 2013; Santra et al. 2013). Rice is an efficient accumulator of arsenic than any other cereal crops (Su et al. 2010) and consumption of rice has been termed as an important source of inorganic arsenic intake to human body (Meharg et al. 2009).

Greenhouse pot experiments conducted with Bangladeshi rice varieties have showed significant differences in the accumulation of arsenic (Azad et al. 2009, 2013; Norton et al. 2009). Delowar et al. (2005) reported that the accumulation of arsenic in rice grain was in the range 0–0.14 mg kg^{-1} which was cultivated with 0–20 mg l^{-1} of arsenic containing water. Analyzing two widely cultivated rice varieties

---

P. Bhattacharya (✉)
Department of Environmental Science, University of Kalyani,
Kalyani, West Bengal 741235, India

Department of Environmental Science, Kanchrapara College,
Kanchrapara, West Bengal 743145, India
e-mail: piyal_green@yahoo.co.in

A.C. Samal • S.C. Santra
Department of Environmental Science, University of Kalyani,
Kalyani, West Bengal 741235, India

in Bangladesh, Rahman et al. (2007) reported that the BRRI *dhan* 28 and BRRI hybrid *dhan* 1 had difference in the amount of arsenic accumulation (0.5 ± 0.0 and 0.6 ± 0.2 mg kg^{-1} dry weight of arsenic, respectively). Rahman et al. (2008) by studying five different hybrid as well as non-hybrid rice samples concluded that the arsenic translocation from root to shoot (straw) and husk was higher in the hybrid variety (BRRI hybrid *dhan* 1) as compared to those of non-hybrid varieties (BRRI *dhan* 28, BRRI *dhan* 29, BRRI *dhan* 35 and BRRI *dhan* 36). Azad et al. (2009) observed an increase in the grain arsenic uptake of transplanted aman rice with the increase of arsenic treatment in soil. Abedin et al. (2002b) found that 30–50 mg kg^{-1} arsenic containing soil produced rice grains with arsenic levels exceeding the WHO recommended permissible limit of 1 mg kg^{-1}. In our previously conducted study on some other rice varieties, all the studied high yielding and hybrid varieties (*Ratna*, IET 4094, IR 50 and *Gangakaveri*) were found to be higher accumulator of arsenic as compared to all but one local rice variety, *Kerala Sundari* (Bhattacharya et al. 2001). Azad et al. (2013) have recently reported the accumulation of arsenic in the range 0.06–0.47 mg kg^{-1} through a greenhouse pot experiment conducted in Bangladesh.

Thus, a greenhouse pot experiment was conducted to investigate the accumulation and distribution of arsenic in the different fractions of rice plant with increasing soil arsenic treatments (5, 10, 20 and 30 mg kg^{-1} dry weights) on six selected rice varieties (four high yielding varieties MTU 7029, IET 5656, MTU 1010 and CNHR 3, and two local varieties *Nayanmani* and *Danaguri*). The major objective of the present study was to identify the rice varieties that are resistant to arsenic phytotoxicity. The findings would have significant impacts on agriculture and public health of arsenic-contaminated 14 districts of Gangetic Bengal.

16.2 Materials and Methods

16.2.1 Experimental Condition

The pot culture experiment on different rice (*Oryza sativa* L.) varieties was carried out in a greenhouse at the Department of Environmental Science, University of Kalyani. The experimental site was selected on the basis of having good sunshine throughout the day. Although the experiment was conducted in a greenhouse, the environmental conditions inside the greenhouse were not controlled. The greenhouse was only used to protect the experiment from natural calamities (such as heavy rainfall, northwester wind, etc.) and disturbances by animals.

16.2.2 Soil Collection and Pot Preparation

Soil was collected from the campus of University of Kalyani at a depth of 0–15 cm. The physico-chemical properties of soils used for pot experiments are given in Table 16.1. Initial arsenic content of the soils prior to treatment was 2.3 ± 0.07 mg kg^{-1}

Table 16.1 Physico-chemical properties of experimental pot soil

Soil parameters	Range
Clay (%)	66–69
Sand (%)	9–15
Silt (%)	17–24
Texture	Clay loam
pH	7.8 ± 0.18
Organic carbon (%)	0.92 ± 0.13
Total nitrogen (%)	0.14 ± 0.02
Available phosphorous (mg kg^{-1})	14 ± 1
Total arsenic (mg kg^{-1})	2.3 ± 0.07

dry weights. After collection, the soil was air dried for 7 days and aggregates were broken by gentle crushing. The materials such as dry roots, grasses, stones and plastics were removed and the soil was thoroughly mixed to homogenize. Earthen pots (40 × 40 cm) were used for rice cultivation. The pots were designed to prevent the loss of water soluble arsenic from pots (Rahman et al. 2007). About 10 kg of soil was taken in total 90 pots comprising four different arsenic treatments (5, 10, 20 and 30 mg kg^{-1} dry weights) along with one control treatment (no arsenic dosing), each with three replications for the six different rice plant varieties. The arsenic was applied in the form of sodium arsenate (Na$_2$HAsO$_4$), which can easily convert to arsenite under reducing and submerged condition of paddy soil (Abedin et al. 2002a). Chemical fertilizers or nutritional solutions were not added to pot soil.

The tap water, used for irrigation, contained arsenic below the detection limit (<0.03 μg l^{-1}). Thus, there was no chance of arsenic input from the tap water to the pot soil. After the application of arsenic, soils were left in the pots for 2 days without irrigation. Then tap water was used to irrigate the pots to make the soil clay suitable for rice seedling transplantation. About 3–4 cm water level above the soil surface was maintained in the pots before and after seedling transplantation. The water level was maintained in each pot throughout the growth period. Irrigation was stopped before 10 days of harvest (Azad et al. 2009).

16.2.3 Selection of Rice Varieties and Seedling Transplantation

Four high yielding rice varieties MTU 7029 (*Swarna*), IET 5656, MTU 1010 and CNHR 3 and two local varieties *Nayanmani* and *Danaguri* were selected through germination test for this greenhouse pot experiment. Rice seedlings of 21 days old were carefully uprooted from nursery-bed and transplanted to pots under flooded condition. Eight seedlings, six inches apart from each other, were transplanted to each pot. The seedlings, which died within 7 days of transplantation, were discarded and replaced by new seedlings.

16.2.4 Sample Collection, Preservation and Digestion

The full-grown rice plants were carefully uprooted at their maturity (90–120 days after transplantation). Then the collected samples were separated into different parts and washed thoroughly with arsenic-free water to remove soil and other contaminants, followed by rinsing with de-ionized water with continuous shaking for several minutes. Finally, the samples were dried in the hot air oven at 60 °C for 72 h and stored in airtight polyethylene bags at room temperature with proper labeling. Proper care was taken at each step to minimize any contamination.

The samples were digested following the heating block digestion procedure (Rahman et al. 2007), diluted to 25 ml with de-ionized water and filtered through Whatman No. 41 filter papers and finally stored in polyethylene bottles. Prior to sample digestion all glass apparatus were washed with 2 % HNO_3 followed by rinsing with de-ionized water and drying.

16.2.5 Analysis of Total Arsenic

The total arsenic was analyzed by flow injection hydride generation atomic absorption spectrometer (FI-HG-AAS, Perkin Elmer AAnalyst 400) using external calibration (Welsch et al. 1990). The optimum HCl concentration was 10 % (v/v) and 0.4 % $NaBH_4$ (Merck, Germany; synthesis grade; 96 %) produced the maximum sensitivity. For each sample three replicates were taken and the mean values were obtained on the basis of calculation of those three replicates. Standard Reference Material (SRM) from National Institute of Standards and Technology (NIST), USA was analyzed in the same procedure at the start, during and at the end of the measurements to ensure continued accuracy. The observed arsenic concentrations of SRM Rice Flour (1568A) showed >97 % recovery.

16.3 Results and Discussion

The experimental soil belonged to clay loam type (Table 16.1). The soil was found to be slightly basic in nature (pH 7.8 ± 0.18) and with 2.3 ± 0.07 mg kg^{-1} initial arsenic content. The background value of arsenic in the non-irrigated soils of the study area was reported to be 2.3–3.1 mg kg^{-1} dry weights (Bhattacharya et al. 2010b).

The impact of soil arsenic treatments on accumulation of arsenic in different fractions of the six selected rice varieties are shown in Fig. 16.1. The uptake of arsenic in rice plants was observed to vary with the different local and high yielding rice varieties. This finding is concurrent with the earlier observations by Delowar et al. (2005), Williams et al. (2006) and Bhattacharya et al. (2013). With gradual increase in concentrations of arsenic treatments in pot soil, the accumulation of

16 A Greenhouse Pot Experiment to Study Arsenic Accumulation

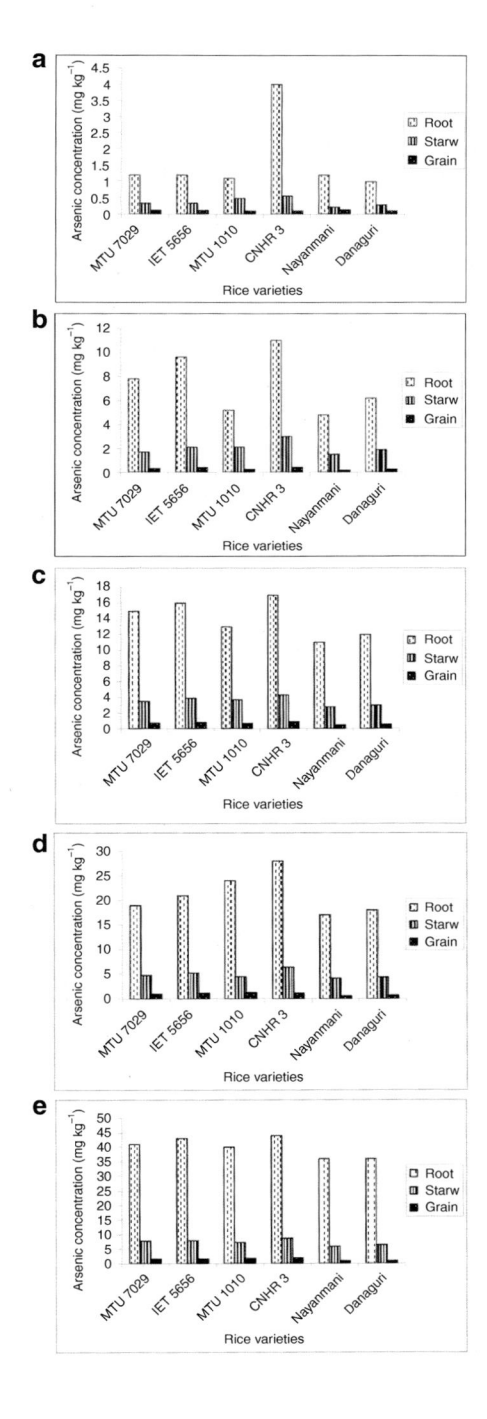

Fig. 16.1 Effect of soil arsenic treatments in pot soil [(**a**) control, (**b**) 5, (**c**) 10, (**d**) 20 and (**e**) 30 mg kg^{-1}] on arsenic accumulation in various parts of the six selected rice (*O. sativa* L.) varieties of Gangetic West Bengal, India

arsenic in different fractions of rice plant was found to increase at dissimilar rates in different rice plant varieties. It is also evident from the results that arsenic accumulated predominantly in root of the rice plant, irrespective of its variety. Iron plaques are commonly formed on the root surfaces of aquatic plants including rice by releasing oxygen to their rhizosphere through aerenchyma. This results in the oxidation of ferrous iron to ferric iron and the precipitation of iron oxides on the root surfaces (Armstrong 1964). Composition of iron oxides were later reported to be dominantly of ferrihydrite (63 %), followed by goethite (32 %) and siderite (5 %) (Hansel et al. 2001). All these precipitated iron oxides have strong adsorptive capacity for arsenate. According to Liu et al. (2004a), the formation of iron plaques around root surfaces of the rice plant has a significant influence on binding arsenic and reducing its translocation to the above ground tissues (straw, husk and grain) of the plant. The presence of iron plaque was found to sequester arsenic and form a buffer zone that alters the entry of arsenic into plants (Liu et al. 2004b). For example at 10 mg kg^{-1} arsenic dosing in pot soil the accumulation of arsenic in root was in the range 11 ± 1.2–017 ± 3.1 mg kg^{-1} dry weights. It was followed by the accumulation in the straw (2.8 ± 0.52–4.3 ± 0.85 mg kg^{-1} dry weight of arsenic) and grain (0.48 ± 0.15–0.90 ± 0.15 mg kg^{-1} dry weight of arsenic) parts of rice plant. The decreasing trend of accumulation of arsenic in rice plant parts (root > straw > grain) as detected in the present study is in good agreement with the previous findings by Rahman et al. (2007) and Bhattacharya et al. (2010a, 2013).

The results clearly show that rice straw is a moderate accumulator of arsenic. The rate of arsenic accumulation in rice straw was noticed to be concurrent with increasing soil arsenic treatments (Fig. 16.1). Previously, a significant correlation ($r = 0.961$) had been observed by us between average arsenic contents in straw part of different rice varieties and arsenic doses in pot soil (Bhattacharya et al. 2013). In rural West Bengal, rice straw is the most favoured and economical food given to cattle. Thus, accumulation of arsenic in rice straw induces additional risk of arsenic entry to human through cattle milk (Ulman et al. 1998; Datta et al. 2010) and meat (Rana et al. 2012; Bundschuh et al. 2012). Much higher arsenic accumulation ability in rice straw by hybrid rice varieties as compared to non-hybrid varieties had been also reported by Abedin et al. (2002a); Rahman et al. (2007).

The average arsenic concentration in the paddy field soil of West Bengal was reported to be just below 10 mg kg^{-1}, the global average arsenic level in agricultural soil (Das et al. 2002; Bhattacharya et al. 2010a, b; Samal et al. 2011). The comparison of arsenic accumulation in grain of the six rice varieties in the present study at 10 mg kg^{-1} arsenic dosing showed that CNHR 3, a high yielding rice variety was the highest accumulator of arsenic (0.90 ± 0.15 mg kg^{-1} dry weight) while *Nayanmani*, a local rice variety was the lowest accumulator (0.48 ± 0.15 mg kg^{-1} dry weight). At 10 mg kg^{-1} of arsenic dosing in pot soil the accumulation of arsenic in rice grain in any of the studied sample did not exceed 1 mg kg^{-1} (WHO permissible limit). But, with the increasing concentration of arsenic added to the pot soil, the accumulation of arsenic in rice grain was found to increase, but at dissimilar rate (Fig. 16.1). At the maximum level of arsenic dosing in pot soil (30 mg kg^{-1}), comparison of arsenic accumulation in grain of the different rice varieties showed that CNHR 3 still remains as the highest accumulator of arsenic (1.9 ± 0.53 mg kg^{-1} dry weight) as compared to the *Nayanmani* rice variety with accumulation as low as 0.84 ± 0.18 mg kg^{-1} dry weight of arsenic. Figure 16.1 shows that the high yielding rice varieties (CNHR 3, MTU 1010, MTU 7029 and IET 5656) are on an average higher accumulator of arsenic as compared to the studied two local rice varieties, *Nayanmani* and *Danaguri*. Uptake of arsenic upto 2 mg kg^{-1} by an *Aman* rice variety had been reported by Huq et al. (2011). Table 16.2 describes the comparison among the previous works on the accumulation of arsenic in rice grain using a greenhouse pot experiment with the present findings.

Apart from *Nayanmani* and *Danaguri* the accumulation of arsenic in rice grain was found to exceed the WHO recommended permissible limit in rice (1 mg kg^{-1}) at 20 mg kg^{-1} arsenic dosing in pot soil, which is very much close to the reported highest content of arsenic (19.4 mg kg^{-1}) in soil of West Bengal (Roychowdhury et al. 2005). This surpassing of the 1 mg kg^{-1} limit by the four out of six studied rice varieties at the 20 mg kg^{-1} arsenic dosing is considerably alarming. The arsenic content of the paddy field soil of West Bengal and that of irrigation water was previously accounted to be significantly correlated (Bhattacharya et al. 2010b). Thus, an eminent possibility of increase of arsenic concentration in the paddy field soils of the entire arsenic-contaminated areas of West Bengal can be hypothesized from the present study. Moreover, if the situation is not immediately taken care of, it can be predicted that consumption of arsenic-contaminated rice may become the potent route for arsenic entry into human body along with the drinking water pathway.

Table 16.2 Accumulations of arsenic in rice grain in the present study compared with that reported from other green house pot experiments

Arsenic in soil (mg kg^{-1})	Arsenic in rice grain (mg kg^{-1})	Reference
Control-8	0.15–0.42	Abedin et al. (2002a)
0–20	0–0.14	Delowar et al. (2005)
Control-30	0.24–0.75	Rahman et al. (2007)
6.4–80	0.2–0.3	Rahman et al. (2008)
0–50	0.69–1.6	Azad et al. (2009)
0–20	0.22–0.81	Khan et al. (2010)
Control-0.5 mg l^{-1} (in irrigation water)	0.01–2	Huq et al. (2011)
1.9–40	0–2.6	Bhattacharya et al. (2013)
3.7–14.6	0.06–0.47	Azad et al. (2013)
2.3–30	0.1–1.9	Present study

16.4 Conclusions

The potentiality of arsenic contamination in groundwater of Gangetic Bengal is increasing day by day and enhancing the human health risk from arsenic toxicity via water-soil-plant-human pathway. Arsenic was found to accumulate in the range 0.10–1.9 mg kg^{-1} dry weight in rice grain with 2.3–30 mg kg^{-1} dry weight arsenic treatment in soil. Thus, prompt management strategy needs to be taken by the Government in encouraging cultivation of less arsenic accumulating rice varieties (e.g., *Nayanmani* and *Danaguri*) in arsenic-contaminated areas of West Bengal. Along with it, rice varieties that require huge irrigation water are found to accumulate higher amount of arsenic (e.g., CNHR 3, MTU 1010, MTU 7029 and IET 5656) which should be avoided. More emphasis is to be given for cultivation of crops accumulating very low amount of arsenic. This will support the economy of the farmers and also reduce the potential entry of arsenic into human food chain.

Acknowledgements The authors are thankful to the Department of Environment, Government of West Bengal, India for providing funding to carry out the investigation and to the Department of Environmental Science, University of Kalyani, West Bengal for providing the laboratory facilities. The authors are also thankful to the critical comments of the anonymous reviewer that helped to improve the manuscript considerably.

References

Abedin MJ, Feldmann J, Meharg AA (2002a) Uptake kinetics of arsenic species in rice plants. Plant Physiol 128(3):1120–1128

Abedin MJ, Cresser MS, Meharg AA, Feldmann J, Cotter-Howells J (2002b) Arsenic accumulation and metabolism in rice (*Oryza sativa* L.). Environ Sci Technol 36:962–968

Armstrong W (1964) Oxygen diffusion from the roots of some British bog plants. Nature 204:801–802

Azad MAK, Islam MN, Alam A, Mahmud H, Islam MA, Karim MR, Rahman M (2009) Arsenic uptake and phytotoxicity of T-aman rice (*Oryza sativa* L.) grown in the As-amended soil of Bangladesh. Environmentalist 29:436–440

Azad MAK, Monda AHMFK, Hossain I, Moniruzzaman M (2013) Experiment for arsenic accumulation into rice cultivated with arsenic enriched irrigation water in Bangladesh. Am J Environ Prot 1(3):54–58

Banerjee M, Banerjee N, Bhattacharjee P, Mondal D, Lythgoe PR, Martinez M, Pan J, Polya DA, Giri AK (2013) High arsenic in rice is associated with elevated genotoxic effects in humans. Sci Rep 3, Article number: 2195. doi:10.1038/srep02195

Bhattacharya P, Samal AC, Majumdar J, Santra SC (2010a) Accumulation of arsenic and its distribution in rice plant (*Oryza sativa* L.) in Gangetic West Bengal, India. Paddy Water Environ 8(1):63–70

Bhattacharya P, Samal AC, Majumdar J, Santra SC (2010b) Arsenic contamination in rice, wheat, pulses and vegetables: a study in an arsenic affected area of West Bengal, India. Water Air Soil Pollut 213:3–13

Bhattacharya P, Samal AC, Majumdar J, Banerjee S, Santra SC (2013) In-vitro assessment on the impact of soil arsenic in the eight rice varieties of West Bengal, India. J Hazard Mater 262:1091–1097

Bundschuh J, Nath B, Bhattacharya P, Liu CW, Armienta MA, López MVM, Lopez DL, Jean JS, Cornejo L, Macedo LFL, Filho AT (2012) Arsenic in the human food chain: the Latin American perspectives. Sci Total Environ 429:92–106

Chakraborti D, Das B, Nayak B, Pal A, Rahman MM, Sengupta MK, Hossain MA, Ahamed S, Sahu M, Saha KC, Mukherjee SC, Pati S, Dutta RN, Quamruzzaman Q (2008) Groundwater arsenic contamination in Ganga-Meghna-Brahmaputra plain, its health effects and an approach for mitigation. In: UNESCO UCI groundwater conference proceedings. http://www.groundwater-conference.uci.edu/proceedings.html#chapter1

Chakraborti D, Das B, Rahman MM, Chowdhury UK, Biswas B, Goswami AB, Nayak B, Pal A, Sengupta MK, Ahamed S, Hossain A, Basu G, Roychowdhury T, Das D (2009) Status of groundwater arsenic contamination in the state of West Bengal, India: a 20 years study report. Mol Nutr Food Res 53(5):542–551

Das HK, Sengupta PK, Hossain A, Islam M, Islam F (2002) Diversity of environmental arsenic pollution in Bangladesh. In: Ahmed MF, Tanveer SA, Badruzzaman ABM (eds) Bangladesh environment, vol 1. Bangladesh Paribesh Andolon, Dhaka

Datta BK, Mishra A, Singh A, Sar TK, Sarkar S, Bhatacharya A, Chakraborty AK, Mandal TK (2010) Chronic arsenicosis in cattle with special reference to its metabolism in arsenic endemic village of Nadia district, West Bengal, India. Sci Total Environ 409(2):284–288

Delowar HKM, Yoshida I, Harada M, Sarkar AA, Miah MNH, Razzaque AHM, Uddin MI, Adhana K, Perveen MF (2005) Growth and uptake of arsenic by rice irrigated with As-contaminated water. J Food Agric Environ 3(2):287–291

Hansel CM, Fendorf S, Sutton S, Newville M (2001) Characterization of Fe plaque and associated metals on the roots of mine-waste impacted aquatic plants. Environ Sci Technol 35:3863–3868

Huq SMI, Sultana S, Chakraborty G, Chowdhury MTA (2011) A mitigation approach to alleviate arsenic accumulation in rice through balanced fertilization. Appl Environ Soil Sci 2011:1–8

Khan MA, Islam MR, Panaullah GM, Duxbury JM, Jahiruddin M, Loeppert RH (2010) Accumulation of arsenic in soil and rice under wetland condition in Bangladesh. Plant Soil 333(1–2):263–274

Liu WJ, Zhu YG, Smith A, Smith SE (2004a) Do iron plaque and genotypes affect arsenate uptake and translocation by rice seedlings (*Oryza sativa* L.) grown in solution culture. J Exp Bot 55(403):1707–1713

Liu WJ, Zhu YG, Smith FA, Smith SE (2004b) Do phosphorus nutrition and iron plaque alter arsenate (As) uptake by rice seedlings in hydroponic culture? New Phytol 162:481–488

Meharg AA (2004) Arsenic in rice – understanding a new disaster for South-East Asia. Trends Plant Sci 9:415–417

Meharg AA, Williams PN, Adomako E, Lawgali YY, Deacon C, Villada A, Sun G, Zhu YG, Feldmann J, Raab A, Zhao FJ, Islam R, Hossain S, Yanai J (2009) Geographical variation in total and inorganic arsenic content of polished (white) rice. Environ Sci Technol 43(5):1612–1617

Norton GJ, Islam MR, Deacon CM, Zhao FJ, Stroud JL, McGrath SP, Islam S, Jahiruddin M, Feldmann J, Price AH, Meharg AA (2009) Identification of low inorganic and total grain arsenic rice cultivars from Bangladesh. Environ Sci Technol 43(15):6070–6075

Rahman MA, Hasegawa H, Rahman MM, Rahman MA, Miah MAM (2007) Accumulation of arsenic in tissues of rice plant (*Oryza sativa* L.) and its distribution in fractions of rice grain. Chemosphere 69:942–948

Rahman MA, Hasegawa H, Rahman MM, Miah MAM, Tasmin A (2008) Arsenic accumulation in rice (*Oryza sativa* L.): human exposure through food chain. Ecotoxicol Environ Saf 69:317–324

Rana T, Bera AK, Bhattacharya D, Das S, Pan D, Das SK (2012) Chronic arsenicosis in goats with special reference to its exposure, excretion and deposition in an arsenic contaminated zone. Environ Toxicol Pharmacol 33(2):372–376

Roychowdhury T, Tokunaga H, Uchino T, Ando M (2005) Effect of arsenic-contaminated irrigation water on agricultural land, soil and plants in West Bengal, India. Chemosphere 58:799–810

Samal AC, Kar S, Bhattacharya P, Santra SC (2011) Human exposure to arsenic through foodstuffs cultivated using arsenic contaminated groundwater in areas of West Bengal, India. J Environ Sci Health Part A: Environ Sci Eng 46:1259–1265

Santra SC, Samal AC, Bhattacharya P, Banerjee S, Biswas A, Majumdar J (2013) Arsenic in foodchain and community health risk: a study in Gangetic West Bengal. Proc Environ Sci 18:2–13

Signes-Pastor AJ, Mitra K, Sarkhel S, Hobbes M, Burló F, de Groot WT, Carbonell-Barrachina AA (2008) Arsenic speciation in food and estimation of dietary intake of inorganic arsenic in a rural village of West Bengal, India. J Agric Food Chem 56(20):9469–9474

Smith AH, Lingas EO, Rahman M (2000) Contamination of drinking water by arsenic in Bangladesh: a public health emergency. Bull World Health Organ 78(9):1093–1103

Su YH, McGrath SP, Zhao FJ (2010) Rice is more efficient in arsenite uptake and translocation than wheat and barley. Plant Soil 328:27–34

Ulman C, Gezer S, Anal Ö, Töre IR, Kirca U (1998) Arsenic in human and cow's milk: a reflection of environmental pollution. Water Air Soil Pollut 101(1–4):411–416

Welsch EP, Crock JG, Sanzolone R (1990) Trace level determination of arsenic and selenium using continuous flow hydride generation atomic absorption spectrophotometry (HG-AAS). In: Arbogast BF (ed) Quality assurance manual for the branch of geochemistry, Open-File Rep 90–0668. US Geological Survey, Reston

Williams PN, Islam MR, Raab A, Hossain SA, Meharg AA (2006) Increase in rice grain arsenic for regions of Bangladesh irrigating paddies with elevated arsenic in groundwater. Environ Sci Technol 40:4903–4908

Chapter 17
Status of Arsenic Contamination Along the Gangetic Plain of Ballia and Kanpur Districts, Uttar Pradesh, India and Possible Remedial Measures

N. Sankararamakrishnan, A. Gupta, and V.S. Chauhan

17.1 Introduction

The presence of arsenic in ground water has been reported from many parts of the world, particularly in the Bengal delta of India and Bangladesh (Berg et al. 2001; Rahaman et al. 2013), China (Kinniburgh and Smedley 2000), Vietnam (UNESCAP-UNICEF-WHO 2001) and Nepal (Tandukar et al. 2001). In India, apart from West Bengal, arsenic contamination has been reported in other states like Ballia (Chauhan et al. 2009); Ghazipur and Varanasi districts of UP (Ahamed et al. 2006), middle Gangetic plain of Bihar (Chakraborti et al. 2003), Sahibganj district of Jharkand (Bhattacharjee et al. 2005), and Northeastern Karnataka (Chakraborti et al. 2013). Till date, the data from various reports and research papers indicate that arsenic in ground water used for drinking occurs in more than 250 blocks in approximately 55 districts of India.

It is well known that consumption of arsenic contaminated groundwater leads to chronic health effects. Signs of chronic arsenicalism include dermal lesions, peripheral neuropathy, and skin cancer. These clinical symptoms, especially dermal lesions (the most commonly observed symptom) generally occur after a minimum of 5 years of consumption of arsenic contaminated drinking water. WHO guideline for

---

N. Sankararamakrishnan (✉)
Centre for Environmental Science and Engineering, Indian Institute of Technology Kanpur, Kanpur 208016, Uttar Pradesh, India
e-mail: nalini@iitk.ac.in

A. Gupta
Babasaheb Bhimrao Ambedkar University, Lucknow 226025, Uttar Pradesh, India

V.S. Chauhan
Drinking Water and Sanitation Department, Jharkhand, Ranchi, Jharkhand 834002, India

arsenic in drinking water is 10 µg L^{-1}. BIS permissible limit is presently set at 50 µg L^{-1} in the absence of other sources of water supply.

The removal of arsenic by various methods has been adequately reviewed (Mohan and Pittman 2007). Although co-precipitation, flotation, ion-exchange, ultrafiltration, and reverse osmosis have been used for arsenic removal, the adsorption from solution has received more attention due to its high concentration efficiency. It is well known that there exists a high affinity between inorganic arsenic species and iron. This behaviour was put to advantage to develop Fe(III)-bearing materials like goethite and hematite (Gimenez et al. 2007), ferrihydrite (Raven et al. 1998), Fe(III)-loaded resins (Katsoyiannis and Zouboulis 2002), and iron-oxide-coated sand (Thirunavukkarasu et al. 2005). However, the adsorption was found to be very low in iron coated sand and for iron bearing resins its applicability is limited due to its cost. It is also noteworthy that most of the adsorbents mentioned above was applied to the removal of either As(III) or As(V) ions. It is well established that most of the ground waters in West Bengal and Bangladesh contained both species of arsenic. Thus, the adsorbent should have the capability to remove both inorganic forms of arsenic at neutral pH. Chitosan is a natural polysaccharide with many useful features such as hydrophilicity, biocompatibility, biodegradability and antibacterial properties. Chitosan is also capable of adsorbing several metal ions (Guibal 2004; Chauhan and Sankararamakrishnan 2008; Gupta et al. 2009) because its amino groups can serve as chelation sites. Thus an inexpensive novel adsorbent was synthesized namely chitosan coated sand and its ability to remove total inorganic arsenic at neutral pH have been demonstrated. Also, the present paper discusses the status of arsenic contamination in Ballia and Kanpur districts of U.P.

17.2 Site Description, Materials and Methods

17.2.1 Site Description

In the map of India, location of Uttar Pradesh state, Kanpur and Balli districts are shown in Fig. 17.1. Our study areas are Ballia district (area 3,168 km^2, population 2.75 million) which stretches from 83°38′ to 84°39′ East longitudes and 25°33′ to 26°11′ North latitudes (Fig. 17.2), and Kanpur-Unnao region (26° 05′–27° 02′ N and 80° 03′–81° 04′ E) which covers an area of 4,558 km^2 (Fig. 17.3). Ballia district is located between Ghaghra and Ganga rivers. The blocks (Rewti, Bansidh and Murli Chhapra) were sampled in Ballia district. Around 88 samples were collected in these blocks.

Kanpur district along the Ganges river was divided into four zones (Bithore, Kanpur city, Shuklaganj and Beyond Jajmau) according to geographical as well as land use patterns: Upstream sampling station (agricultural), Central sampling station (industrial), and Downstream sampling station (agricultural). A total of 154 tube-wells were sampled, which included India mark II (installed by government,

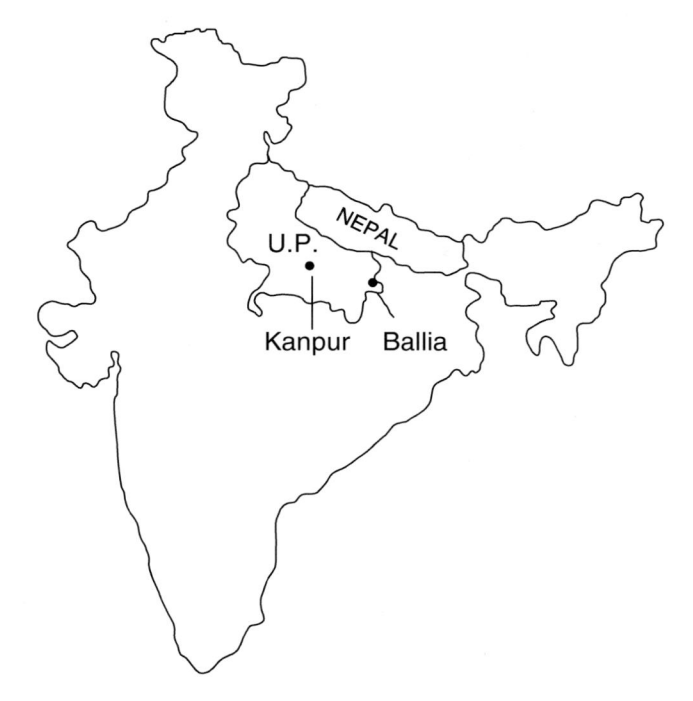

17 Arsenic Contamination Along the Gangetic Plain of U.P. (India)

Fig. 17.1 Map of India showing U.P., Ballia and Kanpur districts

Fig. 17.2 Sampling locations in Ballia district shown by *shaded* blocks

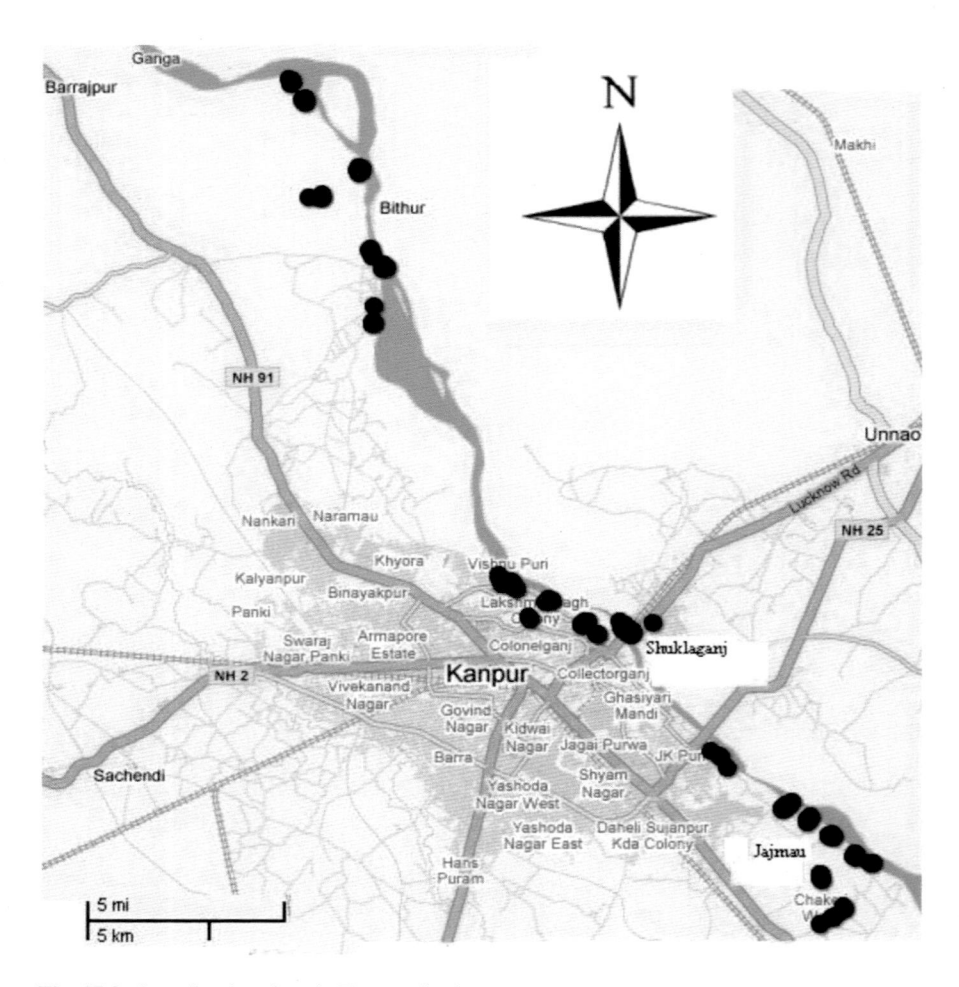

Fig. 17.3 Sampling locations in Kanpur district

90–110 ft depth), as well as privately owned hand pumps (~40 ft depth). Except few, all India mark II tube wells had well constructed concrete apron. However, the ceiling was not satisfactory in the privately owned shallow hand pumps of Shuklaganj zone.

17.2.2 Geomorphology of Ballia and Kanpur Districts

Ahamed et al. (2006) and Acharya and Shah (2007) have detailed the geography, geomorphology and quaternary stratigraphy of the study area from whose work the account of the Ballia district is summarized as follows: The large-scale features of the Ganga plain correspond to major climate changes in the late Quaternary (Singh 2004). The geomorphic surfaces identified in the regional mapping of the Quaternary deposits of the Ganga plain are upland interfluves surface (T_2), marginal fan upland surface (MP), megafan surface (MF), piedmont fan surface (PF), river valley terrace surface (T_1) and active flood plain surface (T_0). A significant aspect of these surfaces is that all of them are depositional surfaces, having a succession of overlying sediments. The Ganga plain foreland basin is a repository of sediments derived from the Himalayas and from Peninsular Craton. The weathered material brought from the Himalaya is deposited in the alluvial plain where they undergo further chemical weathering, mobilizing several anions and cations.

Kanpur district (26.4670°N, 80.3500°E, and 120–130 m above mean sea level) is a part of the Central Ganga Plain with its characteristic landforms (Ansari et al. 1999). The area under study is traversed by the Ganga, Pandu and Loni rivers. The study (lowland or younger alluvial plain) has been identified as flat to gently sloping and slightly undulating terrain of large aerial extent, formed by river deposition; and is limited along river Ganga with the breadth not exceeding 5 km. The sediments comprise recent unconsolidated alluvial material of varying lithology (http://localbodies.up.nic.in/dist30.pdf).

17.2.3 Sampling Methodology and Measurements

Samples from Ballia district were collected between April and June 2007 and samples from Kanpur district were collected during March-April 2010. A GPS device was used for the positioning of each location. Prior to sampling, the hand pumps were flushed with 30–40 L of water. Depth of each tube well ranged from 80 to 100 ft. During sampling, water was filled to the brim of the bottle without any air bubbles. Those for the analysis of cations and sulfate were acidified to 1 % (v/v) HNO_3. Samples were acidified with 2 % (v/v) HCl for the analysis of total As (AsT), As(III), and iron. Samples for anion analysis were left unacidified. In the field, after collection samples were immediately analyzed for ORP and pH measurements by Wagtech ORP/pH electrode. Total inorganic arsenic analysis was carried out using silver diethyldithiocarbamate (APHA 1998). The lower limit of detection was found to be 4 µg L^{-1}. All chemicals and solvents used were of Analytical Reagents (AR) grade. The SDDC was used from Sigma Aldrich and E Merck. Each sample was analyzed twice.

17.2.4 Preparation of Chitosan Coated Sand (CCS)

Chitosan coated sand (CCS) was prepared by using Gangetic belt sand near Kanpur district. Initially, the sand was sieved to a geometric mean size of 0.3 mm and washed twice with deionized distilled water and 1 M HCl to remove the adsorbed

```
┌─────────────────────────────┐
│  Gangetic Sand (<0.3 mm)    │
└──────────────┬──────────────┘
               │   Washed twice with Distilled water
               │   and 1 M HCl
               ▼   Heated to 90°C for 20 h
┌─────────────────────────────┐
│       Activated Sand        │
└──────────────┬──────────────┘
               │   Added 0.5% chitosan
               │   dissolved in Acetic acid and
               ▼   stirred overnight
┌─────────────────────────────┐
│  Chitosan coated sand (CCS) │
└─────────────────────────────┘
```

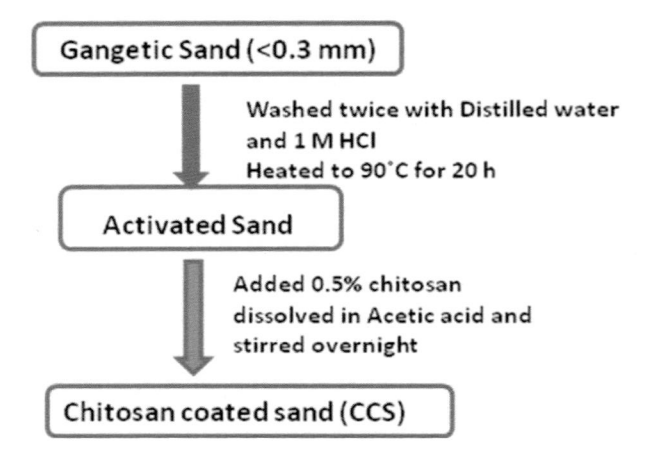

Fig. 17.4 Schematic representation of the preparation of CCS

metal ions and dried at 90 °C for 20 h to activate the sites. Then, activated sand was mixed with the dissolved chitosan solution (0.5 %) and stirred overnight. The coated sand was washed with deionized distilled water and dried at room temperature for further experiments. The entire methodology is depicted in Fig. 17.4.

17.2.5 Batch Adsorption Experiments

Batch equilibrium adsorption isotherm studies were conducted with aqueous solutions of As(III) and As(V) varying the concentration from 1 to 10 mg L^{-1} in 125 mL Erlenmeyer flasks. Amount of adsorbent used was 0.1 g and the solution volume was maintained at 20 mL. The pH of the solution was adjusted to 7. The experiments were carried out at constant temperature of 25 °C and equilibration time was maintained at 4 h at an agitation speed of 200 rpm. After the isothermal equilibration, the sorbent was separated by filtration with Whatman 41 filter paper. The filtrate was analyzed for arsenic. The amount of the arsenic adsorbed (mg) per unit mass of CCS (g), q_e, was obtained by mass balance using the equation (17.1)

$$q_e = \frac{C_i - C_e}{m} \cdot V \qquad (17.1)$$

where C_i and C_e are initial and equilibrium concentrations of the metal ion (mg/L), m is dry mass of chitosan (g) and V is the volume of the solution (L).

17.3 Results and Discussion

17.3.1 Status of Arsenic Contamination in Ballia District

Results of analysis of the water samples for arsenic contamination in Ballia district are presented in Table 17.1. Three blocks namely Murli Chapra, Bansdih and Rewti were sampled. The sampled ground waters were mildly alkaline (pH > 7) and reducing in nature, as evidenced by negative Oxidation Reduction Potential (ORPs). Arsenic was detected in 81 out of 88 samples from the study area with concentrations ranging from 2.5 to 158 µg/L (Table 17.1). The spatial distribution of the total As concentrations is not homogenous over the study area. Approximately 82 % of all groundwater samples exceed 10 µg/L arsenic, which is the WHO guideline value. At 14 wells, As is higher than 100 µg/L. Concentrations higher than 100 µgL^{-1} were observed in Dalon Chhapra, Bhopapur and Vishauli villages of Murli Chhapra and Rewti blocks. Nine deep bore hand pumps (66–75 m) were sampled and total inorganic arsenic concentrations in all these samples exceeded 10 µg L^{-1}, but were below 20 µg L^{-1}. The corresponding shallow well hand pumps contained arsenic at much greater concentrations, up to 158 µg L^{-1} in one case.

Distribution of As(III) and As(V) is depicted in Fig. 17.5. It is evident from the figure that most of the samples contained both As(III) and As(V). The concentration of arsenic(III) was found to be equal/higher than arsenic(V) in most of the samples, although in few samples from Rewti block, concentration of arsenic(V) was higher than arsenic(III).

Table 17.1 Status of arsenic contamination in Ballia district

Blocks	N	<10 µg L^{-1}	<50 µg L^{-1}	50–100 µg L^{-1}	>100 µg L^{-1}
Murli Chapra	48	16	27	9	12
Bansdih	30	2	11	17	2
Rewti	9	1	6	3	–

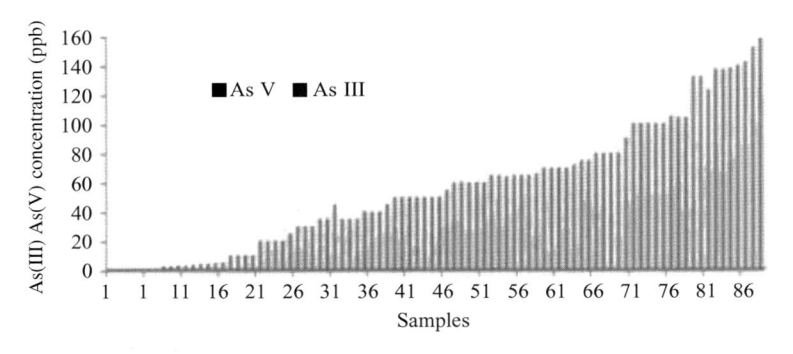

Fig. 17.5 Distribution of As(III) and As(V) in Ballia district

17.3.2 Comparison of Arsenic Concentration in Deep Tube Wells and Its Comparison with Shallow Tube Well

The results of the studies were published earlier (Chauhan et al. 2009).

- Nine deep tube wells (66–75 m) were sampled in Ballia district. The concentration of arsenic was less than 20 µgL^{-1} in all samples.
- The ORP values ranged from −3 to −173 mV suggesting that water is reducing in nature.
- Deep, pre-Holocene, aquifers (>150 m) are generally known to have low-As (BGS/DPHE 2001) and offer a possible alternative source of As-safe drinking water. However, drilling to depths more than 150 m is costly, and may not always be possible with the local drilling techniques.
- The data presented in this study suggest that deeper aquifers (66–75 m) are much lower in arsenic than nearby shallow aquifers and could be a viable and cost-effective source of drinking water in the short to medium term.
- However, the samples did contain arsenic in the range of 12–20 µgL^{-1}, which is higher than 10 µgL^{-1} i.e. WHO Guideline Value for arsenic in drinking water (Table 17.2).

17.3.3 Status of Arsenic in Kanpur-Unnao District

Out of 154 samples included 32 samples from Bithore zone, 30 samples from Kanpur city zone, 33 samples from Jajmau zone and 59 samples from Shuklaganj area (Table 17.3). The data show the arsenic concentrations varying from below detectable limits to concentrations well above those of typical provisional drinking water standards, i.e. Not Detected (nd) – 448 µgL^{-1} As. The pH of the groundwater samples

Table 17.2 Comparison of the parameters in shallow and deep tube well

Parameter	Shallow tube well			Deep tube well		
	Min	Avg	Max	Min	Avg	Max
Arsenic total (µgL^{-1})	n.d.	56.1	158	12	13	18
pH	6.54	7.47	8.14	7.46	7.56	7.87
ORP (mV)	−173	−91	−3	−159	−111	−98

Table 17.3 Status of arsenic contamination in Kanpur district

Zones	N	<10 µg L^{-1}	<50 µg L^{-1}	50–100 µg L^{-1}	>100 µg L^{-1}
Bithore	32	28	32	–	–
Kanpur city	29	26	29	–	–
Jajmau	31	28	32	–	–
Shuklaganj	59	15	31	21	7

ranged from 7.02 to 8.55 and ORP values ranged from 33.7 to −166.9 mV. Negative ORP values in general indicated the anaerobic reducing condition of the ground water.

In Bithore zone, out of 32 samples, 28 samples contain arsenic <10 µg L^{-1} and 31 samples contain arsenic concentration <50 µg L^{-1}. The samples in Kanpur city zone and Beyond Jajmau zone (Table 17.3) contain arsenic below 50 µg L^{-1}. Three samples in Kanpur city and Beyond Jajmau zone exceeded WHO limit of 10 µg L^{-1}.

In Shuklaganj Zone (Chauhan et al. 2012), samples were drawn from both India Mark II and privately owned shallow hand pumps. In this zone, groundwater arsenic concentrations as high as 440 µg L^{-1} were found. It was also observed that most of the private shallow hand pumps contained high levels of arsenic. Out of 59 samples in Shuklaganj, 44 samples are exceeding the WHO permission limit; and 21 samples exceed the BIS permission limit. The seven samples even show the worst condition of drinking water arsenic contamination having more than 100 ppb arsenic in groundwater.

17.3.4 Arsenic Remediation Using Chitosan Coated Sand (CCS)

17.3.4.1 Effect of pH

The solution pH is an important factor for all water and waste water treatment processes because it affects the speciation of metal in water. It is evident from Fig. 17.6 that adsorption increases with increase in pH and reaches a maximum, further

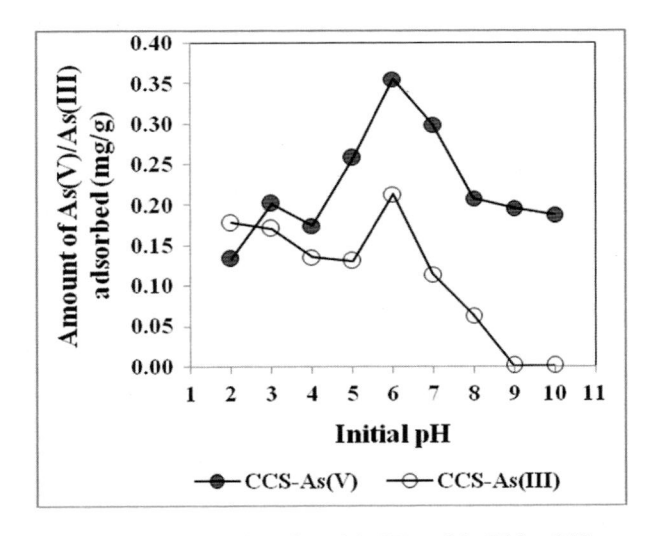

Fig. 17.6 Effect of initial pH on the adsorption of As(III) and As(V) by CCS

increase in pH above 7.5 results in decreased adsorption. Maximum adsorption is found to be at pH 6.0 for both As(V) and As(III) using CCS. The observed difference between pH's could be due to the pH-dependent As speciation. Due to practical reasons, all further experiments were carried out at pH 7.

17.3.4.2 Equilibrium Isotherm

The isotherm models of Langmuir and Freundlich were applied to fit the adsorption equilibrium data of As(III) and As(V) on CCS. Linearized forms of Langmuir and Freundlich isotherms are represented by

$$\frac{1}{qe} = \frac{1}{QbCe} + \frac{1}{Ce\ Q} \quad (17.2)$$

$$\log qe = \frac{1}{n}\log Ce + \log K_f \quad (17.3)$$

where Ce and qe are equilibrium concentration and the amount adsorbed at equilibrium expressed in mg L^{-1}, Q (mg g^{-1}) represents the maximum adsorption capacity and b (mL mg^{-1}) is the Langmuir constant, which represents the affinity between the solute and the adsorbent. The n and K_f are the Freundlich parameters, for values range $1 < n < 10$, which indicated adsorption is considered to be favourable.

Equilibrium adsorption isotherm studies were conducted with aqueous solutions of As(III) and As(V) varying the concentration from 100 to 1,000 μg L^{-1} at pH 7 with a dose rate of 2.5 mg L^{-1} of solution for time period for 2 h.

Linearized Langmuir adsorption plots are shown in Fig. 17.7. The values obtained for the various parameters of Langmuir and Freundlich isotherm parameters are shown in Table 17.4. The adsorption data could be described well by the Langmuir model $(R^2 > 0.977)$ and Freundlich model $(R^2 = 1)$, which interpreted the adsorption process and were fitted to a monolayer adsorption on a homogeneous surface. The Freundlich parameters for n value were both higher than 2.4, indicating that arsenic was favourable to adsorption on adsorbents. A comparison of the various adsorbents reported for arsenic removal is presented in Table 17.5. From the table it is evident that CCS perform efficiently with higher capacity at neutral pH values.

17.4 Conclusions

A systematic study has been conducted on the status of arsenic contamination in few blocks of Ballia and Kanpur districts along the Gangetic plain. In both Ballia and Kanpur districts both species of arsenic are prevalent. The concentration of arsenic in deep bore well of Ballia district was within 20 ppb. A mitigation strategy

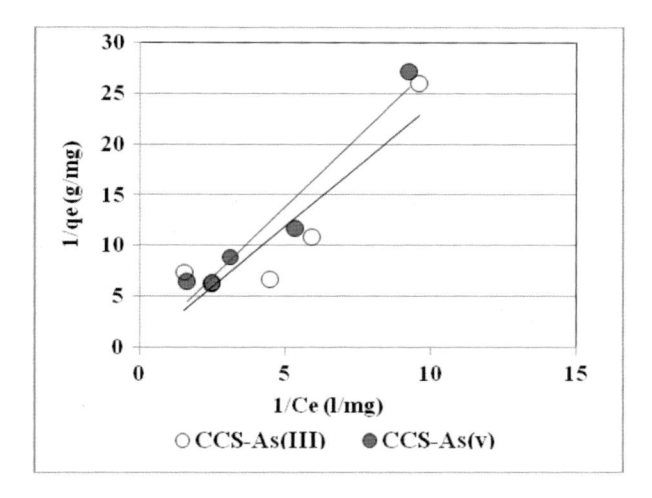

Fig. 17.7 Linearized Langmuir isotherm

Table 17.4 Langmuir and Freundlich model constants

Arsenic species	Langmuir model			Freundlich model		
	Qmax (mg/g)	b (ml/mg)	R^2	1/n	K_f	R^2
As(III)	17	1.144	0.8172	1	2.499	1
As(V)	23	1.016	0.9465	1	2.499	1

Table 17.5 Comparison of adsorption capacities of some adsorbents towards arsenic removal

Adsorbents	Capacity from Langmuir model (mg/g)		References
	As(III)	As(V)	
Goethite	10.1	12.1	Javier et al. 2007
Hematite	10.0	31.3	Javier et al. 2007
Ferrihydride	0.58	0.16	Raven et al. 1998
Fe and Mn oxide coated sand	0.129	–	Chang et al. 2008
FeS-coated sand	10.7	–	Han et al. 2011
Iron oxide coated sand	0.136	–	Hsu et al. 2008
Zerovalent iron coated sand	70.4	–	Wan et al. 2010
Chitosan coated sand	17	23	Present work

using chitosan coated sand has been demonstrated. The prepared adsorbent showed capacity of 17 and 23 mg/g both As(III) and As(V) at neutral pH. This suggests that this adsorbent could be used as an effective filter candidate for arsenic removal.

Acknowledgements NS is thankful to Council of Scientific and Industrial Research (Scheme no. 24(306)09-EMR-II) for financial support to carry out this work. AG is thankful to CSIR for Junior research fellowship.

References

Acharyya SK, Shah BA (2007) Groundwater arsenic contamination affecting different geologic domains in India – a review: influence of geological setting, fluvial geomorphology and quaternary stratigraphy. J Environ Sci Health A 42:1795–1805

Ahamed S, Kumar Sengupta M, Mukherjee A, Amir Hossain M, Das B, Nayak B, Pal A, Chakraborti D (2006) Arsenic groundwater contamination and its health effects in the state of Uttar Pradesh (UP) in upper and middle Ganga plain, India: a severe danger. Sci Total Environ 370(2–3):310–322

Ansari AA, Singh IB, Tobschall HJ (1999) Status of anthropogenically induced metal pollution in the Kanpur–Unnao industrial region of the Ganga Plain, India. Environ Geol 38:25–33

APHA (1998) Standard methods for water and wastewater, 20th edn. American Public Health Association, Washington, DC

Berg M, Tran HC, Nguyen TC, Pham H, Schertenleib R, Giger W (2001) Arsenic contamination in groundwater and drinking water in Vietnam: a human health threat. Environ Sci Technol 35(13):2621–2626

BGS/DPHE (2001) In arsenic contamination of groundwater in Bangladesh. In: Kinniburgh DG, Smedley PL (eds) Final report, BGS technical report WC/00/19. British Geological Survey, Keyworth

Bhattacharjee S, Chakravarty S, Maity S, Dureja V, Gupta KK (2005) Metal contents in the groundwater of Sahebgunj district, Jharkhand, India, with special reference to arsenic. Chemosphere 58:1203–1217

Chakraborti D, Mukherjee SC, Pati S, Sengupta MK, Rahman MM, Chowdhury UK (2003) Arsenic groundwater contamination in Middle Ganga Plain, Bihar, India: a future danger. Environ Health Perspect 111(9):1194–1201

Chakraborti D, Rahman MM, Murrill M, Das R, Siddayya, Patil SG, Sarkar A, Das KK (2013) Environmental arsenic contamination and its health effects in a historic gold mining area of the Mangalur greenstone belt of Northeastern Karnataka, India. J Hazard Mater 262:1048–1055

Chang Y, Song K, Yang J (2008) Removal of As(III) in a column reactor packed with iron-coated sand and manganese-coated sand. J Hazard Mater 150:565–572

Chauhan D, Sankararamakrishnan N (2008) Highly enhanced adsorption for decontamination of lead from battery effluents using chitosan functionalized with xanthate. Bioresour Technol 99:9021–9024

Chauhan VS, Nickson RT, Chauhan D, Iyengar L, Sankararamakrishnan N (2009) Ground water geochemistry of Ballia district, Uttar Pradesh, India and mechanism of arsenic release. Chemosphere 75(1):83–91

Chauhan VS, Yunus M, Sankararamakrishnan N (2012) Geochemistry and mobilization of arsenic in Shuklaganj area of Kanpur-Unnao district, Uttar Pradesh, India. Environ Monit Assess 194:4884–4901

Giménez J, Martínez M, de Pablo J, Rovira M, Duro L (2007) Arsenic sorption onto natural hematite, magnetite, and goethite. J Hazard Mater 141(3):575–580

Guibal E (2004) Interactions of metal ions with chitosan-based sorbents: a review. Sep Purif Technol 38(1):43–74

Gupta A, Chauhan VS, Sankararamakrishnan N (2009) Preparation and evaluation of iron-chitosan composites for removal of As(III) and As(V) from arsenic contaminated real life groundwater. Water Res 43(15):3862–3870

Han Y, Gallegos T, Demond AH, Hayes KF (2011) FeS-coated sand for removal of arsenic(III) under anaerobic conditions in permeable reactive barriers. Water Res 45:593–604

Hsu J, Lin C, Liao C, Chen S (2008) Evaluation of the multiple-ion competition in the adsorption of As(V) onto reclaimed iron-oxide coated sands by fractional factorial design. Chemosphere 72:1049–1055

Javier G, Martinez M, Pablo JD, Rovira M, Duro L (2007) Arsenic sorption onto natural hematite, magnetite, and goethite. J Hazard Mater 141:575–580

Katsoyiannis IA, Zouboulis AI (2002) Removal of arsenic from contaminated water sources by sorption onto iron-oxide coated polymeric materials. Water Res 36:5141–5155

Kinniburgh DG, Smedley PL (2000) Arsenic contamination in ground water of Bangladesh. Final report summary, Bangladesh Department of Public Health Engineering, British Geological Society, Keyworth. http://www.bgs.ac.uk/arsenic

Mohan D, Pittman CU Jr (2007) Arsenic removal from water/wastewater using adsorbents – a critical review. J Hazard Mater 142:1–53

Rahaman S, Sinha AC, Pati R, Mukhopadhyay D (2013) Arsenic contamination: a potential hazard to the affected areas of West Bengal, India. Environ Geochem Health 35(1):119–132

Raven KP, Jain A, Loeppert RH (1998) Arsenite and arsenate adsorption on ferrihydrite: kinetics, equilibrium, and adsorption envelopes. Environ Sci Technol 32(3):344–349

Singh B (2004) Late quaternary history of the Ganga Plain. J Geol Soc India 64:431–454

Tandukar N, Bhattacharya P, Mukherjee AB (2001) Managing arsenic for future. In: Proceedings of international conference on arsenic in Asia-Pacific Region, Adelaide

Thirunavukkarasu OS, Viraraghavan T, Subramanian KS, Chaalal O, Islam MR (2005) Arsenic removal in drinking water – impacts and novel removal technologies. Energy Sources 27(1–2):209–219

UNESCAP-UNICEF-WHO (2001) United Nations Economic and Social Commission for Asia and the Pacific, geology and health: solving arsenic crisis in Asia and Pacific region. UNESCAP-UNICEF-WHO expert group meeting, Bangkok

Wan J, Klein J, Simon S, Joulian C, Dictor M, De luchat V, Dagot C (2010) AsIII oxidation by Thiomonas arsenivorans in up-flow fixed-bed reactors coupled to As sequestration onto zero-valent iron-coated sand. Water Res 44:5098–5108

Chapter 18
A Low-Cost Arsenic Removal Method for Application in the Brahmaputra-Ganga Plains: Arsiron Nilogon

Shreemoyee Bordoloi, Sweety Gogoi, and Robin K. Dutta

18.1 Introduction

18.1.1 The Gangetic Plains vs. The Brahmaputra Plains

There is much similarity between the geological formation of the Gangetic and the Brahmaputra plains as both are formed by sedimentation of soils from the Himalayas. The shallow aquifers of both the plains are also expected to be similar except for some variations caused by difference in rainfall and occurrence of floods. Hence, the findings of arsenicosis and arsenic (As) contamination of groundwater in Bangladesh and the Gangetic plains of India (Chakraborti et al. 2002; Chatterjee et al. 1995) had indicated the possibility of similar fate of the Brahmaputra plains and the Borak plains of the north-eastern region of India also, which was soon proven to be true.

18.1.2 Arsenic Contamination in the Groundwater in Brahmaputra Plains

Contamination of the groundwater of almost the entire north-eastern states of India, in general and in the plains on both sides of the Brahmaputra in Assam, in particular, with high level of As has come to light in 2004 (Nickson et al. 2007; Singh 2004). The As contaminated area extends to the small valleys of the Brahmaputra and its tributaries in Arunachal Pradesh and Nagaland also. Arsenic above 50 µg/L has been detected in the groundwater in 23 out of 27 districts in Assam, 6 out of 13

S. Bordoloi • S. Gogoi • R.K. Dutta (✉)
Department of Chemical Sciences, Tezpur University, Napaam, Tezpur, Assam 784028, India
e-mail: robind@tezu.ernet.in

districts in Arunachal Pradesh and 2 out of 8 districts in Nagaland. The highest concentrations of arsenic was reported in four districts of Assam viz., Jorhat, Lakhimpur, Nalbari and Nagaon in the range of 112–601 µg/L. Other districts where arsenic concentration in groundwater was found in between 100 and 300 µg/L are Baksa, Barpeta, Darrang, Dhemaji, Dhubri, Golaghat, Sibsagar and Sonitpur. Arsenic concentration in groundwater of remaining 11 districts of Assam is reported to be in between 50 and 100 µg/L.

18.1.3 Arsenicosis in the Affected Areas of Assam

Symptoms of arsenicosis were first noticed in some residents in Assam in 2011 during a field trial of the arsenic removal method described in this article (*The Assam Tribune* 2010). Since then, these authors themselves have seen a large number of people with similar symptoms in many other places including the own village of the first author of this article, called Totoya, located in the largest human-inhabited river island, Majuli in the Brahmaputra. This village with a population of about 500 has seen 28 cancer deaths since 1974, which may also have a correlation with groundwater arsenic (Gogoi and Dutta 2014, unpublished data).

18.1.4 Relevance of Application of Arsenic Removal Methods

Though surface water is the best alternative option for arsenic affected areas, it is unlikely for a developing country to switch the source of water from groundwater to surface water in a short period of time, particularly for sparse rural populations. Therefore, one must consider arsenic removal techniques and the economic feasibility and simplicity of the systems in making arsenic remediation strategy. Thus, low-cost user-friendly removal methods appear to be the immediate option for arsenic removal from groundwater in such areas.

18.1.5 Suitable Removal Methods for the Affected Areas of Assam

Various arsenic removal techniques, e.g., ion-exchange, reverse osmosis (RO), ultrafiltration, adsorption, etc., are unsuitable for rural applications in developing countries for one or more drawbacks of high cost, requirement of power, and disposal of large sludge (Mohan and Pittman 2007). The RO method, which is gaining popularity among the middle-class families, removes most of the essential minerals from already low-TDS water of the flood-prone areas of the Brahmaputra plains.

This is a serious problem with the RO apart from its being costly and producing large quantity of contaminated reject water. Some techniques based on coagulation-adsorption are of low-cost, simple and environment-friendly (Jiang 2013). However, the efficiency in removal of arsenic is a very important factor. The detection limit in the atomic adsorption method is 2 mg/L. On the other hand, it is reported that a presence of arsenic in drinking water in concentration as low as 0.17 mg/L may also cause cancer (WHO 2011). The guideline value of 10 mg/L of WHO is only a provisional guideline value because the safe limit is actually below 10 mg/L whereas it is difficult to remove and measure arsenic below 10 mg/L (WHO 2011). Thus, arsenic in drinking water should be as low as possible or undetectable. Most of the arsenic removal methods based on oxidation-coagulation have low efficiency because of inadequate selection of chemicals, viz., the pH-conditioner, the oxidant and the coagulant and their doses.

18.2 The Arsiron Nilogon Method

18.2.1 pH-Conditioning for Arsenic Removal

A particularly important factor leading to lower efficiency of the existing coagulation-adsorption methods is using inappropriate pH conditioning for efficient adsorption of arsenic on the coagulations. Various pH-conditioners, viz., lime, ash, carbonate and bicarbonate salts of sodium and potassium are used for pH conditioning to facilitate removal of arsenic. Choosing the suitable pH-conditioner for removal of arsenic with the optimum dose is necessary to ensure maximum removal of arsenic without adding undesired residual ions arising from the dose and to avoid readjustment of the final pH of the treated water.

18.2.2 pH-Conditioning for Simultaneous Removal of Arsenic and Iron

Iron usually coexists with arsenic and also influences the removal of arsenic. Therefore, a proper pH-conditioning for iron removal (Das et al. 2007; Bordoloi et al. 2011) also pertains to removal of arsenic. We have reported a detail study of pH-conditioning for simultaneous removal of arsenate and iron by using various pH-conditioners, viz., lime, banana ash, bicarbonate and carbonate salts of sodium and potassium and their 1:1 binary mixture (Bordoloi et al. 2013a). The arsenic removal in presence of the pH conditioners has been found to increase in the order: banana ash < carbonates < bicarbonates < lime. However, only the bicarbonate salts provide the suitable pH condition for simultaneous removal of arsenate and iron ions. The potassium salts are more efficient than the corresponding sodium salts.

However, sodium bicarbonate has also been found to be useful. Lime is disadvantageous because it requires post-treatment correction of highly alkaline pH. The pH of the treated water remains in the acceptable range for drinking when the bicarbonates are used.

18.2.3 The Most Suitable Oxidant for Coagulation Method

The oxidation of As(III) to As(V) is critical for achieving optimal performance for most arsenic removal processes. The feasibility of employing ozone, potassium permanganate and sodium hypochlorite for As(III) oxidation and subsequent enhanced arsenic removal have been demonstrated (Dodd et al. 2006; Kim and Nriagu 2000; Li et al. 2007). Potassium permanganate is preferable over other oxidizing agents for the oxidation of As(III) as it is cost-effective, easy to preserve and to handle in practice, especially for small systems in rural areas (Lihua et al. 2009; Sorlini and Gialdini 2010). On the other hand, several studies have reported that arsenic removal from drinking water by coagulation with $FeCl_3$ is more effective than other coagulants such as $Fe_2(SO_4)_3$, $FeSO_4$ and $Al_2(SO_4)_3$ (Qiao et al. 2012; Wickramasinghe et al. 2004). $FeCl_3$ is preferable over its main competitor, viz., alum [$Al_2(SO_4)_3$] also because of the suspected ill effect of the later causing Alzheimer's disease (Simate et al. 2012). Considering the above facts, we have carried out a detailed systematic study of arsenic removal by oxidation-coagulation using $KMnO_4$ and $FeCl_3$ at pH optimized by using $NaHCO_3$ as the pH-conditioner.

18.2.4 The Oxidation-Coagulation at Optimised pH (OCOP) (Bordoloi et al. 2013b)

The method removes both arsenic and iron simultaneously, hence the name – Arsiron Nilogon, where '*Arsiron*' stands for '***Arsenic** and **iron***', and '*Nilogon*' means removal. The simplest arrangement requires only a bucket, a sand-gravel filter and small quantities of baking soda, potassium permanganate and aqueous ferric chloride solution. For large scale community application it needs large treatment chamber and sand-gravel filter. A schematic diagram of the simple arrangement of the system has been shown in Fig. 18.1. The thickness of the sand layer can be 20 cm or more depending upon the sand grain size. Larger grain size will need thicker media. The method involves a three-step addition of the chemicals which takes 3 min, 1 h residence time and a subsequent filtration. The three chemicals are added in the three steps as follows:

Step 1. A specified quantity of baking soda has to be added to a certain quantity of the arsenic containing water in a bucket or any other container and thoroughly mixed by stirring with a stick for a while.

Fig. 18.1 Schematic diagram of the OCOP system

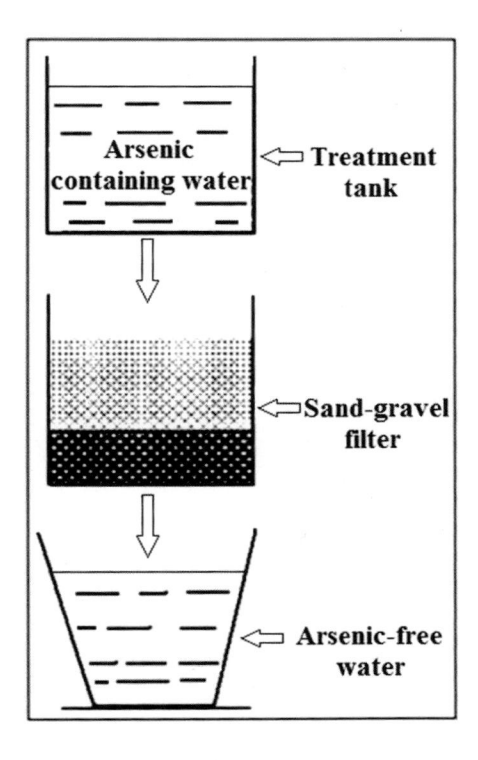

Step 2. A specified quantity of potassium permanganate is added to the water and with thorough mixing by stirring. It mixes immediately.

Step 3. A specified quantity of $FeCl_3$ as coagulant is then added to the water with thorough mixing by stirring for at least a minute.

The method has to be modified if the water already contains soluble iron. In presence of appreciable concentration of dissolved Fe(II) ions along with arsenic, an additional amount of potassium permanganate is necessary. The doses of the chemicals can be seen in Table 18.1.

Coagulation in the form of reddish brown particles will be visible within minutes after addition of the chemicals. The water is then allowed to settle for at least an hour. It can be decanted and filtered through sand-gravel filters.

18.2.5 The Chemical Tricks of Arsiron Nilogon

The three chemicals used in the present method helps in various ways in removal of arsenic and also iron making the method highly efficient and safe (Bordoloi et al. 2011, 2013a, b):

(a) $NaHCO_3$ maintains the pH of the water in a range in which $FeCl_3$ works most efficiently. Arsenate ions are removed more efficiently in the pH range provided

Table 18.1 Doses of the chemicals for iron-free and iron-containing water with arsenic concentration in the range of 100–500 μg/L

Groundwater type	Baking soda in mg/L	Potassium permanganate		FeCl$_3$	
		in mg/L	in ml (of 5 % aq. solution)/10 L of water	in mg/L	in Ml (of 25 % aq. solution)/10 L of water
Having <1 mg/L dissolved iron	100	0.5	0.1	25	1
Having 1–5 mg/L dissolved iron	100	4	0.8	25	1
Having >5 mg/L dissolved iron	100		Till it imparts a light purple colour[a]	25	1

[a]For the water containing iron concentration above 5 mg/L, potassium permanganate was added until it imparts a light purple colour to the water. The colour disappears after coagulation

by NaHCO$_3$ than at pH higher or lower than that. Arsenate adsorption decreases in more alkaline solutions due to competition with OH$^-$ ions. The arsenate adsorption decreases in acidic conditions due to protonation of arsenate ions.

(b) NaHCO$_3$ provides the optimum pH condition for precipitation of Fe(II) if soluble Fe(II) ion is present along with arsenic in the water. The precipitation of iron is slow at lower pH. On the other hand, there remains some residual iron in the water at higher pH.

(c) KMnO$_4$ oxidises difficult-to-remove arsenite ions to easy-to-remove arsenate ions. KMnO$_4$, a popular oxidizing agent used in water treatment, has another edge over its competitors as it exists as stable solid with high water solubility. KMnO$_4$ also oxidizes dissolved ferrous iron to insoluble ferric iron if ferrous iron is present in the water along with arsenic.

(d) In acidic medium, Mn(VII) of KMnO$_4$ is itself reduced to soluble Mn(II) state which is unwanted in drinking water. In the mild alkaline condition provided by NaHCO$_3$, Mn(VII) itself is reduced to Mn(IV) state and separates out as insoluble MnO$_2$ without leaving any residual dissolved manganous ions in the treated water.

(e) The arsenite oxidation efficiency for the Mn(VII)-As(III) system is higher under basic condition provided by baking soda than that under acidic conditions. The carbonate ions of NaHCO$_3$ accelerate oxidation of arsenite to arsenate (Kim et al. 2006).

(f) FeCl$_3$ is a Lewis acid. The charges of Fe^{3+} attract negatively charged ions including arsenate and other particulates, grow in size and settle down quickly. FeCl$_3$ is more efficient than alum as a coagulant and does not leave any residual toxic substance in the water unlike alum which leaves aluminum ions released in water suspected of being responsible for Alzheimer's diseases. FeCl$_3$ coagulates sufficiently in the mild pH range provided by NaHCO$_3$.

(g) The pH of about 8.3 initially provided by NaHCO$_3$ is brought to 7.3 by FeCl$_3$, producing H$^+$ ions by hydrolysis. The pH 7.3 is in the middle of the acceptable pH range of drinking water, which is 6.5–8.5. Though FeCl$_3$ is acidic and corrosive in aqueous solution it is safe in the presence of NaHCO$_3$ in the chosen dose.

(h) The mild disinfectant KMnO$_4$ also kills the bacteria present in the water.

18.2.6 Field Experience of the Arsiron Nilogon Method

User trial of the oxidation-coagulation at optimized pH (OCOP) method with optimized doses of pH-conditioner, oxidant and coagulant was carried out in different arsenic affected areas in Assam (Bordoloi et al. 2013b). User trainings were conducted for application of the method as well as for collection of samples of treated water for further analysis. The field trial was conducted at 16 spots which included 10 households with 10 L, 5 schools with 25 L and 1 school with 200 L capacity units. The chemicals were supplied in small kits containing solid $NaHCO_3$, 5 % aqueous $KMnO_4$ solution in a small plastic bottle along with a dropper and 25 % $FeCl_3$ in a larger plastic bottle with a marked measure. The water before and after the treatment by the OCOP method were analysed by atomic absorption method and the results of the field trial are shown in Table 18.2.

It can be seen from the results in Table 18.2 that arsenic was removed to below 5 µg/L when the initial Fe ion concentration in the water was less than 1.0 mg/L. Concentrations of Fe were also reduced to below 0.1 mg/L in all cases of initial iron of less than 1.0 mg/L. The pH of the treated water remained within 7.4–8.2, which is within the acceptable range for drinking, i.e., 6.5–8.5. Thus, the results show that over 95 % arsenic removal can be achieved with the used doses of the pH conditioner,

Table 18.2 Concentrations of As and Fe before and after treatment by the OCOP method in some field water from tube wells at households (H) and schools (S) in Jorhat, Sonitpur and Nalbari districts of Assam

Source[a] sl. no.	User sl. no.	Unit capacity (L)	[As]/(µg/L)[b] Before	After	[Fe]/(mg/L)[b] Before	After	pH[b] Before	After
	H1			3.7		0.06		7.7
1	H2	10	196.4	5.3	0.19	0.06	7.5	7.7
	H3			4.5		0.06		7.7
	H4			4.1		0.08		8.1
2	H5	10	208.5	5.2	0.28	0.08	7.8	8.2
	H6			6.1		0.08		8.1
	H7			3.9		0.08		7.9
3	H8	10	211.2	3.6	0.14	0.08	7.6	7.9
	H9			5.2		0.08		7.8
4	H10	10	220.2	4.3	0.25	0.07	7.7	7.9
5	S1	25	204.7	3.9	0.41	0.08	7.5	7.8
6	S2	25	238.1	7.6	2.61	0.14	7.3	7.5
7	S3	25	229.1	7.4	3.15	0.17	7.0	7.4
8	S4	25	127.4	6.6	2.35	0.15	7.6	7.9
9	S5	25	185.4	8.2	5.73	0.20	7.5	7.7
10	S6	200	106.5	9.2	16.25	0.36	7.4	7.8

[a]Sources 1–5 were at Tatigaon village and 6 & 7 at Kharikotia village in Jorhat district; sources 8 & 9 were at Kakila and Kutumgaon villages, respectively, in Sonitpur district and source 10 was at Kaithalkuchi village in Nalbari district

[b]Error limits: $\Delta[As] = \pm 0.4$ µg/L, $\Delta[Fe] = \pm 0.05$ mg/L and $\Delta pH = \pm 0.03$

oxidant and coagulants from arsenic-containing water with low dissolved iron. Despite addition of higher dose of $KMnO_4$, arsenic was lowered only to below 8 µg/L when the water contained iron above 1.0 mg/L, which is slightly poorer than that in the presence of lower dissolved iron concentrations. This indicates that although initially present iron can remove arsenic to some extent, there is a negative effect of initially present iron on the arsenic removal to a very low level even if the initial ferrous ions are oxidized in the process. However, with the chosen dose of $KMnO_4$, the OCOP process could remove both arsenic and iron to below their respective WHO guideline values.

18.3 Suitability Analysis

18.3.1 User-Safety

The concentrations of Fe, Mn, Na and Cl, which are present in the materials used in Arsiron Nilogon, remain within levels prescribed by the WHO for drinking water. Some heavy metals are also removed in Arsiron Nilogon. The method partially removes the bacteria if present in the water as $KMnO_4$ is a mild disinfectant. The final pH of water after treatment is about 7.3, which is in the middle of the acceptable range of pH for drinking water.

18.3.2 Environmental Impact

The solid sludge passes the toxicity characteristic leaching procedure (TCLP) test of the United State Environmental Protection Agency (US-EPA) for dumping even in land-fill (USEPA 1992). The test showed the arsenic concentration in the TCLP leachate of the sludge sample to be less than 10 µg/L whereas the maximum permissible TCLP limit of the US-EPA is 5 mg/L. Thus, the arsenic concentration in the TCLP extract of the solid sludge of the present OCOP method is about 500 times lower than the US-EPA TCLP limit for disposing in land-fill. The solid sludge can therefore be disposed in land-fill safely. The solid sludge can be collected in an earthen pot containing sand and a small hole at the bottom. The water coming out from the earthen pot also is arsenic-free. The solid sludge collected in the pot, smaller in size than that of a pee, can be disposed conveniently in such a way that it does not contaminate drinking water or agricultural land.

18.3.3 User-Friendliness

The recurring cost of the present Arsiron Nilogon technique has been estimated to be about INR 1 per 100 L of water. Thus, Arsiron Nilogon is a low-cost method. In addition to high efficiency and low-cost, an option of using without electricity, safety and simplicity of operation make Arsiron Nilogon suitable for poor villagers in sparsely populated arsenic affected areas.

18.4 Conclusions

No doubt, all these chemicals have been already in use for arsenic removal but the present Arsiron Nilogon method uses them in a more scientific way. Removal of arsenic and iron by the OCOP method using $NaHCO_3$, $KMnO_4$ and $FeCl_3$ as pH-conditioner, oxidant and coagulant, respectively, is highly efficient for simultaneous removal of both arsenic and iron from groundwater. The results of the field trial including the potability of the treated water are promising. This together with high efficiency, low-cost, simplicity of operation, safety and environment-friendliness and option of non-requirement of electricity suggest that the OCOP method has a great potential for arsenic removal in rural areas where alternate arsenic-free water is not available. The method is being continued by the users except where alternate water has been made available later by the government and is gaining popularity. The method is expected to be applicable in the arsenic affected areas of both the Brahmaputra and the Gangetic plains as they both have similar geological conditions.

References

Bordoloi S, Nath SK, Dutta RK (2011) Iron ion removal from groundwater using banana ash, carbonates and bicarbonates of Na and K, and their mixtures. Desalination 281:190–198

Bordoloi S, Nath M, Dutta RK (2013a) pH-conditioning for simultaneous removal of arsenic and iron ions from groundwater. Process Saf Environ Prot 91:405–414

Bordoloi S, Nath SK, Gogoi S, Dutta RK (2013b) Arsenic and iron removal from groundwater by oxidation-coagulation at optimized pH: laboratory and field studies. J Hazard Mater 260:618–626

Chakraborti D, Rahman MM, Paul K, Chowdhury UK, Sengupta MK, Lodh D, Chanda CR, Saha KC, Mukherjee SC (2002) Arsenic calamity in the Indian subcontinent: what lessons have been learned? Talanta 58:3–22

Chatterjee A, Das D, Mandal BK, Chowdhury TR, Samanta G, Chakraborti D (1995) Arsenic in ground-water in six districts of West-Bengal, India – the biggest arsenic calamity in the world. I. Arsenic species in drinking-water and urine of the affected people. Analyst 120:643–650

Das B, Hazarika P, Saikia G, Kalita H, Goswami DC, Das HB, Dube SN, Dutta RK (2007) Removal of iron from groundwater by ash: a systematic study of a traditional method. J Hazard Mater 141:834–841

Dodd MC, Vu ND, Ammann A, Le VC, Kissner R, Pham HV, Cao TH, Berg M, Gunten UV (2006) Kinetics and mechanistic aspects of As(III) oxidation by aqueous chlorine, chloramines, and ozone: relevance to drinking water treatment. Environ Sci Technol 40:3285–3292

Jiang JQ (2013) Removing arsenic from groundwater for the developing world – a review. Water Sci Technol 44:89–98

Kim MJ, Nriagu J (2000) Oxidation of arsenite in groundwater using ozone and oxygen. Sci Total Environ 247:71–79

Kim J, Korshin GV, Frenkel AI, Velichenk AB (2006) Electrochemical and XAFS studies of effects of carbonate on oxidation of arsenite. Environ Sci Technol 40:228–234

Li N, Fan M, Leeuwen JV, Saha B, Yang H, Huang CP (2007) Oxidation of As(III) by potassium permanganate. J Environ Sci 19:783–786

Lihua S, Sun Lihua S, Liu Ruiping L, Xia Shengji X, Yang Yanling Y, Guibai L (2009) Enhanced As(III) removal with permanganate oxidation, ferric chloride precipitation and sand filtration as pretreatment of ultrafiltration. Desalination 243:122–131

Mohan D, Pittman CU Jr (2007) Arsenic removal from water/wastewater using adsorbents – a critical review. J Hazard Mater 142:1–53

Nickson R, Sengupta C, Mitra P, Dave SN, Banerjee AK, Bhattacharya A, Basu S, Kakoti N, Moorthy NS, Wasuja M, Kumar M, Mishra DS, Ghosh A, Vaish DP, Srivastava AK, Tripathi RM, Singh SN, Prasad R, Bhattacharya S, Deverill P (2007) Current knowledge on the distribution of arsenic in groundwater in five states of India. J Environ Sci Health Part A 42:1707–1718

Qiao J, Jiang Z, Sun B, Sun Y, Wang Q, Guan XH (2012) Arsenate and arsenite removal by $FeCl_3$: effects of pH, As/Fe ratio, initial As concentration and co-existing solutes. Sep Purif Technol 92:106–114

Simate GS, Iyuke SE, Ndlovu S, Heydenrych M, Walubita LF (2012) Human health effects of residual carbon nanotubes and traditional water treatment chemicals in drinking water. Environ Int 39:38–49

Singh AK (2004) Arsenic contamination of groundwater of North Eastern India. In: Proceedings of national seminar on hydrology with local theme on water quality. National Institute of Hydrology, Roorkee, 22–23 Nov 2004

Sorlini S, Gialdini F (2010) Conventional oxidation treatments for the removal of arsenic with chlorine dioxide, hypochlorite, potassium permanganate and monochloramine. Water Res 44:5653–5659

The Assam Tribune (2010) Symptoms of Arsenicosis detected in some Tatigaon residents, 4 Aug 2010

U.S. Environmental Protection Agency (USEPA) (1992) Method 1311: toxicity characteristic leaching procedure. Washington, DC

WHO (2011) Guidelines for drinking-water quality, recommendations, 4th edn. World Health Organization, Geneva

Wickramasinghe SR, Han B, Zimbron J, Shen Z, Karim MN (2004) Arsenic removal by coagulation and filtration: comparison of groundwaters from the United States and Bangladesh. Desalination 169:231–244

Index

A
Acute, 97
Adsorption, 11, 12, 49, 64, 66, 76, 131, 135, 137, 138, 151, 192, 194, 202, 206, 207
Aeolian, 97
Aerobic, 205
Alkaline, 3, 9, 101, 129, 133, 291
Alluvial fan, 4, 59
Alluvium, alluvial, 4, 5, 7, 10, 12, 13, 20, 23, 24, 28, 33, 34, 38, 39, 41, 42, 45, 47, 50, 57, 59, 66, 68, 69, 71, 75, 98, 112, 121, 126, 128, 143, 146, 149, 152, 157, 162, 197, 200, 255
Amphiboles, 20, 127
Anaerobic, 49, 137, 205, 255
Anthropogenic, 173
Aquifer, 3–5, 7, 10, 12, 24, 27, 29, 33, 34, 42, 43, 47, 48, 58, 68, 75, 144, 147, 151, 152, 154, 156, 157, 161, 162, 164, 166–168, 171, 202, 205, 230, 257, 263, 282, 289
Aquitard, 128, 147
Aragonite, 121
Argillaceous, 38, 47, 49, 69, 75, 127, 146
Arsenicosis, 63, 76, 90, 205, 216, 240, 241, 245, 290
Arsenopyrite, 173
Atmospheric deposition, 150

B
Bank scour, 46
Bewildering, 168
Biocide, 131
Biomarker, 238
Biotite, 3
Birnessite, 121
Body burden, 77
Boulder bed, 84
Bowen's gangrene, 240, 244
Braided, 59
Buffer zone, 98

C
Calcareous, 28, 38, 39, 41, 44, 47
Calcite, 25, 119, 122, 134, 135, 152, 172, 174
Ca-plagioclase, 3
Carbonates, 3, 8, 9, 12, 101–103, 116, 119, 130, 134, 135, 138, 149, 151, 152, 291, 294
Carbonate weathering, 9, 12, 102, 130, 135, 138, 152, 157
Cardiovascular, 65
Catchment, 92, 136
Cation exchange, 10
Cenozoic, 59
Chlorite, 20, 46, 102, 127
Chronic arsenicalism, 275
Clay lens, 202
Clay patches, 188
Competitive, 12
Confined aquifer, 42, 68, 127, 146
Conglomerate, 38, 69
Consumption, 45
Cratonic, 4, 12
Cretaceous, 12

D

Decomposition, 101, 138, 150, 180, 184, 186, 188
Delta, 144, 161, 230
Deposition, 71, 74, 127, 137
Dietary intake, 264
Dissolution, 3, 12, 66, 72, 74, 81, 103, 114, 119, 121, 125, 128, 134, 138
Dolomites, 25, 119, 122, 134
Downstream, 10, 12, 83, 85, 90
Drainage, 47
Drill, 22, 93

E

Earthquake, 69
Electromyography, 245
Endemic, 66
Epidemiological survey, 65, 216, 244
Erosion, 46, 49, 71
Eutrophic lake, 98
Evaporation, 24, 28, 29, 134
Exploitation, 83, 95

F

Factor analysis, 171
Fault, 38, 200–201
Feldspar, 20, 25, 46, 127, 134
Ferruginous, 38, 41, 44, 47, 69, 70
Fissure, 18
Flood plain, 4, 12, 24, 28, 34, 38, 39, 47, 59, 66, 68, 72, 75, 90, 99, 101, 127, 136, 146, 202, 205, 207, 230, 279
Florigen, 263
Fluorosis, 150
Fluvial dynamics, 63
Fluvio-deltaic, 4, 10, 74, 126, 163
Fluvio-lacustrine, 39, 146
Fold, 38
Foliation, 68
Fracture, 18, 22, 68, 202

G

Gangetic plain, 3, 4, 8, 12, 13, 17–20, 23, 33, 34, 38, 39, 45, 47, 66, 72, 82, 88, 90, 98, 103, 112, 114, 116, 121, 126, 144, 145, 197, 206, 216, 255, 297
Gastrointestinal, 65
Geochemical, 27, 28, 63, 103, 125, 134, 138, 157, 162, 164, 172, 174, 200
Geogenic, 12, 66, 112, 144, 197, 258, 263

Geology, 33, 68, 84, 98, 112, 125, 134, 138
Geomorphology, 5, 19, 33, 34, 38, 39, 47, 63, 98, 173, 230, 263, 278
Geophysical, 27
Gneisses, 59
Goethite, 20, 25, 26, 46, 119, 122, 127
Grain size, 184
Grain-size sediment, 45, 63
Greenalite, 122
Guttate melanosis, 240

H

Hand percussion, 22
Hardness, 134, 139
Hematite, 172
Hematology, 33, 65
Herbicide, 207
Heterogeneity, 33, 162
Himalayan foothill, 3, 7, 12, 13, 38, 45, 66, 69, 102
Holocene deposits, 17, 20, 34, 38, 39, 47, 69, 75, 92, 127
Humus, 24, 28
Hydraulic conductivity, 18, 20
Hydraulic head, 47, 50, 206
Hydrobiotite, 3
Hydrocarbon, 180
Hydrochemical, 173
Hydrogeochemical facies, 12, 102, 116, 131, 133, 139, 167, 202, 207
Hydrogeology, 4, 33, 71
Hyper pigmentation, 18, 33, 126

I

Igneous rock, 4
Incongruent dissolution, 102
Infiltration, 99
Interactions, 174
Interfluve, 12, 20, 47, 98, 127, 128, 144, 145, 147
Iron, 26, 28, 276

J

Joint, 18, 68
Jute decomposing, 181

K

Kaolinite, 46, 127
Keratosis, 18, 126

L

Lacustrine, 97
Landfill, 202, 206, 209
Landform, 39, 90
Landscape, 90
Leaching, 66, 153, 186, 191, 192
Leucomelanosis, 240
Lithology, 4, 10, 22, 28, 90, 98, 149, 202, 279

M

Marshes, 48
Mass flooding, 74
Matsatung glacier, 33
Meander, 41, 47, 49, 85, 90, 98, 127
Meteoric, 28, 29
Microbial activity, 26, 66, 92, 136
Microbial reduction, 4, 81, 116, 125, 137
Mineralogy, 125, 138, 162, 174, 188
Mitigation, 72, 158, 198
Mobilization, 10, 11, 13, 17, 26, 27, 29, 49, 62, 66, 71, 72, 93, 111, 125, 126, 136, 138, 156, 157, 171, 174, 180, 186, 190, 192, 194, 198, 200, 206
Montmorillonite, 46
Mottled clay, 69
Mudstone, 69
Muscovite, 20, 46, 127

N

Natural levee, 39, 90
Neurological, 65
Neutralization, 130
Nilogon, 292, 296
Nitrate, 164
Nsutite, 121

O

Ophiolitic rock, 48
Ox-bow lake, 44, 85, 112, 127
Oxidation, 180
Oxidation-reduction potential (ORP), 21, 99, 172, 174, 180, 188, 190, 205

P

Paddy field, 181
Palaeo-channel, 44
Paleozoic, 59
Particulate organic carbon (POC), 181
Peninsular India, 4, 38, 48, 197, 263

Perched aquifer, 24
Permanent hardness, 134
Permissible, 179
Phosphate, 180
Photo-oxidation, 183
Phreatic, 84
Phytotoxicity, 266
Piedmont, 4, 7, 10, 12, 20
Plagioclase feldspar, 101
Pleistocene, 34, 38, 41, 43, 44, 47, 49, 50, 68, 75, 127
Polycarbonate, 183
Pond, 33
Porosity, 22
Potentiometric titration, 147
Precambrian, 7, 12, 71
Pre-Cenozoic, 4, 12, 13
Precipitation, 26, 59, 138, 174
Pulmonary, 65
Pyrites, 24, 48
Pyrolusite, 121
Pyroxenes, 102

Q

Quartz, 20, 25, 38, 46, 48, 119, 127, 134
Quaternary alluvium, 4, 20, 33, 34, 38, 39, 66, 68, 112

R

Rajmahal basalt, 7
Redox sensitive element, 166
Residence-time, 205
Resistivity, 18, 22–24, 26, 28
Rhodochrosite, 172

S

Salinity, 23–24, 28, 70, 139
Saltwater intrusion, 207
Sandstone, 22, 24, 38, 68, 69, 84, 90
Sanitation, 170, 173, 174
Saturation, 18, 25, 28, 174, 206
Schist, 59, 69
Schlumberger method, 22
Sedimentology, 4
Seepage, 180
Seismic zone, 68
Serpentine, 48
Shale, 48, 58–59, 69, 84
Silicate, 3, 4, 8, 9, 103, 134
Silicate weathering, 4, 8, 9, 102, 130, 134, 138

Siltstone, 69
Smectite, 3, 102
Solubility, 150, 205
Solution cavity, 68
Spatial distribution, 17–18, 34, 38, 130
Spatio-vertical heterogeneity, 173
Spatial variation, 92, 105
Stable isotopes, 27, 29
Straighthead disease, 261
Stratigraphy, 34, 46, 75, 199
Subsidence, 38
Sulfate, 24, 130, 134
Sulphide, 3
Supersaturated, 119
Suspended sediments, 46, 50
Swamp, 33, 49, 90
Synclinal valley, 69

T
Temporal variation, 45, 105
Terrain, 4, 8, 10–11, 13
Texture, 18, 28
Thermodynamic, 174
Topography, 33, 90, 92
Transportation, 49, 111, 188
Tributary, 3, 4, 7, 12, 25, 39, 44, 46, 59, 67, 84, 85, 90, 126

Tropical, 163
Tropical monsoon, 67
Tsangpo suture, 48

U
Unconfined aquifer, 68, 128
Unconsolidated, 18, 68, 84
Unsaturated, 101
Uplifting, 68–69
Upstream, 4, 10, 13, 83, 85

V
Variegated clay, 38
Vermiculite, 3, 102
Vulnerability, 93

W
Water-rock interaction, 12, 103, 116
Weathering, 3–4, 8, 22, 72, 101, 116, 135
Wetland, 49, 82

X
X-ray diffraction, 183

Printed by Printforce, the Netherlands